Springer Texts in Statistics

For further volumes:
http://www.springer.com/series/417

Jean-Michel Marin • Christian P. Robert

Bayesian Essentials with R

Second Edition

Jean-Michel Marin
Université Montpellier 2
Montpellier, France

Christian P. Robert
Université Paris-Dauphine
Paris, France

ISSN 1431-875X
ISSN 2197-4136 (electronic)
ISBN 978-1-4614-8686-2
ISBN 978-1-4614-8687-9 (eBook)
DOI 10.1007/978-1-4614-8687-9
Springer New York Heidelberg Dordrecht London

Library of Congress Control Number: 2013950378

Printed on acid-free paper

Springer is part of Springer Science+Business Media (www.springer.com)

To our most rewarding case studies,
Chloé & Lucas, Joachim & Rachel

Preface

After that, it was down to attitude.
—**Ian Rankin**, ***Black & Blue***.—

The purpose of this book is to provide a self-contained entry into practical and computational Bayesian statistics using generic examples from the most common models for a class duration of about seven blocks that roughly correspond to 13–15 weeks of teaching (with three hours of lectures per week), depending on the intended level and the prerequisites imposed on the students. (That estimate does not include practice—i.e., R programming labs, writing data reports—since those may have a variable duration, also depending on the students' involvement and their programming abilities.) The emphasis on *practice* is a strong commitment of this book in that its primary audience consists of graduate students who need to use (Bayesian) statistics as a tool to analyze their experiments and/or datasets. The book should also appeal to scientists in all fields who want to engage into Bayesian statistics, given the versatility of the Bayesian tools. Bayesian essentials can also be used for a more classical statistics audience when aimed at teaching a quick entry to Bayesian statistics at the end of an undergraduate program, for instance. (Obviously, it can supplement another textbook on data analysis at the graduate level.)

This book is an extensive revision of our previous book, *Bayesian Core*, which appeared in 2007, aiming at the same goals. (Glancing at this earlier version will show the filiation to most readers.) However, after publishing *Bayesian Core* and teaching from it to different audiences, we soon realized that the level of mathematics therein was actually more involved than the one expected by those audiences. Students were also asking for more advice and more R code than what was then available. We thus decided upon a major revision, producing a manual that cut the mathematics and expanded the R code, changing as well some chapters and replacing some datasets. We had at first even larger ambitions in terms of contents, but had eventually to sacrifice new chapters for the sake of completing the book before we came to blows! To stress further the changes from the 2007 version, we also decided on a new title, *Bayesian Essentials*, that was actually suggested by Andrew Gelman during a visit to Paris.

The current format of the book is one of a quick coverage of the topics, always backed by a motivated problem and a corresponding dataset (available in the associated R package, *bayess*), and a detailed resolution of the inference procedures pertaining to this problem, always including commented R programs or relevant parts of R programs. Special attention is paid to the derivation of prior distributions, and operational reference solutions are proposed for each model under study. Additional cases are proposed as exercises. The spirit is not unrelated to that of Nolan and Speed (2000), with more emphasis on the methodological backgrounds. While the datasets are inspired by real cases, we also cut on their description and the motivations for their analysis. The current format thus serves as a unique textbook for a service course for scientists aimed at analyzing data the Bayesian way or as an introductory course on Bayesian statistics.

Note that we have not included any BUGS-oriented hierarchical analysis in this edition. This choice is deliberate: We have instead focussed on the Bayesian processing of mostly standard statistical models, notably in terms of prior specification and of the stochastic algorithms that are required to handle Bayesian estimation and model choice questions. We plainly expect that the readers of our book will have no difficulty in assimilating the BUGS philosophy, relying, for instance, on the highly relevant books by Lunn et al. (2012) and Gelman et al. (2013).

A course corresponding to the book has now been taught by both of us for several years in a second year master's program for students aiming at a professional degree in data processing and statistics (at Université Paris Dauphine, France) as well as in several US and Canadian universities. In Paris Dauphine the first half of the book was used in a 6-week (intensive) program, and students were tested on both the exercises (meaning all exercises) and their (practical) mastery of the datasets, the stated expectation being that they should go beyond a mere reproduction of the R outputs presented in the book. While the students found that the amount of work required by this course was rather beyond their usual standards (!), we observed that their understanding and mastery of Bayesian techniques were much deeper and more ingrained than in the more formal courses their counterparts had in the years before. In short, they started to think about the purpose of a Bayesian statistical analysis rather than on the contents of the final test and they ended up building a true intuition about what the results should look like, intuition that, for instance, helped them to detect modeling and programming errors! In most subjects, working on Bayesian statistics from this perspective created a genuine interest in the approach and several students continued to use this approach in later courses or, even better, on the job.

Exercises are now focussed on solving problems rather than addressing finer theoretical points. Solutions to about half of the exercises are freely available on our webpages. We insist upon the point that the developments contained in those exercises are often relevant for fully understanding in the chapter.

Thanks

We are immensely grateful to colleagues and friends for their help with this book and its previous version, *Bayesian Core*, in particular, to the following people: François Perron somehow started thinking about this book and did a thorough editing of it during a second visit to Dauphine, helping us to adapt it more closely to North American audiences. He also adopted *Bayesian Core* as a textbook in Montréal as soon as it appeared. George Casella made helpful suggestions on the format of the book. Jérôme Dupuis provided capture–recapture slides that have been recycled in Chap. 5. Arnaud Doucet taught from the book at the University of British Columbia, Vancouver. Jean-Dominique Lebreton provided the European dipper dataset of Chap. 5. Gaelle Lefol pointed out the Eurostoxx series as a versatile dataset for Chap. 7. Kerrie Mengersen collaborated with both of us on a review paper about mixtures that is related to Chap. 6, Jim Kay introduced us to the Lake of Menteith dataset. Mike Titterington is thanked for collaborative friendship over the years and for a detailed set of comments on the book (quite in tune with his dedicated editorship of *Biometrika*). Jean-Louis Foulley provided us with some dataset and with extensive comments on their Bayesian processing. Even though we did not use those examples in the end, in connection with the strategy not to include BUGS-oriented materials, we are indebted to Jean-Louis for this help. Gilles Celeux carefully read the manuscript of the first edition and made numerous suggestions on both content and style. Darren Wraith, Julyan Arbel, Marco Banterle, Robin Ryder, and Sophie Donnet all reviewed some chapters or some R code and provided highly relevant comments, which clearly contributed to the final output. The picture of the caterpillar nest at the beginning of Chapter 3 was taken by Brigitte Plessis, Christian P. Robert's spouse, near his great-grand-mother's house in Brittany.

We are also grateful to the numerous readers who sent us queries about potential typos, as there were indeed many typos and if not unclear statements. Thanks in particular to Jarrett Barber, Hossein Gholami, we thus encourage all new readers of *Bayesian Essentials* to do the same!

The second edition of *Bayesian Core* was started, thanks to the support of the *Centre International de Rencontres* Mathématiques (CIRM), sponsored by both the Centre National de la Recherche Scientifique (CNRS) and the Société Mathématique de France (SMF), located on the Luminy campus near Marseille. Being able to work "in pair" in the center for 2 weeks was an invaluable opportunity, boosted by the lovely surroundings of the Calanques, where mountain and sea meet! The help provided by the CIRM staff during the stay is also most gratefully acknowledged.

Montpellier, France — Jean-Michel Marin
Paris, France — Christian P. Robert
September 19, 2013

Contents

1

User's Manual

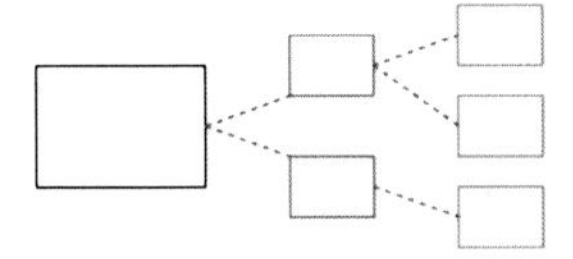

The bare essentials, in other words.
—**Ian Rankin**, ***Tooth & Nail.***—

Roadmap

The **Roadmap** is a section that will start each chapter by providing a commented table of contents. It also usually contains indications on the purpose of the chapter.

For instance, in this initial chapter, we explain the typographical notations that we adopted to distinguish between the different semantic levels of the course. We also try to detail how one should work with this book and how one could best benefit from this work. This chapter is to be understood as a user's (or instructor's) manual that details our pedagogical choices. It also seems the right place to introduce the programming language R, which we use to illustrate all the introduced concepts.

In each chapter, both Ian Rankin's quotation and the figure on top of the title page are (at best) vaguely related to the topic of the chapter, and one should not waste too much time pondering their implications and multiple meanings. The similarity with the introductory chapter of *Introducing Monte Carlo Methods with R* is not coincidental, as Robert and Casella (2009) used the same skeleton as in *Bayesian Core* and as we restarted from their version.

J.-M. Marin and C.P. Robert, *Bayesian Essentials with R*, Springer Texts in Statistics, DOI 10.1007/978-1-4614-8687-9_1,

1.1 Expectations

The key word associated with this book is *modeling*, that is, the ability to build up a probabilistic interpretation of an observed phenomenon and the "story" that goes with it. The "grand scheme" is to get anyone involved in analyzing data to process a dataset within this coherent methodology. This means picking a parameterized probability distribution, denoted by f_θ, and extracting information about (shortened in "estimating") the unknown parameter θ of this probability distribution in order to provide a convincing interpretation of the reasons that led to the phenomenon at the basis of the dataset (and/or to be able to draw predictions about upcoming phenomena of the same nature). Before starting the description of the probability distributions, we want to impose on the reader the essential feature that a model is an *interpretation* of a real phenomenon that fits its characteristics up to some degree of approximation rather than an *explanation* that would require the model to be "true". In short, there is no such thing as a "true model", even though some models are more appropriate than others!

In this book, we chose to describe the use of "classical" probability models for several reasons: First, it is often better to start a trip on well-traveled paths because they are less likely to give rise to unexpected surprises and misinterpretations. Second, they can serve as references for more advanced modelings: Quantities that appear in both simple and advanced modelings should get comparable estimators or, if not, the more advanced modeling should account for that difference. At last, the deliberate choice of an artificial model should give a clearer meaning to the motto that *all models are false* in that it illustrates the fact that a model is not necessarily justified by the theory beyond the modeled phenomenon but that its corresponding inference can nonetheless be exploited *as if* it were a true model. By the end of the book, the reader should also be in a position to assess the relevance of a particular model for a given dataset.

Working with this book should not appear as a major endeavor: The datasets are described along with the methods that are relevant for the corresponding model, and the statistical analysis is provided with detailed comments. The R code that backs up this analysis is included and commented throughout the text. If there is a difficulty with this scheme, it actually starts at this point: Once the reader has seen the analysis, it should be possible for her or him to repeat this analysis or a similar analysis with no further assistance. Even better, the reader should try to read as little as possible of the analysis proposed in this book and on the opposite hand should try to conduct the following stage of the analysis *before* reading the proposed (but not unique) solution. The ultimate lesson here is that there are indeed many ways to analyze a dataset and to propose modeling scenarios and inferential schemes. It is beyond the purpose of this book to provide all of those analyses, and the reader (or the instructor) is supposed to look for alternatives on her or his own.

We thus expect readers to place themselves in a realistic situation to conduct this analysis in life-threatening (or job-threatening) situations. As detailed in the preface, the course was originally intended for students in the last year of study toward a professional degree, and it seems quite reasonable to insist that they face similar situations before entering their incoming job!

1.2 Prerequisites and Further Reading

This being a textbook about statistical modeling, the students are supposed to have a background in both probability and statistics, at the level, for instance, of Casella and Berger (2001). In particular, a knowledge of standard sampling distributions and their properties is desirable. Lab work in the spirit of Nolan and Speed (2000) is also a plus. (One should read in particular their Appendix A on "How to write lab reports?") Further knowledge about Bayesian statistics is not a requirement, although using Robert (2007) or Hoff (2009) as further references would bring a better insight into the topics treated here.

Similarly, we expect students to be able to understand the bits of R programs provided in the analysis, mostly because the syntax of R is very simple. We include an introduction to this language in this chapter and we refer to Dalgaard (2002) for a deeper entry and also to Venables and Ripley (2002).

Besides Robert (2007), the philosophy of which is obviously reflected in this book, other reference books pertaining to applied Bayesian statistics include Gelman et al. (2013), Carlin and Louis (1996), and Congdon (2001, 2003). More specific books that cover parts of the topics of a given chapter are mentioned (with moderation) in the corresponding chapter, but we can quote here the relevant books of Holmes et al. (2002), Pole et al. (1994), and Gill (2002). We want to stress that the citations are limited for efficiency purposes: There is no extensive coverage of the literature as in, e.g., Robert (2007) or Gelman et al. (2013), because the prime purpose of the book is to provide a working methodology, for which incremental improvements and historical perspectives are not directly relevant.

While we also cover simulation-based techniques in a self-contained perspective, and thus do not assume prior knowledge of Monte Carlo methods, detailed references are Robert and Casella (2004, 2009) and Chen et al. (2000).

Although we had at some stage intended to write a new chapter about hierarchical Bayes analysis, we ended up not including this chapter in the current edition and this for several reasons. First, we were not completely convinced about the relevance of a specific hierarchical chapter, given that the hierarchical theme is somehow transversal to the book and pops in the mixture (Chap. 6), dynamic (Chap. 7) and image (Chap. 8) chapters. Second, the revision took already too long and creating a brand new chapter did not sound a manageable goal. Third, managing realistic hierarchical models meant relying on codes written in JAGS and BUGS, which clashed with the philosophy of backing the whole book on R codes. This was subsumed by the recent and highly relevant publication of *The BUGS Book* (Lunn et al., 2012) and by the incoming new edition of *Bayesian Data Analysis* (Gelman et al., 2013).

1.3 Styles and Fonts

Presentation often matters almost as much as content towards a better understanding, and this is particularly true for data analyzes, since they aim to reproduce a realistic situation of a consultancy job where the consultant must report to a customer the results of an analysis. An equilibrated use of graphics, tables, itemized comments, and short paragraphs is, for instance, quite important for providing an analysis that stresses the different conclusions of the work, as well as the points that are yet unclear and those that could be expanded.

In particular, because this book is doing several things at once (that is, to introduce theoretical and computational concepts and to implement them in realistic situations), it needs to differentiate between the purposes and the levels of the parts of the text so that it is as obvious as possible to the reader. To this effect, we take advantage of the many possibilities of modern computer editing, and in particular of LaTeX, as follows.

First, a minimal amount of theoretical bases is required for dealing with the model introduced in each chapter, either for Bayesian statistics or for Monte Carlo theory. This aspect of the material is necessarily part of the main text, but it is also kept to a minimum—just enough for the book to be self-contained—and therefore occasional references to more detailed books such as Robert (2007) and Robert and Casella (2004) are necessary. These sections need be well-understood before handling the following applications or realistic cases. This book is primarily intended for those without a strong background in the theory of Bayesian statistics or computational methods, and "theoretical" sections are essential for them, hence the need to keep those sections within the main text.

Statistics is as much about data processing as about mathematical and probabilistic modeling. To enforce this principle, we center each chapter around one or two specific realistic datasets that are described early enough in the chapter to be used extensively throughout the chapter. These datasets are available on the book's Website (`http://www.ceremade.dauphine.fr/~xian/BCS/`) and are part of the corresponding R package `bayess`, as **normaldata**, **capturedata**, and so on, the name being chosen in reference to the case/chapter heading. (Some of these datasets are already available as `datasets` in the R language.) In particular, we explain the "how and why" of the corresponding dataset in a separate paragraph in this shaded format. This style is also used for illustrating theoretical developments for the corresponding dataset and for specific computations related to this dataset. For typographical convenience, large graphs and tables may appear outside these sections, in subsequent pages, but are obviously mentioned and identified within them.

Example 1.1. There may also be a need for detailed examples in addition to the main datasets, although we strived to keep them to a minimum and only for very specific issues where the reference dataset was not appropriate. They follow this numbered style, the sideways triangle indicating the end of the example. ◀

↯ The last style used in the book is the warning, represented by a lightning ↯ symbol in the margin: This entry is intended to signal major warnings about things that can (and do) go wrong "otherwise"; that is, if the warning is not taken into account. Needless to say, these paragraphs must be given the utmost attention!

A diverse collection of exercises is proposed at the end of each chapter, with solutions to all those exercises freely available on Springer-Verlag webpage.

1.4 An Introduction to R

This section attempts at introducing R to newcomers in a few pages and, as such, it should not be considered as a proper introduction to R. Entire volumes, such as the monumental *R Book* by Crawley (2007), and the introduction by Dalgaard (2002), are dedicated to the practice of this language, and therefore additional efforts (besides reading this chapter) will be required from the reader to sufficiently master the language.[1] However, before discouraging anyone, let us comfort you with the fact that:

(a) The syntax of R is simple and logical enough to quickly allow for a basic understanding of simple R programs, as should become obvious in a few paragraphs.
(b) The best, and in a sense the only, way to learn R is through trial-and-error on simple and then more complex examples. Reading the book with a computer available nearby is therefore the best way of implementing this recommendation.

In particular, the embedded help commands `help()` and `help.search()` are very good starting points to gather information about a specific function or a general issue, even though more detailed manuals are available both locally and on-line. Note that `help.start()` opens a Web browser linked to the local manual pages.

One may first wonder why we support using R as the programming interface for this introduction to Monte Carlo methods, since there exist other

[1]If you decide to skip this chapter, be sure to at least print the handy R Reference Card available at http://cran.r-project.org/doc/contrib/Short-refcard.pdf that summarizes, in four pages, the major commands of R.

languages, most (all?) of them faster than R, like Matlab, and some even free, like C or Python. We obviously have no partisan or commercial involvement in this language.[2] Rather, besides the ease of presentation, our main reason for this choice is that the language combines a sufficiently high power (for an interpreted language) with a very clear syntax both for statistical computation and graphics. R is a flexible language that is *object-oriented* and thus allows the manipulation of complex data structures in a condensed and efficient manner. Its graphical abilities are also remarkable. R provides a powerful *interface* that can integrate programs written in other languages such as C, C++, Fortran, Perl, Python, and Java. At last, it is increasingly common to see people who develop new methodology simultaneously producing an R package in support of their approach and to back up introductory statistics courses with illustrations in R.

One choice we have *not* addressed above is "why R and not BUGS?" BUGS (which stands for Bayesian inference Using Gibbs Sampling) is a Bayesian analysis software developed since the early 1990s, mostly by researchers from the Medical Research Council (MRC) at Cambridge University. The most common version is WinBugs, working under Windows, but there also exists an open-source version called OpenBugs. So, to return to the initial question, we are not addressing the possible links and advantages of BUGS simply because the purpose is different. While access to Monte Carlo specifications is possible in BUGS, most computing operations are handled by the software itself, with the possible outcome that the user does not bother about this side of the problem and instead concentrates on Bayesian modeling. Thus, while R can be easily linked with BUGS and simulation can be done via BUGS, we think that a lower-level language such as R is more effective in bringing you in touch. However, more advanced models like the hierarchical models cannot be easily handled by basic R programming and packages are not necessarily available to handle the variety of those models and call for other programming languages like JAGS. (JAGS standing for *Just Another Gibbs Sampler* and being dedicated to the study of Bayesian hierarchical models. This program is also freely available and distributed under the GNU Licence, the current version being JAGS 3.3.0.)

1.4.1 Getting Started

The R language is straightforward to install: it can be downloaded (obviously free) from one of the numerous CRAN (Comprehensive R Archive Network) mirror Websites around the world.[3]

At this stage, we refrain from covering the installation of the R package and thus assume that (a) R is installed on the machine you want to work with and (b) that you have managed to launch it (in most cases, you simply have

---

[2] Once again, R is a freely distributed and open-source language.

[3] The main CRAN Website is http://cran.r-project.org/.

to click on the proper icon). In the event you use a friendly (GUI) interface like RKWard, the interface opens several windows whose use should be self-explanatory (along with a proper on-line help). Otherwise, you should then obtain a terminal window whose first lines resemble the following, most likely with a more recent version:

```
R version 2.14.1 (2011-12-22)
Copyright (C) 2011 The R Foundation for Statistical Computing
ISBN 3-900051-07-0
Platform: i686-pc-linux-gnu (32-bit)

R is free software and comes with ABSOLUTELY NO WARRANTY.
You are welcome to redistribute it under certain conditions.
Type 'license()' or 'licence()' for distribution details.

R is a collaborative project with many contributors.
Type 'contributors()' for more information and
'citation()' on how to cite R or R packages in publications.

Type 'demo()' for some demos, 'help()' for on-line help, or
'help.start()' for an HTML browser interface to help.
Type 'q()' to quit R.

>
```

Neither this austere beginning nor the prospect of using a line editor should put you off, though, as there are many other ways of inputting and outputting commands and data, as we shall soon see! The final line above with the symbol > means that the R software is waiting for a command from the user. This character > at the beginning of each line in the executable window is called the *prompt* and precedes the line command, which is terminated by pressing the `RETURN` key. At this early stage, all commands will be passed as line commands, and you should thus spot commands thanks to this symbol.

Commands and programs that need to be stopped during their execution, for instance because they take too long or too much memory to complete or because they involve a programming mistake such as an infinite loop, can be stopped by the Control-C double-key action without exiting the R session.

For memory and efficiency reasons, R does not install all the available functions and programs when launched but only the basic *packages* that it requires to run properly. Additional packages can be loaded via the `library` command, as in

```
> library(mnormt) # Multivariate Normal and t Distributions
```

and the entire list of available packages is provided by `library()`. (The symbol `#` in the prompt lines above indicates a comment: All characters following `#` until the end of the command line are ignored. Comments are recommended to improve the readability of your programs.) There exist hundreds of packages available on the Web.[4] Installing a new package such as the package `mnormt` is done by downloading the file from the Web depository and calling

```
> install.package("mnormt")
```

For a given package, the `install.package` command obviously needs to be executed only once, while the `library` call is required each time R is launched (as the corresponding package is not kept as part of the `.RData` file). Thus, it is good practice to include calls to required libraries within your R programs in order to avoid error messages when launching them.

1.4.2 R Objects

As with many advanced programming languages, R distinguishes between several types of *objects*. Those types include scalar, vector, matrix, time series, data frames, functions, or graphics. An R object is mostly characterized by a *mode* that describes its contents and a *class* that describes its structure. The R function `str` applied to any R object, including R functions, will show its structure. For instance,

```
> str(log)
function (x, base = exp(1))
```

The different modes are

- `null` (empty object),
- `logical` (`TRUE` or `FALSE`),
- `numeric` (such as `3`, `0.14159`, or `2+sqrt(3)`),
- `complex`, (such as `3-2i` or `complex(1,4,-2)`), and
- `character` (such as `''Blue''`, `''binomial''`, `''male''`, or `''y=a+bx''`),

and the main classes are `vector`, `matrix`, `array`, `factor`, `time-series`, `data.frame`, and `list`. Heterogeneous objects such as those of the `list` class can include elements with various modes. Manual entries about those classes can be obtained via the help commands `help(data.frame)` or `?matrix` for instance.

R can operate on most of those types as a regular function would operate on a scalar, which is a feature that should be exploited as much as possible for compact and efficient programming. The fact that R is interpreted rather than compiled involves many subtle differences, but a major issue is that all variables in the system are evaluated and stored at every step of R programs. This means that loops in R are enormously time-consuming and should be avoided at all costs! Therefore, using the shortcuts offered by R in the manipulation of vectors, matrices, and other structures is a must.

---

[4]Packages that have been validated and tested by the R core team are listed at http://cran.r-project.org/src/contrib/PACKAGES.html.

The vector *class*

As indicated logically by its name, the `vector` object corresponds to a mathematical vector of elements of the same type, such as `(TRUE,TRUE,FALSE)` or `(1,2,3,5,7,11)`. Creating small vectors can be done using the R command `c()` as in

```
> a=c(2,6,-4,9,18)
```

This fundamental function combines or concatenates terms together. For instance,

```
> d=c(a,b)
```

concatenates the two vectors `a` and `b` into a new vector `d`. Note that decimal numbers should be encoded with a dot, character strings in quotes `" "`, and logical values with the character strings `TRUE` and `FALSE` or with their respective abbreviations `T` and `F`. Missing values are encoded with the character string `NA`.

In Fig. 1.1, we give a few illustrations of the use of vectors in R. The character `+` indicates that the console is waiting for a supplementary instruction, which is useful when typing long expressions. The assignment operator is `=`, not to be confused with `==`, which is the Boolean operator for equality. An older assignment operator is `<-`, as in

```
> x <- c(3,6,9)
```

and, at least for compatibility reasons, it still remains functional in current versions of R, but we prefer using the equality sign. (As pointed out by Spector (2009), an exception is when using `system.time`, briefly described in Fig. 1.8, since `=` is then used to identify keywords, although `=` can preserve its initial purpose if curly brackets `{` and `}` delimit the allocation commands.)

↯ A misleading feature of the assignment operator `<-` is found in Boolean expressions such as

```
> if (x[1]<-2) ...
```

which is supposed to test whether or not `x[1]` is less than -2 but ends up allocating 2 to `x[1]`, erasing its current value! Adding a space in the expression is sufficient to solve the problem: `if (x[1] < -2)`.
Note also that using

```
> if (x[1]=-2) ...
```

mistakenly instead of `(x[1]==-2)` has the same consequence.

New R objects are simply defined by assigning them a value, as in the first line of Fig. 1.1, without a preliminary declaration of type (as in the C language).

`> a=c(5,5.6,1,4,-5)`	build the object a containing a numeric vector of dimension 5 with elements 5, 5.6, 1, 4, –5
`> a[1]`	display the first element of a
`> b=a[2:4]`	build the numeric vector b of dimension 3 with elements 5.6, 1, 4
`> d=a[c(1,3,5)]`	build the numeric vector d of dimension 3 with elements 5, 1, –5
`> 2*a`	multiply each element of a by 2 and display the result
`> b%%3`	provides each element of b modulo 3
`> d%/%2.4`	computes the integer division of each element of d by 2.4
`> e=3/d`	build the numeric vector[5] e of dimension 3 and elements 3/5, 3, –3/5
`> log(d*e)`	multiply the vectors d and e term by term and transform each term into its natural logarithm
`> sum(d)`	calculate the sum of d
`> length(d)`	display the length of d
`> t(d)`	transpose d, the result is a row vector
`> t(d)%*%e`	scalar product between the row vector t(b) and the column vector e with identical length
`> t(d)*e`	element-wise product between two vectors with identical lengths
`> g=c(sqrt(2),log(10))`	build the numeric vector g of dimension 2 and elements $\sqrt{2}$, log(10)
`> e[d==5]`	build the subvector of e that contains the components e[i] such that d[i]=5
`> a[-3]`	create the subvector of a that contains all components of a but the third.
`> is.vector(d)`	display the logical expression TRUE if a vector and FALSE else

Fig. 1.1. Illustrations of the processing of vectors in R

Note[6] in Table 1.1 the convenient use of Boolean expressions to extract subvectors from a vector without having to resort to a component-by-component test (and hence a loop). The quantity `d==5` is itself a vector of Booleans, while the number of components satisfying the constraint can be computed by `sum(d==5)`. The ability to apply scalar functions to vectors as a whole is also a major advantage of R. In the event the function depends on a parameter or an option, this quantity can be entered as in

[5] The variable e is not predefined in R as exp(1).

[6] Positive and negative indices cannot be used simultaneously.

```
> e=lgamma(e^2) #warning: this is not the exponential basis,
  exp(1)
```

which returns the vector with components $\log \Gamma(e_i^2)$. Functions that are specially designed for vectors include, for instance, `sample`, `order`, `sort` and `rank`, which all have to do with manipulating the order in which the components of the vector occur.

Besides their numeric and logical indexes, the components of a vector can also be identified by names. For a given vector `x`, `names(x)` is a vector of characters of the same length as `x`. This additional attribute is most useful when dealing with real data, where the components have a meaning such as `"unemployed"` or `"democrat"`. Those names can also be erased by the command

```
> names(x)=NULL
```

↯ The : operator found in Fig. 1.1 is a very useful device that defines a consecutive sequence, but it is also fragile in that sequences do not always produce what is expected. For instance, `1:2*n` corresponds to `(1:2)*n` rather than `1:(2*n)`.

The matrix, array, *and* factor *classes*

The `matrix` class provides the R representation of matrices. A typical entry is, for instance,

```
> x=matrix(vec,nrow=n,ncol=p)
```

which creates an $n \times p$ matrix whose elements are those of the vector `vec`, assuming this vector is of dimension np. An important feature of this entry is that, in a somewhat unusual way, the components of `vec` are stored by column, which means that `x[1,1]` is equal to `vec[1]`, `x[2,1]` is equal to `vec[2]`, and so on, except if the option `byrow=T` is used in `matrix`. (Because of this choice of storage convention, working on R matrices column-wise is faster then working row-wise.) Note also that, if `vec` is of dimension $n \times p$, it is not necessary to specify both the `nrow=n` and `ncol=p` options in `matrix`. One of those two parameters is sufficient to define the matrix. On the other hand, if `vec` is *not* of dimension $n \times p$, `matrix(vec,nrow=n,ncol=p)` will create an $n \times p$ matrix with the components of `vec` repeated the appropriate number of times. For instance,

```
> matrix(1:4,ncol=3)
     [,1] [,2] [,3]
[1,]    1    3    1
[2,]    2    4    2
Warning message:
data length [4] is not a submultiple or multiple of the
  number
of columns [3] in matrix in: matrix(1:4, ncol = 3)
```

produces a 2×3 matrix along with a warning message that something may be missing in the call to `matrix`. Note again that $1, 2, 3, 4$ are entered consecutively when following the column (or *lexicographic*) order. Names can be given to the rows and columns of a matrix using the `rownames` and `colnames` functions.

Note that, in some situations, it is useful to remember that an R matrix can also be used as a vector. If `x` is an $n \times p$ matrix, `x[i+p*(j-1)]` is equal to `x[i,j]`, i.e., `x` can also be manipulated as a vector made of the columns of `vec` piled on top of one another. For instance, `x[x>5]` is a vector, while `x[x>5]=0` modifies the right entries in the matrix `x`. Conversely, vectors can be turned into $p \times 1$ matrices by the command `as.matrix`. Note that `x[1,]` produces the first row of `x` as a vector rather than as a $p \times 1$ matrix.

R allows for a wide range of manipulations on matrices, both termwise and in the classical matrix algebra perspective. For instance, the standard matrix product is denoted by `%*%`, while `*` represents the term-by-term product. (Note that taking the product `a%*%b` when the number of columns of `a` differs from the number of rows of `b` produces an error message.) Figure 1.2 gives a few examples of matrix-related commands. The `apply` function is particularly easy to use for functions operating on matrices by row or column.

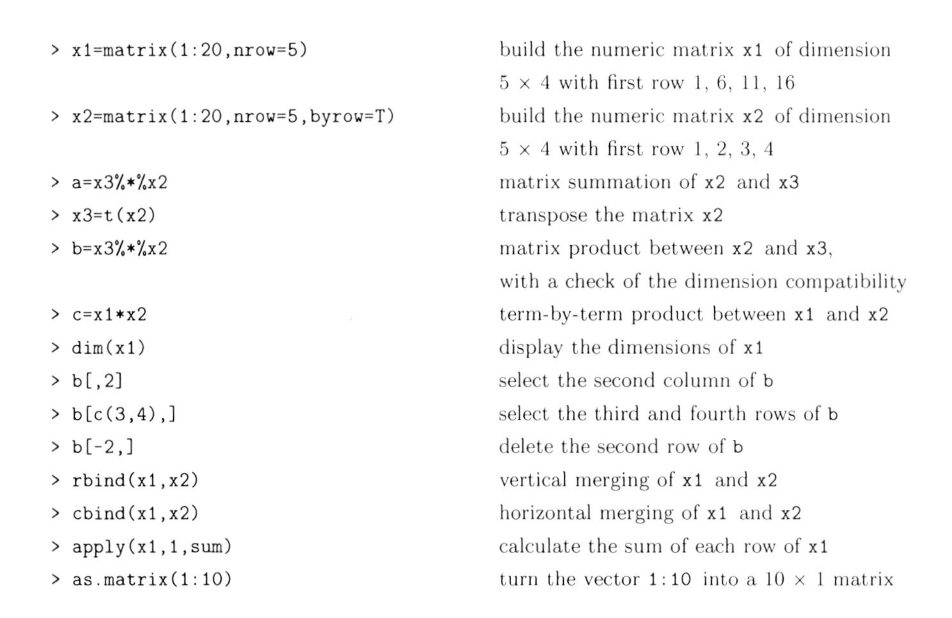

`> x1=matrix(1:20,nrow=5)`	build the numeric matrix `x1` of dimension 5×4 with first row 1, 6, 11, 16
`> x2=matrix(1:20,nrow=5,byrow=T)`	build the numeric matrix `x2` of dimension 5×4 with first row 1, 2, 3, 4
`> a=x3%*%x2`	matrix summation of `x2` and `x3`
`> x3=t(x2)`	transpose the matrix `x2`
`> b=x3%*%x2`	matrix product between `x2` and `x3`, with a check of the dimension compatibility
`> c=x1*x2`	term-by-term product between `x1` and `x2`
`> dim(x1)`	display the dimensions of `x1`
`> b[,2]`	select the second column of `b`
`> b[c(3,4),]`	select the third and fourth rows of `b`
`> b[-2,]`	delete the second row of `b`
`> rbind(x1,x2)`	vertical merging of `x1` and `x2`
`> cbind(x1,x2)`	horizontal merging of `x1` and `x2`
`> apply(x1,1,sum)`	calculate the sum of each row of `x1`
`> as.matrix(1:10)`	turn the vector `1:10` into a 10×1 matrix

Fig. 1.2. Illustrations of the processing of matrices in R

The function `diag` can be used to extract the vector of the diagonal elements of a matrix, as in `diag(a)`, or to create a diagonal matrix with a given diagonal, as in `diag(1:10)`. Since matrix algebra is central to good programming in R, as matrix programming allows for the elimination of time-consuming loops, it is important to be familiar with matrix manipulation. For instance, the function crossprod replaces the product t(x)%*%y on either vectors or matrices by crossprod(x,y) more efficiently:

```
> system.time(crossprod(1:10^6,1:10^6))
   user  system elapsed
  0.016   0.048   0.066
> system.time(t(1:10^6)%*%(1:10^6))
   user  system elapsed
  0.084   0.036   0.121
```

Eigen-analysis of square matrices is also included in the base package. For instance, chol(m) returns the upper triangular factor of the Choleski decomposition of m; that is, the matrix R such that $R^{\mathrm{T}}R$ is equal to m. Similarly, eigen(m) returns a list that contains the eigenvalues of m (some of which can be complex numbers) as well as the corresponding eigenvectors (some of which are complex if there are complex eigenvalues). Related functions are svd and qr, which provide the singular values and the QR decomposition of their argument, respectively. Note that the inverse M^{-1} of a matrix M can be found either by solve(M) (recommended) or ginv(M), which requires downloading the library MASS and also produces generalized inverses (which may be a mixed blessing since the fact that a matrix is not invertible is not signaled by ginv). Special versions of solve are backsolve and forwardsolve, which are restricted to upper and lower diagonal triangular systems, respectively. Note also the alternative of using chol2inv which returns the inverse of a matrix m when provided by the Choleski decomposition chol(m).

Structures with more than two indices are represented by *arrays* and can also be processed by R commands, for instance x=array(1:50,c(2,5,5)), which gives a three-entry table of 50 terms. Once again, they can also be interpreted as vectors.

The apply function used in Fig. 1.2 is a very powerful device that operates on arrays and, in particular, matrices. Since it can return arrays, it bypasses calls to multiple loops and makes for (sometimes) quicker and (always) cleaner programs. It should not be considered as a panacea, however, as apply hides calls to loops inside a single command. For instance, a comparison of apply(A, 1,mean) with rowMeans(A) shows the second version is about 200 times faster. Using linear algebra whenever possible is therefore a more efficient solution. Spector (2009, Sect. 8.7) gives a detailed analysis of the limitations of apply and the advantages of vectorization in R.

A factor is a vector of characters or integers used to specify a discrete classification of the components of other vectors with the same length. Its main difference from a standard vector is that it comes with a level attribute used to specify the possible values of the factor. This structure is therefore appropriate to represent qualitative variables. R provides both ordered and unordered factors, whose major appeal lies within model formulas, as illustrated in Fig. 1.3. Note the subtle difference between apply and tapply.

`> state=c("tas","tas","sa","sa","wa")`	create a vector with five values
`> statef=factor(state)`	distinguish entries by group
`> levels(statef)`	give the groups
`> incomes=c(60,59,40,42,23)`	create a vector of incomes
`> tapply(incomes,statef,mean)`	average the incomes for each group
`> statef=factor(state,`	define a new level with one more
`+ levels=c("tas","sa","wa","yo"))`	group than observed
`> table(statef)`	return statistics for all levels

Fig. 1.3. Illustrations of the factor class

The list *and* data.frame *classes*

A list in R is a rather loose object made of a collection of other arbitrary objects known as its *components*.[7] For instance, a list can be derived from n existing objects using the function list:

```
a=list(name_1=object_1,...,name_n=object_n)
```

This command creates a list with n arguments using object_1,...,object_n for the components, each being associated with the argument's name, name_i. For instance, a$name_1 will be equal to object_1. (It can also be represented as a[[1]], but this is less practical, as it requires some bookkeeping of the order of the objects contained in the list.) Lists are very useful in preserving information about the values of variables used within R functions in the sense that all relevant values can be put within a list that is the output of the corresponding function (see Sect. 1.4.5 for details about the construction of functions in R). Most standard functions in R, for instance eigen in Fig. 1.4, return a list as their output. Note the use of the abbreviations vec and val in the last line of Fig. 1.4. Such abbreviations are acceptable as long as they do not induce confusion. (Using res$v would not work!)

The local version of apply is lapply, which computes a function for each argument of the list

```
> x = list(a = 1:10, beta = exp(-3:3),
+ logic = c(TRUE,FALSE,FALSE,TRUE))
> lapply(x,mean) #compute the empirical means
$a
[1] 5.5
```

[7] Lists can contain lists as elements.

`> li=list(num=1:5,y="color",a=T)`	create a list with three arguments
`> a=matrix(c(6,2,0,2,6,0,0,0,36),nrow=3)`	create a (3,3) matrix
`> res=eigen(a,symmetric=T)`	diagonalize `a` and
`> names(res)`	produce a list with two arguments: `vectors and values`
`> res$vectors`	`vectors` arguments of `res`
`> diag(res$values)`	create the diagonal matrix of eigenvalues
`> res$vec%*%diag(res$val)%*%t(res$vec)`	recover `a`

Fig. 1.4. Chosen features of the list class

```
$beta
[1] 4.535125
$logic
[1] 0.5
```

provided each argument is of a mode that is compatible with the function argument (i.e., is numeric in this case). A "user-friendly" version of `lapply` is `sapply`, as in

```
> sapply(x,mean)
       a     beta    logic
5.500000 4.535125 0.500000
```

The last class we briefly mention here is the `data.frame`. A data frame is a list whose elements are possibly made of differing modes and attributes but have the same length, as in the example provided in Fig. 1.5. A data frame can be displayed in matrix form, and its rows and columns can be extracted using matrix indexing conventions. A list whose components satisfy the restrictions imposed on a data frame can be coerced into a data frame using the function `as.data.frame`. The main purpose of this object is to import data from an external file by using the `read.table` function.

1.4.3 Probability Distributions in R

R is primarily a statistical language. It is therefore well-equipped with probability distributions. As described in Table 1.1, all standard distributions are available, with a clever programming shortcut: A "core" name, such as `norm`, is associated with each distribution and the four basic associated functions, namely the cdf, the pdf, the quantile function, and the simulation procedure, are defined by appending the prefixes `d, p, q, r` to the core name, such as `dnorm()`, `pnorm()`, `qnorm()`, and `rnorm()`. Obviously, each function requires

`> v1=sample(1:12,30,rep=T)`	simulate 30 independent uniform random variables on {1, 2, ..., 12}
`> v2=sample(LETTERS[1:10],30,rep=T)`	simulate 30 independent uniform random variables on $\{a, b, ..., j\}$
`> v3=runif(30)`	simulate 30 independent uniform random variables on [0, 1]
`> v4=rnorm(30)`	simulate 30 independent realizations from a standard normal distribution
`> xx=data.frame(v1,v2,v3,v4)`	create a data frame

Fig. 1.5. Definition of a data.frame

additional entries, as in `pnorm(1.96)` or `rnorm(10,mean=3,sd=3)`. Recall that `pnorm()` and `qnorm()` are inverses of one another.

Table 1.1. Standard distributions with R core name

Distribution	Core	Parameters	Default values
Beta	`beta`	`shape1, shape2`	
Binomial	`binom`	`size, prob`	
Cauchy	`cauchy`	`location, scale`	0, 1
Chi-square	`chisq`	`df`	
Exponential	`exp`	`1/mean`	1
Fisher	`f`	`df1, df2`	
Gamma	`gamma`	`shape,1/scale`	NA, 1
Geometric	`geom`	`prob`	
Hypergeometric	`hyper`	`m, n, k`	
Log-Normal	`lnorm`	`mean, sd`	0, 1
Logistic	`logis`	`location, scale`	0, 1
Normal	`norm`	`mean, sd`	0, 1
Poisson	`pois`	`lambda`	
Student	`t`	`df`	
Uniform	`unif`	`min, max`	0, 1
Weibull	`weibull`	`shape`	

In addition to these probability functions, R also provides a battery of (classical) statistical tools, ranging from descriptive statistics to nonparametric tests and generalized linear models. A description of these abilities is not possible in this section but we refer the reader to, e.g., Dalgaard (2002) or Venables and Ripley (2002) for a complete entry.

1.4.4 Graphical Facilities

Another clear advantage of using the R language is that it allows a very rich range of graphical possibilities. Functions such as `plot` and `image` can be customized to a large extent, as described in Venables and Ripley (2002) or Murrell (2005) (the latter being entirely dedicated to the R graphic abilities). Even though the default output of `plot` as for instance in

```
> plot(faithful)
```

is not highly most enticing, `plot` is incredibly flexible: To see the number of parameters involved, you can type `par()` that delivers the default values of all those parameters.

↯ The wealth of graphical possibilities offered by R should be taken advantage of cautiously! That is, good design avoids clutter, small fonts, unreadable scale, etc. The recommendations found in Tufte (2001) are thus worth following to avoid horrid outputs like those often found in some periodicals! In addition, graphs produced by R usually tend to look nicer on the current device than when printed or included in a slide presentation. Colors may change, font sizes may turn awkward, separate curves may end up overlapping, and so on.

Before covering the most standard graphic commands, we start by describing the notion of device that is at the core of those graphic commands. Each graphical operation sends its outcome to a *device*, which can be a graphical window (like the one that automatically appears when calling a graphical command for the first time as in the example above) or a file where the graphical outcome is stored for printing or other uses. Under Unix, Linux and mac OS, launching a new graphical window can be done via `X11()`, with many possibilities for customization (such as size, positions, color, etc.). Once a graphical window is created, it is given a device number and can be managed by functions that start with `dev.`, such as `dev.list`, `dev.set`, and others. An important command is `dev.off`, which closes the current graphical window. When the device is a file, it is created by a function that is named after its driver. There are therefore a `postscript`, a `pdf`, a `jpeg`, and a `png` function. When printing to a file, as in the following example,

```
> pdf(file="faith.pdf")
> par(mfrow=c(1,2),mar=c(4,2,2,1))
> hist(faithful[,1],nclass=21,col="grey",main="",
+ xlab=names(faithful)[1])
> hist(faithful[,2],nclass=21,col="wheat",main="",
+ xlab=names(faithful)[2])
> dev.off()
```

closing the sequence with `dev.off()` is compulsory since it completes the file, which is then saved. If the command `pdf(file="faith.pdf")` is repeated, the earlier version of the `pdf` file is erased.

Of course, using a line command interface for controlling graphics may seem antiquated, but this is the consequence of the R object-oriented philosophy. In addition, current graphs can be saved to a postscript file using the `dev.copy` and `dev.print` functions. Note that R-produced graphs tend to be large objects, in part because the graphs are not pictures of the current state but instead preserve every action ever taken.

As already stressed above, plot is a highly versatile tool that can be used to represent functional curves and two-dimensional datasets. Colors (chosen by colors() or colours() out of 650 hues), widths, and types can be calibrated at will and LaTeX-like formulas can be included within the graphs using expression. Text and legends can be included at a specific point with locator (see also identify) and legend. An example of (relatively simple) output is

```
> plot(as.vector(time(mdeaths)),as.vector(mdeaths),cex=.6,
+ pch=19,xlab="",ylab="Monthly deaths from bronchitis")
> lines(spline(mdeaths),lwd=2,col="red",lty=3)
> ar=arima(mdeaths,order=c(1,0,0))$coef
> lines(as.vector(time(mdeaths))[-1], ar[2]+ar[1]*
+ (mdeaths[-length(mdeaths)]-ar[2]),col="blue",lwd=2,lty=2)
> title("Splines versus AR(1) predictor")
> legend(1974,2800,legend=c("spline","AR(1)"),col=c("red",
+ "blue"),lty=c(3,2),lwd=c(2,2),cex=.5)
```

represented in Fig. 1.6, which compares spline fitting to an AR(1) predictor and to an SAR(1,12) predictor. Note that the seasonal model is doing worse.

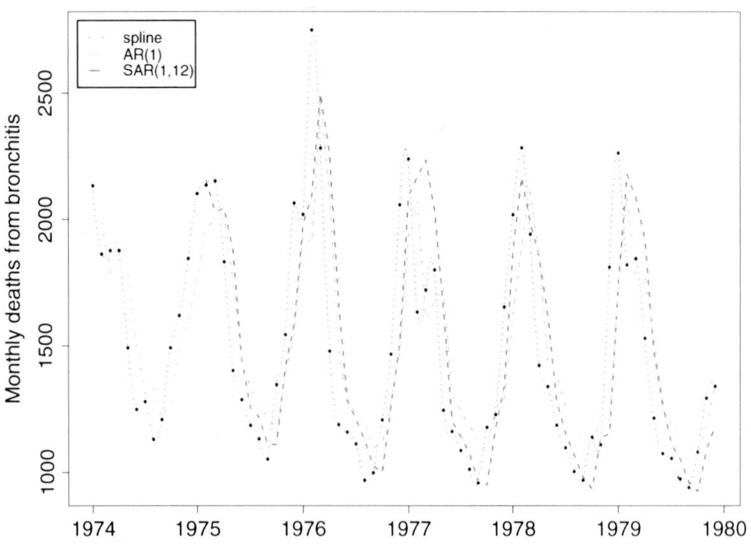

Fig. 1.6. Monthly deaths from bronchitis in the UK over the period 1974–1980 and fits by a spline approximation and an AR predictor

Useful graphical functions include

- `hist` for constructing and optionally plotting histograms of datasets;
- `points` for adding points on an existing graph;
- `lines` for linking points together on an existing graph;
- `polygon` for filling the area between two sets of points;
- `barplot` for creating barplots;
- `boxplot` for creating boxplots.

The two-dimensional representations offered by `image` and `contour` are quite handy when providing likelihood or posterior surfaces. Figure 1.7 gives some of the most usual graphical commands.

```
> x=rnorm(100)
> hist(x,nclass=10, prob=T)                  compute and plot an histogram
                                             of x
> curve(dnorm(x),add=T)                      draw the normal density on top
> y=2*x+rnorm(100,0,2)
> plot(x,y,xlim=c(-5,5),ylim=c(-10,10))      draw a scatterplot of x against y
> lines(c(0,0),c(1,2),col="sienna3")
> boxplot(x)                                 compute and plot
                                             a box-and-whiskers plot of x
> state=c("tas","tas","sa","sa","wa","sa")
> statef=factor(state)
> barplot(table(statef))                     draw a bar diagram of x
```

Fig. 1.7. Some standard plotting commands

1.4.5 Writing New R Functions

One of the strengths of R is that new functions and libraries can be created by anyone and then added to Web depositories to continuously enrich the language. These new functions are not distinguishable from the core functions of R, such as `median()` or `var()`, because those are also written in R. This means their code can be accessed and potentially modified, although it is safer to define new functions. (A few functions are written in C, though, for efficiency.) Learning how to write functions designed for one's own problems is paramount for their resolution, even though the huge collection of available R functions may often contain a function already written for that purpose.

A function is defined in R by an assignment of the form

```
name=function(arg1[=expr1],arg2[=expr2],...) {
expression
...
expression
value
}
```

where expression denotes an R command that uses some of the arguments arg1, arg2, ... to calculate a value, value, that is the outcome of the function. The braces indicate the beginning and the end of the function and the brackets some possible default values for the arguments. Note that producing a value at the end of a function is essential because anything done within a function is local and temporary, and therefore lost once the function has been exited, unless saved in value (hence, again, the appeal of list()). For instance, the following function, named sqrnt, implements a version of Newton's method for calculating the square root of y:

```
sqrnt=function(y) {
x=y/2
while (abs(x*x-y) > 1e-10) x=(x+y/x)/2
x
}
```

When designing a new R function, it is more convenient to use an external text editor and to store the function under development in an external file, say myfunction.R, which can be executed in R as source("myfunction.R"). Note also that some external commands can be launched within an R function via the very handy command system(). This is, for instance, the easiest way to incorporate programs written in other languages (e.g., Fortran, C, Matlab) within R programs.

Without getting deeply into R programming, let us note a distinction between global and local variables: the former are defined in the core of the R code and are recognized everywhere, while the later are only defined within a specific function. This means in particular that a local variable, locax say, initialized within a function, myfunc say, will not be recognized outside myfunc. (It will not even be recognized in a function defined within myfunc.)

The expressions used in a function rely on a syntax that is quite similar to those of other programming languages, with conditional statements such as

```
if (expres1) expres2 else expres3
```

where expres1 is a logical value, and loops such as

```
for (name in expres1) expres2
```

and

```
while (name in expres1) expres2
```

where expres1 is a collection of values, as illustrated in Fig. 1.8. In particular, Boolean operators can be used within those expressions, including == for testing equality, != for testing inequality, & for the logical and, | for the logical or, and ! for the logical contradiction.

Since R is an interpreted language, avoiding loops is generally a good idea, but this may render programs much harder to read. It is therefore extremely useful to include comments within the programs by using the symbol #.

```
> bool=T;i=0                                  separate commands by semicolons
> while(bool==T) {i=i+1; bool=(i<10)}         stop at i = 11
> s=0;x=rnorm(10000)
> system.time(for (i in 1:length(x)){         output sum(x) and
+ s=s+x[i]})[3]                               provide computing time
> system.time(t(rep(1,10000))%*%x)[3]         compare with vector product
> system.time(sum(x))[3]                      compare with sum() efficiency
```

Fig. 1.8. Some artificial loops in R

1.4.6 Input and Output in R

Large data objects need to be read as values from external files rather than entered during an R session at the keyboard (or by cut-and-paste). Input facilities are simple, but their requirements are fairly strict. In fact, there is a clear presumption that it is possible to modify input files using other tools outside R.

An entire data frame can be read directly with the `read.table()` function. Plain files containing rows of values with a single mode can be downloaded using the `scan()` function, as in

```
> a=matrix(scan("myfile"),nrow=5,byrow=T)
```

When data frames have been produced by another statistical software, the library `foreign` can be used to input those frames in R. For example, the function `read.spss()` allows ones to read SPSS data frames.

Conversely, the generic function `save()` can be used to store all R objects in a given file, either in binary or ASCII format. (The alternative function `dump()` is more rudimentary but also useful.) The function `write.table()` is used to export R data frames as ASCII files.

1.4.7 Administration of R Objects

During an R session, objects are created and stored by name. The command `objects()` (or, alternatively, `ls()`) can be used to display, within a directory called the *workspace*, the names of the objects that are currently stored. Individual objects can be deleted with the function `rm()`. Removing all objects created so far is done by `rm(list=ls())`.

All objects created during an R session (including functions) can be stored permanently in a file in provision of future R sessions. At the end of each R session, obtained by the command `quit()` (which can be abbreviated as `q()`), the user is given the opportunity to save all the currently available objects, as in

```
>q()
Save workspace image? [y/n/c]:
```

If the user answers y, the object created during the current session and those saved from earlier sessions are saved in a file called .RData and located in the working directory. When R is called again, it reloads the workspace from this file, which means that the user starts the new session exactly where the old one had stopped. In addition, the entire past command history is stored in the file .Rhistory and can be used in the current or in later sessions by using the command history().

1.5 The bayess Package

Since this is originally a paper book, copying by hand the R code represented on the following pages to your computer terminal would be both tedious and time-wasting. We have therefore gathered all the programs and codes of this book within an R package called bayess that you should download from CRAN before proceeding to the next chapter. Once downloaded on your computer following the instructions provided on the CRAN Webpage, the package bayess is loaded into your current R session by library(bayess). All the functions defined inside the package are then available, and so is a step-by-step reproduction of the examples provided in the book, using the demo command:

```
> demo(Chapter.1)

        demo(Chapter.1)
        ---- ~~~~~~~~~


Type  <Return>   to start :

> # Chapter 1 R commands
>
> # Section 1.4.2
>
> str(log)
function (x, base = exp(1))

> a=c(2,6,-4,9,18)

> x <- c(3,6,9)

> d=a[c(1,3,5)]

> e=3/d

> e=lgamma(e^2)
```

```
> S=readline(prompt="Type  <Return>   to continue : ")
Type  <Return>   to continue :
```

and similarly for the following chapters. Obviously, all commands contained in the demonstrations and all functions defined in the package can be accessed and modified.

↯ Although most steps of the demonstrations are short, some may require longer execution times. If you need to interrupt the demonstration, recall that `Ctrl-C` is an interruption command.

2

Normal Models

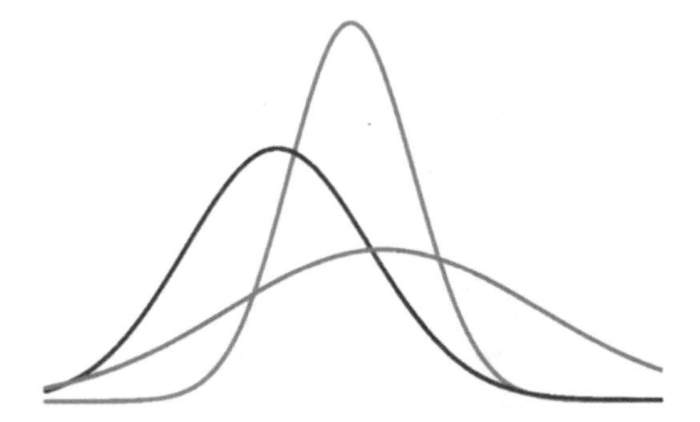

This was where the work really took place.
—**Ian Rankin**, ***Knots & Crosses***.—

Roadmap

This chapter uses the standard normal $\mathcal{N}(\mu, \sigma^2)$ distribution as an easy entry to generic Bayesian inferential methods. As in every subsequent chapter, we start with a description of the data used as a chapter benchmark for illustrating new methods and for testing assimilation of the techniques. We then propose a corresponding statistical model centered on the normal distribution and consider specific inferential questions to address at this level, namely parameter estimation, model choice, and outlier detection, once set the description of the Bayesian resolution of inferential problems. As befits a first chapter, we also introduce here general computational techniques known as Monte Carlo methods.

J.-M. Marin and C.P. Robert, *Bayesian Essentials with R*, Springer Texts in Statistics, DOI 10.1007/978-1-4614-8687-9_2,

2.1 Normal Modeling

The normal (or Gaussian) distribution $\mathscr{N}(\mu, \sigma^2)$, with density on $\mathbb{R}$,

$$f(x|\mu, \sigma) = \frac{1}{\sqrt{2\pi}\sigma} \exp\left\{-\frac{1}{2\sigma^2}(x-\mu)^2\right\},$$

is certainly one of the most studied and one of the most used distributions because of its "normality": It appears both as the limit of additive small effects and as a representation of symmetric phenomena without long tails, and it offers many openings in terms of analytical properties and closed-form computations. As such, it is thus the natural opening to a modeling course, even more than discrete and apparently simpler models such as the binomial and Poisson models we will discuss in the following chapters. Note, however, that we do not advocate at this stage the use of the normal distribution as a one-fits-all model: There exist many continuous situations where a normal model is inappropriate for many possible reasons (e.g., skewness, fat tails, dependence, multimodality).

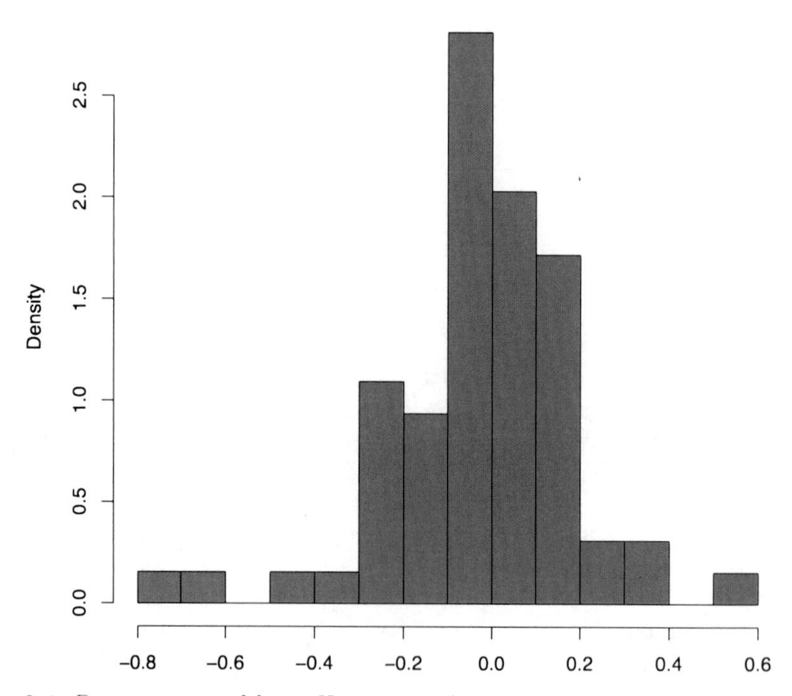

Fig. 2.1. Dataset **normaldata**: Histogram of the observed fringe shifts in Illingworth's 1927 experiment

Our normal dataset, **normaldata**, is linked with the famous Michelson–Morlay experiment that opened the way to Einstein's relativity theory in 1887. The experiment was intended to detect the "æther flow" and hence the existence of æther, this theoretical medium physicists postulated at this epoch was necessary to the transmission of light. Michelson's measuring device consisted in measuring the difference in the speeds of two light beams travelling the same distance in two orthogonal directions. As often in physics, the measurement was done by interferometry and differences in the travelling time inferred from shift in the fringes of the light spectrum. However, the experiment produced very small measurements that were not conclusive for the detection of the æther. Later experiments tried to achieve higher precision, as the one by Illingworth in 1927 used here as **normaldata**, only to obtain smaller and smaller upper bounds on the æther windspeed. While the original dataset is available in R as `morley`, the entries are approximated to the nearest multiple of ten and are therefore difficult to analyze as normal observations.

The 64 data points in **normaldata** are associated with session numbers (first column), corresponding to different times of the day, and the values in the second column represent the averaged fringe displacement due to orientation taken over ten measurements made by Illingworth, who assumed a normal error model. Figure 2.1 produces an histogram of the data by the simple R commands

```
> data(normaldata)
> shift=normaldata[,2]
> hist(shift,nclass=10,col="steelblue",prob=TRUE,main="")
```

This histogram seems compatible with a symmetric unimodal distribution such as the normal distribution. As shown in Fig. 2.2 by a qq-plot obtained by the commands

```
> qqnorm((shift-mean(shift))/sd(shift),pch=19,col="gold2")
> abline(a=0,b=1,lty=2,col="indianred",lwd=2)
```

which compare the empirical cdf with a pluggin normal estimate, The $\mathscr{N}(\mu, \sigma^2)$ fit may not be perfect, though, because of (a) a possible bimodality of the histogram and (b) potential outliers.

As mentioned above, the use of a normal distribution for modeling a given dataset is a convenient device that does not need to correspond to a perfect fit. With some degree of approximation, the normal distribution may agree with the data sufficiently to be used in place of the true distribution (if any). There exist, however, some setups where the normal distribution is thought to be the exact distribution behind the dataset (or where departure from normality has a significance for the theory behind the observations). In Marin and Robert (2007), we introduced a huge dataset related to the astronomical concept of

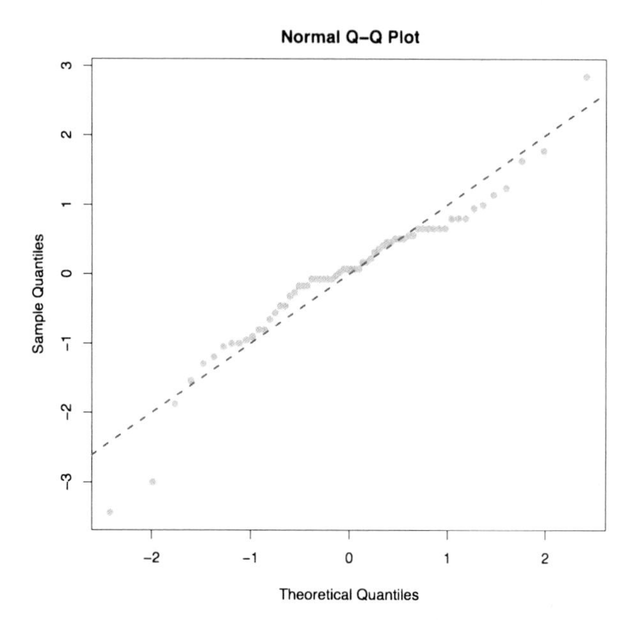

Fig. 2.2. Dataset **normaldata**: qq-plot of the observed fringe shifts against the normal quantiles

the cosmological background noise that illustrated this point, but chose not to reproduce the set in this edition due to the difficulty in handling it.

2.2 The Bayesian Toolkit

2.2.1 Posterior Distribution

Given an independent and identically distributed (later abbreviated as iid) sample $\mathscr{D}_n = (x_1, \ldots, x_n)$ from a density $f(x|\theta)$, depending upon an unknown parameter $\theta \in \Theta$, for instance the mean μ of the benchmark normal distribution, the associated likelihood function is

$$\ell(\theta|\mathscr{D}_n) = \prod_{i=1}^{n} f(x_i|\theta) \,. \tag{2.1}$$

This function of θ is a fundamental entity for the analysis of the information provided about θ by the sample $\mathscr{D}_n$, and Bayesian analysis relies on (2.1) to draw its inference on θ. For instance, when $\mathscr{D}_n$ is a normal $\mathscr{N}(\mu, \sigma^2)$ sample of size n and $\theta = (\mu, \sigma^2)$, we get

$$\ell(\theta|\mathscr{D}_n) = \prod_{i=1}^{n} \exp\{-(x_i - \mu)^2/2\sigma^2\}/\sqrt{2\pi}\sigma$$

$$\propto \exp\left\{-\sum_{i=1}(x_i-\mu)^2/2\sigma^2\right\}/\sigma^n$$

$$\propto \exp\left\{-\left(n\mu^2 - 2n\bar{x}\mu + \sum_{i=1} x_i^2\right)/2\sigma^2\right\}/\sigma^n$$

$$\propto \exp\left\{-\left[n(\mu-\bar{x})^2 + s^2\right]/2\sigma^2\right\}/\sigma^n,$$

where $\bar{x}$ denotes the empirical mean and where s^2 is the sum $\sum_{i=1}^n (x_i - \bar{x})^2$. This shows in particular that $\bar{x}$ and s^2 are sufficient statistics.

↯ In the above display of equations, the sign $\propto$ means *proportional to*. This proportionality is understood for functions of θ, meaning that the discarded constants do not depend on θ but may well depend on the data $\mathscr{D}_n$. This shortcut is both handy in complex Bayesian derivations and fraught with danger when considering several levels of parameters.

The major input of the Bayesian approach, compared with a traditional likelihood approach, is that it modifies the likelihood function into a *posterior* distribution, which is a valid probability distribution on Θ defined by the classical Bayes' formula (or theorem)

$$\pi(\theta|\mathscr{D}_n) = \frac{\ell(\theta|\mathscr{D}_n)\pi(\theta)}{\int \ell(\theta|\mathscr{D}_n)\pi(\theta)\,\mathrm{d}\theta}. \tag{2.2}$$

The factor $\pi(\theta)$ in (2.2) is called the *prior* and it obviously has to be chosen to start the analysis.

↯ The posterior density is a probability density on the parameter, which does not mean the parameter θ need be a genuine random variable. This density is used as an inferential tool, not as a truthful representation.

A first motivation for this approach is that the prior distribution summarizes the *prior information* on θ; that is, the knowledge that is available on θ *prior* to the observation of the sample $\mathscr{D}_n$. However, the choice of $\pi(\theta)$ is often decided on practical grounds rather than strong subjective beliefs or overwhelming prior information. A second motivation for the Bayesian construct is therefore to provide a fully probabilistic framework for the inferential analysis, with respect to a reference measure $\pi(\theta)$.

As an illustration, consider the simplest case of the normal distribution with known variance, $\mathscr{N}(\mu,\sigma^2)$. If the prior distribution on μ, $\pi(\mu)$, is the normal $\mathscr{N}(0,\sigma^2)$, the posterior distribution is easily derived via Bayes' theorem

$$\pi(\mu|\mathscr{D}_n) \propto \pi(\mu)\,\ell(\theta|\mathscr{D}_n)$$
$$\propto \exp\{-\mu^2/2\sigma^2\}\exp\left\{-n(\bar{x}-\mu)^2/2\sigma^2\right\}$$

$$\propto \exp\left\{-(n+1)\mu^2/2\sigma^2 + 2n\mu\bar{x}/2\sigma^2\right\}$$
$$\propto \exp\left\{-(n+1)[\mu - n\bar{x}/(n+1)]^2/2\sigma^2\right\},$$

which means that this posterior distribution in μ is a normal distribution with mean $n\bar{x}/(n+1)$ and variance $\sigma^2/(n+1)$. The mean (and mode) of the posterior is therefore different from the classical estimator $\bar{x}$, which may seem as a paradoxical feature of this Bayesian analysis. The reason for the difference is that the prior information that μ is close enough to zero is taken into account by the posterior distribution, which thus shrinks the original estimate towards zero. If we were given an alternative information that μ was close to ten, the posterior distribution would similarly shrink μ towards ten. The change from a factor n to a factor $(n+1)$ in the (posterior) variance is similarly explained by the prior information, in that accounting for this information reduces the variability of our answer.

For **normaldata**, we can first assume that the value of σ is the variability of the Michelson–Morley apparatus, namely 0.75. In that case, the posterior distribution on the fringe shift average μ is a normal $\mathscr{N}(n\bar{x}/(n+1), \sigma^2/(n+1))$ distribution, hence with mean and variance

```
> n=length(shift)
> mmu=sum(shift)/(n+1); mmu
[1] -0.01461538
> vmu=0.75^2/(n+1); vmu
[1] 0.008653846
```

represented on Fig. 2.3 as a dotted curve.

The case of a normal distribution with a known variance being quite unrealistic, we now consider the general case of an iid sample $\mathscr{D}_n = (x_1, \ldots, x_n)$ from the normal distribution $\mathscr{N}(\mu, \sigma^2)$ and $\theta = (\mu, \sigma^2)$. Keeping the same prior distribution $\mathscr{N}\left(0, \sigma^2\right)$ on μ, which then appears as a conditional distribution of μ given σ^2, *i.e.*, relies on the generic decomposition

$$\pi(\mu, \sigma^2) = \pi(\mu|\sigma^2)\pi(\sigma^2),$$

we have to introduce a further prior distribution on σ^2. To make computations simple at this early stage, we choose an exponential $\mathscr{E}(1)$ distribution on σ^{-2}. This means that the random variable $\omega = \sigma^{-2}$ is distributed from an exponential $\mathscr{E}(1)$ distribution, the distribution on σ^2 being derived by the usual change of variable technique,

$$\pi(\sigma^2) = \exp(-\sigma^{-2}) \left|\frac{\mathrm{d}\sigma^{-2}}{\mathrm{d}\sigma^2}\right| = \exp(-\sigma^{-2})\,(\sigma^2)^{-2}.$$

(This distribution is a special case of an inverse gamma distribution, namely $\mathcal{IG}(1,1)$.) The corresponding posterior density on θ is then given by

$$\begin{aligned}\pi((\mu,\sigma^2)|\mathscr{D}_n) &\propto \pi(\sigma^2)\times\pi(\mu|\sigma^2)\times\ell((\mu,\sigma^2)|\mathscr{D}_n)\\ &\propto (\sigma^{-2})^{1/2+2}\exp\left\{-(\mu^2+2)/2\sigma^2\right\}\\ &\quad\times(\sigma^{-2})^{n/2}\exp\left\{-\left(n(\mu-\overline{x})^2+s^2\right)/2\sigma^2\right\}\\ &\propto (\sigma^2)^{-(n+5)/2}\exp\left\{-\left[(n+1)(\mu-n\bar{x}/(n+1))^2+(2+s^2)\right]/2\sigma^2\right\}\\ &\propto (\sigma^2)^{-1/2}\exp\left\{-(n+1)[\mu-n\bar{x}/(n+1)]^2/2\sigma^2\right\}.\\ &\quad\times(\sigma^2)^{-(n+2)/2-1}\exp\left\{-(2+s^2)/2\sigma^2\right\}.\end{aligned}$$

Therefore, the posterior on θ can be decomposed as the product of an inverse gamma distribution on σ^2, $\mathscr{IG}((n+2)/2,[2+s^2]/2)$—which is the distribution of the inverse of a gamma $\mathscr{G}((n+2)/2,[2+s^2]/2)$ random variable—and, conditionally on σ^2, a normal distribution on μ, $\mathscr{N}(n\bar{x}/(n+1),\sigma^2/(n+1))$. The interpretation of this posterior is quite similar to the case when σ is known, with the difference that the variability in σ induces more variability in μ, the marginal posterior in μ being then a Student's t distribution[1] (Exercise 2.1)

$$\mu|\mathscr{D}_n \sim \mathscr{T}\left(n+2, n\bar{x}/(n+1), (2+s^2)/(n+1)(n+2)\right),$$

with $n+2$ degrees of freedom, a location parameter proportional to $\bar{x}$ and a scale parameter (almost) proportional to s.

For **normaldata**, an $\mathscr{E}xp(1)$ prior on σ^{-2} being compatible with the value observed on the Michelson–Morley experiment, the parameters of the t distribution on μ are therefore $n=64$,

```
> mtmu=sum(shift)/(n+1);mtmu
[1] -0.01461538
> stmu=(2+(n-1)*var(shift))/((n+2)*(n+1));stmu
[1] 0.0010841496
```

We compare the resulting posterior with the one based on the assumption $\sigma=0.75$ on Fig. 2.3, using the `curve` commands (note that the `mnormt` library may require the preliminary installation of the corresponding package by `install.packages("mnormt")`):

```
> library(mnormt)
> curve(dmt(x,mean=mmu,S=stmu,df=n+2),col="chocolate2",lwd=2,
+ xlab="x",ylab="",xlim=c(-.5,.5))
> curve(dnorm(x,mean=mmu,sd=sqrt(vmu)),col="steelblue2",
+ lwd=2,add=TRUE,lty=2)
```

[1] We will omit the reference to Student in the subsequent uses of this distribution, as is the rule in anglo-saxon textbooks.

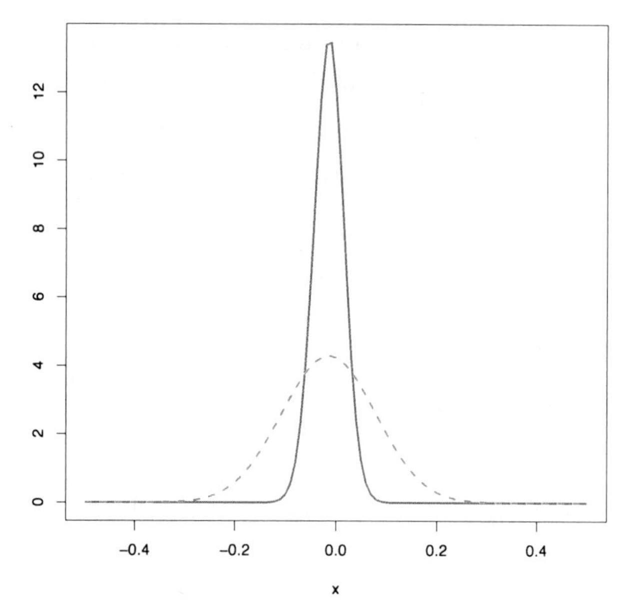

Fig. 2.3. Dataset **normaldata**: Two posterior distributions on μ corresponding to an hypothetical $\sigma = 0.75$ *(dashed lines)* and to an unknown σ^2 under the prior $\sigma^{-2} \sim \mathscr{E}(1)$ *(plain lines)*

Although this may sound counterintuitive, in this very case, estimating the variance produces a reduction in the variability of the posterior distribution on μ. This is because the postulated value of σ^2 is actually inappropriate for Illingworth's experiment, being far too large. Since the posterior distribution on σ^2 is an $\mathscr{IG}(33, 1.82)$ distribution for **normaldata**, the probability that σ is as large as 0.75 can be evaluated as

```
> digmma=function(x,shape,scale){dgamma(1/x,shape,scale)/x^2}
> curve(digmma(x,shape=33,scale=(1+(n+1)*var(shift))/2),
+ xlim=c(0,.2),lwd=2)
> pgamma(1/(.75)^2,shape=33,scale=(1+(n+1)*var(shift))/2)
[1] 8.99453e-39
```

which shows that 0.75 is quite unrealistic, being ten times as large as the mode of the posterior density on σ^2.

The above R command `library(mnormt)` calls the `mnormt` library, which contains useful additional functions related with multivariate normal and t distributions. In particular, `dmt` allows for location and scale parameters in the t distribution. Note also that s^2 is computed as `(n-1)*var(shift)` because R implicitly adopts a classical approach in using the "best unbiased estimator" of σ^2.

2.2.2 Bayesian Estimates

A concept that is at the core of Bayesian analysis is that one should provide an inferential assessment *conditional on the realized value of* $\mathscr{D}_n$. Bayesian analysis gives a proper probabilistic meaning to this conditioning by allocating to θ a probability distribution. Once the prior distribution is selected, Bayesian inference formally is "over"; that is, it is completely determined since the estimation, testing, and evaluation procedures are automatically provided by the prior and the way procedures are compared (or penalized). For instance, if estimations $\hat{\theta}$ of θ are compared via the sum of squared errors,

$$\mathrm{L}(\theta,\hat{\theta}) = ||\theta - \hat{\theta}||^2 ,$$

the corresponding Bayes optimum is the *expected* value of θ under the posterior distribution,[2]

$$\hat{\theta} = \int \theta\, \pi(\theta|\mathscr{D}_n)\, \mathrm{d}\theta = \frac{\int \theta\, \ell(\theta|\mathscr{D}_n)\, \pi(\theta)\, \mathrm{d}\theta}{\int \ell(\theta|\mathscr{D}_n)\, \pi(\theta)\, \mathrm{d}\theta} , \tag{2.3}$$

for a given sample $\mathscr{D}_n$.

When no specific penalty criterion is available, the estimator (2.3) is often used as a default estimator, although alternatives are also available. For instance, the *maximum a posteriori estimator* (MAP) is defined as

$$\hat{\theta} = \arg\max_\theta \pi(\theta|\mathscr{D}_n) = \arg\max_\theta \pi(\theta)\ell(\theta|\mathscr{D}_n), \tag{2.4}$$

where the function to maximize is usually provided in closed form. However, numerical problems often make the optimization involved in finding the MAP far from trivial. Note also here the similarity of (2.4) with the maximum likelihood estimator (MLE): The influence of the prior distribution $\pi(\theta)$ on the estimate progressively disappears as the number of observations n increases, and the MAP estimator often recovers the asymptotic properties of the MLE.

For **normaldata**, since the posterior distribution on σ^{-2} is a $\mathscr{G}(32, 1.82)$ distribution, the posterior expectation of σ^{-2} given Illingworth's experimental data is $32/1.82 = 17.53$. The posterior expectation of σ^2 requires a supplementary effort in order to derive the mean of an inverse gamma distribution (see Exercise 2.2), namely

$$\mathbb{E}^\pi[\sigma^2|\mathscr{D}_n] = 1.82/(33-1) = 0.057 .$$

[2]Estimators are functions of the data $\mathscr{D}_n$, while estimates are values taken by those functions. In most cases, we will denote them with a "hat" symbol, the dependence on $\mathscr{D}_n$ being implicit.

Similarly, the MAP estimate is given here by

$$\arg\max_\theta \pi(\sigma^2|\mathscr{D}_n) = 1.82/(33+1) = 0.054$$

(see also Exercise 2.2). These values therefore reinforce our observation that the Michelson–Morley precision is not appropriate for the Illingworth experiment, which is much more precise indeed.

2.2.3 Conjugate Prior Distributions

The selection of the prior distribution is an important issue in Bayesian statistics. When prior information is available about the data or the model, it can (and must) be used in building the prior, and we will see some implementations of this recommendation in the following chapters. In many situations, however, the selection of the prior distribution is quite delicate, due to the absence of reliable prior information, and default solutions must be chosen instead. Since the choice of the prior distribution has a considerable influence on the resulting inference, this inferential step must be conducted with the utmost care.

From a computational viewpoint, the most convenient choice of prior distributions is to mimic the likelihood structure within the prior. In the most advantageous cases, priors and posteriors remain within the same parameterized family. Such priors are called *conjugate*. While the foundations of this principle are too advanced to be processed here (see, e.g., Robert, 2007, Chap. 3), such priors exist for most usual families, including the normal distribution. Indeed, as seen in Sect. 2.2.1, when the prior on a normal mean is normal, the corresponding posterior is also normal.

Since conjugate priors are such that the prior and posterior densities belong to the same parametric family, using the observations boils down to an update of the parameters of the prior. To avoid confusion, the parameters involved in the prior distribution on the model parameter are usually called *hyperparameters*. (They can themselves be associated with prior distributions, then called *hyperpriors*.)

For most practical purposes, it is sufficient to consider the conjugate priors described in Table 2.1. The derivation of each row is straightforward if painful and proceeds from the same application of Bayes' formula as for the normal case above (Exercise 2.5). For distributions that are not within this table, a conjugate prior may or may not be available (Exercise 2.6).

An important feature of conjugate priors is that one has a priori to select two hyperparameters, e.g., a mean and a variance in the normal case. On the one hand, this is an advantage when using a conjugate prior, namely that one has to select only a few parameters to determine the prior distribution. On the other hand, this is a drawback in that the information known a priori on μ may be either insufficient to determine both parameters or incompatible with the structure imposed by conjugacy.

Table 2.1. Conjugate priors for the most common statistical families

$f(x\|\theta)$	$\pi(\theta)$	$\pi(\theta\|x)$
Normal $\mathscr{N}(\theta,\sigma^2)$	Normal $\mathscr{N}(\mu,\tau^2)$	$\mathscr{N}(\rho(\sigma^2\mu+\tau^2 x),\rho\sigma^2\tau^2)$ $\rho^{-1}=\sigma^2+\tau^2$
Poisson $\mathscr{P}(\theta)$	Gamma $\mathscr{G}(\alpha,\beta)$	$\mathscr{G}(\alpha+x,\beta+1)$
Gamma $\mathscr{G}(\nu,\theta)$	Gamma $\mathscr{G}(\alpha,\beta)$	$\mathscr{G}(\alpha+\nu,\beta+x)$
Binomial $\mathscr{B}(n,\theta)$	Beta $\mathscr{B}e(\alpha,\beta)$	$\mathscr{B}e(\alpha+x,\beta+n-x)$
Negative Binomial $\mathscr{N}eg(m,\theta)$	Beta $\mathscr{B}e(\alpha,\beta)$	$\mathscr{B}e(\alpha+m,\beta+x)$
Multinomial $\mathscr{M}_k(\theta_1,\ldots,\theta_k)$	Dirichlet $\mathscr{D}(\alpha_1,\ldots,\alpha_k)$	$\mathscr{D}(\alpha_1+x_1,\ldots,\alpha_k+x_k)$
Normal $\mathscr{N}(\mu,1/\theta)$	Gamma $\mathscr{G}(\alpha,\beta)$	$\mathscr{G}(\alpha+0.5,\beta+(\mu-x)^2/2)$

2.2.4 Noninformative Priors

There is no compelling reason to choose conjugate priors as our priors, except for their simplicity, but the restrictive aspect of conjugate priors can be attenuated by using *hyperpriors* on the hyperparameters themselves, although we will not deal with this additional level of complexity in the current chapter. The core message is therefore that conjugate priors are nice to work with, but require a hyperparameter determination that may prove awkward in some settings and that may moreover have a lasting impact on the resulting inference.

Instead of using conjugate priors, one can opt for a completely different perspective and rely on so-called *noninformative* priors that aim at attenuating the impact of the prior on the resulting inference. These priors are fundamentally defined as coherent extensions of the uniform distribution. Their purpose is to provide a reference measure that has as little as possible bearing on the inference (relative to the information brought by the likelihood). We first warn the reader that, for unbounded parameter spaces, the densities of noninformative priors actually fail to integrate to a finite number and they are defined instead as positive measures. While this sounds like an invalid extension of the probabilistic framework, it is quite correct to define the corresponding posterior distributions by (2.2), as long as the integral in the denominator is finite (almost surely). A more detailed account is for instance provided in Robert (2007, Sect. 1.5) about this possibility of using σ-finite measures (sometimes called *improper* priors) in settings where true probability prior distributions are too difficult to come by or too subjective to be accepted by all. For instance, *location models*

$$x \sim p(x - \theta)$$

are usually associated with flat priors $\pi(\theta) = 1$ (note that these models include the normal $\mathscr{N}(\theta, 1)$ as a special case), while *scale models*

$$x \sim \frac{1}{\theta} f\left(\frac{x}{\theta}\right)$$

are usually associated with the log-transform of a flat prior, that is,

$$\pi(\theta) = 1/\theta \, .$$

In a more general setting, the (noninformative) prior favored by most Bayesians is the so-called *Jeffreys prior*,[3] which is related to the Fisher information matrix

$$I^F(\theta) = \text{var}_\theta \left(\frac{\partial \log f(X|\theta)}{\partial \theta} \right)$$

by

$$\pi^J(\theta) = \left| I^F(\theta) \right|^{1/2} ,$$

where $|I|$ denotes the determinant of the matrix I.

Since the mean μ of a normal model is a location parameter, when the variance σ^2 is known, the standard choice of noninformative parameter is an arbitrary constant $\pi(\mu)$ (taken to be 1 by default). Given that this flat prior formally corresponds to the limiting case $\tau = \infty$ in the conjugate normal prior, it is easy to verify that this noninformative prior is associated with the posterior distribution $\mathscr{N}(x, 1)$, which happens to be the likelihood function in that case. An interesting consequence of this observation is that the MAP estimator is also the maximum likelihood estimator in that (special) case. For the general case when $\theta = (\mu, \sigma^2)$, the Fisher information matrix leads to the Jeffreys prior $\pi^J(\theta) = 1/\sigma^3$ (Exercise 2.4). The corresponding posterior distribution on (μ, σ^2) is then

$$\begin{aligned} \pi((\mu,\sigma^2)|\mathscr{D}_n) &\propto (\sigma^{-2})^{(3+n)/2} \exp\left\{ - \left(n(\mu - \overline{x})^2 + s^2 \right) / 2\sigma^2 \right\} \\ &\propto \sigma^{-1} \exp\left\{ -n(\mu - \bar{x})^2 / 2\sigma^2 \right\} \times (\sigma^2)^{-(n+2)/2} \exp\left\{ \frac{-s^2}{2\sigma^2} \right\} , \end{aligned}$$

that is,

$$\theta \sim \mathscr{N}\left(\bar{x}, \sigma^2/n\right) \times \mathscr{IG}\left(n/2, s^2/2\right) .$$

a product of a conditional normal on μ by an inverse gamma on σ^2. Therefore the marginal posterior distribution on μ is a t distribution (Exercise 2.1)

$$\mu|\mathscr{D}_n \sim \mathscr{T}\left(n, \bar{x}, s^2/n^2)\right) .$$

---

[3]Harold Jeffreys was an English geophysicist who developed and formalized Bayesian methods in the 1930s in order to analyze geophysical data. He ended up writing an influential treatise on Bayesian statistics entitled *Theory of Probability*.

For **normaldata**, the difference in Fig. 2.3 between the noninformative solution and the conjugate posterior is minor, but it expresses that the prior distribution $\mathscr{E}(1)$ on σ^{-2} is not very appropriate for the Illingworth experiment, since it does not put enough prior weight on the region of importance, i.e. near 0.05. As a result, the most concentrated posterior is (seemingly paradoxically) the one associated with the noninformative prior!

↯ A major (and potentially dangerous) difference between proper and improper priors is that the posterior distribution associated with an improper prior is not necessarily defined, that is, it may happen that

$$\int \pi(\theta)\ell(\theta|\mathscr{D}_n)\,\mathrm{d}\theta < \infty \tag{2.5}$$

does not hold. In some cases, this difficulty disappears when the sample size is large enough. In others (see Chap. 6), it may remain whatever the sample size. But the main thing is that, when using improper priors, condition (2.5) must always be checked.

2.2.5 Bayesian Credible Intervals

One point that must be clear from the beginning is that the Bayesian approach is a complete inferential approach. Therefore, it covers confidence evaluation, testing, prediction, model checking, and point estimation. We will progressively cover the different facets of Bayesian analysis in other chapters of this book, but we address here the issue of confidence intervals because it is rather a straightforward step from point estimation.

As with everything else, the derivation of the confidence intervals (or confidence regions in more general settings) is based on the posterior distribution $\pi(\theta|\mathscr{D}_n)$. Since the Bayesian approach processes θ as a random variable, a natural definition of a confidence region on θ is to determine $C(\mathscr{D}_n)$ such that

$$\pi(\theta \in C(\mathscr{D}_n)|\mathscr{D}_n) = 1 - \alpha \tag{2.6}$$

where α is a predetermined level such as 0.05.[4]

The important difference with a traditional perspective in (2.6) is that the integration is done over the parameter space, rather than over the observation space. The quantity $1 - \alpha$ thus corresponds to the probability that a random θ belongs to this set $C(\mathscr{D}_n)$, rather than to the probability that the random set contains the "true" value of θ. Given this drift in the interpretation of a

[4] There is nothing special about 0.05 when compared with, say, 0.87 or 0.12. It is just that the famous 5 % level is accepted by most as an acceptable level of error. If the context of the analysis tells a different story, another value for α (including one that may even depend on the data) should be chosen!

confidence set (rather called a *credible set* by Bayesians), the determination of the best[5] credible set turns out to be easier than in the classical sense: indeed, this set simply corresponds to the values of θ with the highest posterior values,

$$C(\mathscr{D}_n) = \{\theta;\, \pi(\theta|\mathscr{D}_n) \geq k_\alpha\}\ ,$$

where k_α is determined by the coverage constraint (2.6). This region is called the *highest posterior density* (HPD) region.

For **normaldata**, since the marginal posterior distribution on μ associated with the Jeffreys prior is the t distribution, $\mathscr{T}(n, \bar{x}, s^2/n^2)$,

$$\pi(\mu|\mathscr{D}_n) \propto \left[n(\mu - \bar{x})^2 + s^2\right]^{-(n+1)/2}$$

with $n = 64$ degrees of freedom. Therefore, due to the symmetry properties of the t distribution, the 95 % credible interval on μ is centered at $\bar{x}$ and its range is derived from the `0.975` quantile of the t distribution with n degrees of freedom,

```
> qt(.975,df=n)*sqrt((n-1)*var(shift)/n^2)
[1] 0.05082314
```

since the `mnormt` package does not compute quantiles. The resulting confidence interval is therefore given by

```
> qt(.975,df=n)*sqrt((n-1)*var(shift)/n^2)+mean(shift)
[1] 0.03597939
> -qt(.975,df=n)*sqrt((n-1)*var(shift)/n^2)+mean(shift)
[1] -0.06566689
```

i.e. equal to $[-0.066, 0.036]$. In conclusion, the value 0 belongs to this credible interval on μ and this (noninformative) Bayesian analysis of **normaldata** shows that, indeed, the absence of æther wind is not infirmed by Illingworth's experiment.

↯ While the shape of an optimal Bayesian confidence set is easily derived, the computation of either the bound k_α or the set $C(\mathscr{D}_n)$ may be too challenging to allow an analytic construction outside conjugate setups (see Exercise 2.11).

2.3 Bayesian Model Choice

Deciding the validity of some assumptions or restrictions on the parameter θ is a major part of the statistician's job. In classical statistics, this type of

[5]In the sense of producing the smallest possible volume with a given coverage.

problems goes under the name of *hypothesis testing*, following the framework set by Fisher, Neyman and Pearson in the 1930s. Hypothesis testing considers a decision problem where an hypothesis is either true or false and where the answer provided by the statistician is also a statement whether or not the hypothesis is true. However, we deem this approach to be too formalized—even though it can be directly reproduced from a Bayesian perspective, as shown in Robert (2007, Chap. 5)—, we strongly favour a model choice philosophy, namely that two or more models are proposed in parallel and assessed in terms of their respective fits of the data. This view acknowledges the fact that models are at best approximations of reality and it does not aim at finding a "true model", as hypothesis testing may do. In this book, we will thus follow the later approach and take the stand that inference problems expressed as hypothesis testing by the classical statisticians are in fact comparisons of different models. In terms of numerical outcomes, both perspectives—Bayesian hypothesis testing vs. Bayesian model choice—are exchangeable but we already warn the reader that, while the Bayesian solution is formally very close to a likelihood (ratio) statistic, its numerical values often strongly differ from the classical solutions.

2.3.1 The Model Index as a Parameter

The essential novelty when dealing with the comparison of models is that this issue makes the model itself an unknown quantity of interest. Therefore, if we are comparing two or more models with indices $k = 1, 2, \ldots, J$, we introduce a model indicator $\mathfrak{M}$ taking values in $\{1, 2, \ldots, J\}$ and representing the index of the "true" model. If $\mathfrak{M} = k$, then the data $\mathscr{D}_n$ are generated from a statistical model M_k with likelihood $\ell(\theta_k|\mathscr{D}_n)$ and parameter θ_k taking its value in a parameter space Θ_k. An obvious illustration is when opposing two standard parametric families, e.g., a normal family against a t family, in which case $J = 2$, $\Theta_1 = \mathbb{R} \times \mathbb{R}_+^*$—for mean and variance—and $\Theta_2 = \mathbb{R}_+^* \times \mathbb{R} \times \mathbb{R}_+^*$—for degree of freedom, mean and variance—, but this framework also includes soft or hard constraints on the parameters, as for instance imposing that a mean μ is positive.

In this setting, a natural Bayes procedure associated with a prior distribution π is to consider the posterior probability

$$\delta^\pi(\mathscr{D}_n) = \mathbb{P}^\pi(\mathfrak{M} = k|\mathscr{D}_n)\,,$$

i.e., the posterior probability that the model index is k, and select the index of the model with the highest posterior probability as the model preferred by the data $\mathscr{D}_n$. This representation implies that the prior π is defined over the collection of model indices, $\{1, 2, \ldots, J\}$, and, conditionally on the model index $\mathfrak{M}$, on the corresponding parameter space, Θ_k. This construction may sound both artificial and incomplete, as there is no prior on the parameter θ_k unless $\mathfrak{M} = k$, but it nonetheless perfectly translates the problem at hand:

inference on θ_k is meaningless unless this is the parameter of the correct model. Furthermore, the quantity of interest integrates out the parameter, since

$$\mathbb{P}^\pi(\mathfrak{M}=k|\mathscr{D}_n) = \frac{\mathbb{P}^\pi(\mathfrak{M}=k)\int \ell(\theta_k|\mathscr{D}_n)\pi_k(\theta_k)\,\mathrm{d}\theta_k}{\sum_{j=1}^J \mathbb{P}^\pi(\mathfrak{M}=j)\pi_j(\theta_j)\,\mathrm{d}\theta_j}\,.$$

↯ We believe it is worth emphasizing the above point: A parameter θ_k associated with a model does not have a statistical meaning outside this model. This means in particular that the notion of parameters "common to all models" often found in the literature, including the Bayesian literature, is not acceptable within a model choice perspective. Two models must have distinct parameters, if only because the purpose of the analysis is to end up with a single model.

The choice of the prior π is highly dependent on the value of the prior model probabilities $\mathbb{P}^\pi(\mathfrak{M}=k)$. In some cases, there is experimental or subjective evidence about those probabilities, but in others, we are forced to settle for equal weights $\mathbb{P}^\pi(\mathfrak{M}=k)=1/J$. For instance, given a single observation $x\sim\mathcal{N}(\mu,\sigma^2)$ from a normal model where σ^2 is known, assuming $\mu\sim\mathcal{N}(\xi,\tau^2)$, the posterior distribution $\pi(\mu|x)$ is the normal distribution $\mathcal{N}(\xi(x),\omega^2)$ with

$$\xi(x)=\frac{\sigma^2\xi+\tau^2 x}{\sigma^2+\tau^2} \quad \text{and} \quad \omega^2=\frac{\sigma^2\tau^2}{\sigma^2+\tau^2}.$$

If the question of interest is to decide whether μ is negative or positive, we can directly compute

$$\begin{aligned}\mathbb{P}^\pi(\mu<0|x) &= \mathbb{P}^\pi\left(\frac{\mu-\xi(x)}{\omega}<\frac{-\xi(x)}{\omega}\right)\\ &= \Phi\left(-\xi(x)/\omega\right)\,,\end{aligned} \tag{2.7}$$

where Φ is the normal cdf. This computation does not seem to follow from the principles we just stated but it is only a matter of perspective as we can derive the priors on both models from the original prior. Deriving this posterior probability indeed means that, a priori, μ is negative with probability $\mathbb{P}^\pi(\mu<0)=\Phi(-\xi/\tau)$ and that, in this model, the prior on μ is the truncated normal

$$\pi_1(\mu)=\frac{\exp\{-(\mu-\xi)^2/2\tau^2\}}{\sqrt{2\pi}\tau\Phi(-\xi/\tau)}\,\mathbb{I}_{\mu<0}\,,$$

while μ is positive with probability $\Phi(\xi/\tau)$ and, in this second model, the prior on μ is the truncated normal

$$\pi_2(\mu)=\frac{\exp\{-(\mu-\xi)^2/2\tau^2\}}{\sqrt{2\pi}\tau\Phi(\xi/\tau)}\,\mathbb{I}_{\mu>0}\,.$$

The posterior probability of $\mathbb{P}^\pi(\mathfrak{M}=k|\mathscr{D}_n)$ is the core object in Bayesian model choice and, as indicated above, the default procedure is to select the model with the highest posterior probability. However, in decisional settings where the choice between two models has different consequences depending on the value of k, the boundary in $\mathbb{P}^\pi(\mathfrak{M} = k|\mathscr{D}_n)$ between choosing one model and the other may be far from 0.5. For instance, in a pharmaceutical trial, deciding to start production of a new drug does not have the same financial impact as deciding to run more preliminary tests. Changing the bound away from 0.5 is in fact equivalent to changing the prior probabilities of both models.

2.3.2 The Bayes Factor

A notion central to Bayesian model choice is the *Bayes factor*

$$B_{21}^\pi(\mathscr{D}_n) = \frac{\mathbb{P}^\pi(\mathfrak{M} = 2|\mathscr{D}_n)/\mathbb{P}^\pi(\mathfrak{M} = 1|\mathscr{D}_n)}{\mathbb{P}^\pi(\mathfrak{M} = 2)/\mathbb{P}^\pi(\mathfrak{M} = 1)},$$

which corresponds to the classical odds or likelihood ratio, the difference being that the parameters are integrated rather than maximized under each model. While this quantity is a simple one-to-one transform of the posterior probability, it can be used for Bayesian model choice without first resorting to a determination of the prior weights of both models. Obviously, the Bayes factor depends on prior information through the choice of the model priors π_1 and π_2,

$$B_{21}^\pi(\mathscr{D}_n) = \frac{\int_{\Theta_2} \ell_2(\theta_2|\mathscr{D}_n)\pi_2(\theta_2)\,\mathrm{d}\theta_2}{\int_{\Theta_1} \ell_1(\theta_1|\mathscr{D}_n)\pi_1(\theta_1)\,\mathrm{d}\theta_1} = \frac{m_2(\mathscr{D}_n)}{m_1(\mathscr{D}_n)},$$

and thus it can clearly be perceived as a Bayesian likelihood ratio which replaces the likelihoods with the marginals under both models.

The evidence brought by the data $\mathscr{D}_n$ can be calibrated using for instance Jeffreys' scale of evidence:

- if $\log_{21}(B_{21}^\pi)$ is between 0 and 0.5, the evidence against model M_1 is *weak*,
- if it is between 0.5 and 1, it is *substantial*,
- if it is between 1 and 2, it is *strong*, and
- if it is above 2, it is *decisive*.

While this scale is purely arbitrary, it provides a reference for model assessment in a generic setting.

Consider now the special case when we want to assess whether or not a specific value of one of the parameters is appropriate, for instance $\mu = 0$ in the **normaldata** example. While the classical literature presents this problem as *a point null hypothesis*, we simply interpret it as the comparison of two models, $\mathscr{N}(0, \sigma^2)$ and $\mathscr{N}(\mu, \sigma^2)$, for Illingworth's data. In a more general framework, when the sample $\mathscr{D}_n$ is distributed as $\mathscr{D}_n \sim f(\mathscr{D}_n|\theta)$, if we decompose θ as $\theta = (\delta, \omega)$ and if the restricted model corresponds to the fixed value $\delta = \delta_0$, we define $\pi_1(\omega)$ as the prior under the restricted model (labelled M_1) and $\pi_2(\theta)$

as the prior under the unrestricted model (labelled M_2). The corresponding Bayes factor is then

$$B_{21}^{\pi}(\mathscr{D}_n) = \frac{\int_{\Theta} \ell(\theta|\mathscr{D}_n)\pi_2(\theta)\,\mathrm{d}\theta}{\int_{\Omega} \ell((\delta_0,\omega)|\mathscr{D}_n)\pi_1(\omega)\,\mathrm{d}\omega}$$

Note that, as hypotheses, point null problems often are criticized as artificial and impossible to test (in the sense of *how often can one distinguish* $\theta = 0$ *from* $\theta = 0.0001$ *?!*), but, from a model choice perspective, they simply correspond to more parsimonious models whose fit to the data can be checked against the fit produced by an unconstrained model. While the unconstrained model obviously contains values that produce a better fit, averaging over the whole parameter space Θ may still result in a small integrated likelihood $m_2(\mathscr{D}_n)$. The Bayes factor thus contains an automated penalization for complexity, a feature missed by the classical likelihood ratio statistic.

↯ In the very special case when the whole parameter is constrained to a fixed value, $\theta = \theta_0$, the marginal likelihood under model M_1 coincides with the likelihood $\ell(\theta_0|\mathscr{D}_n) = f(\mathscr{D}_n|\theta_0)$ and the Bayes factor simplifies in

$$B_{21}^{\pi}(\mathscr{D}_n) = \frac{\int_{\Theta} f(\mathscr{D}_n|\theta)\pi_2(\theta)\,\mathrm{d}\theta}{f(\mathscr{D}_n|\theta_0)}\,.$$

For $x \sim \mathscr{N}(\mu, \sigma^2)$ and σ^2 known, consider assessing $\mu = 0$ when $\mu \sim \mathscr{N}(0, \tau^2)$ under the alternative model (labelled M_2). The Bayes factor is the ratio

$$\begin{aligned} B_{21}^{\pi}(\mathscr{D}_n) &= \frac{m_2(x)}{f(x|(0,\sigma^2))} \\ &= \frac{\sigma}{\sqrt{\sigma^2+\tau^2}} \frac{e^{-x^2/2(\sigma^2+\tau^2)}}{e^{-x^2/2\sigma^2}} \\ &= \sqrt{\frac{\sigma^2}{\sigma^2+\tau^2}} \exp\left\{\frac{\tau^2 x^2}{2\sigma^2(\sigma^2+\tau^2)}\right\}\,. \end{aligned}$$

Table 2.2 gives a sample of the values of the Bayes factor when the normalized quantity x/σ varies. They obviously depend on the choice of the prior variance τ^2 and the dependence is actually quite severe, as we will see below with the *Jeffreys–Lindley paradox*.

For **normaldata**, since we saw that setting σ to the Michelson–Morley value of 0.75 was producing a poor outcome compared with the noninformative solution, the comparison between the constrained and the unconstrained models is not very trustworthy, but as an illustration, it gives the following values:

Table 2.2. Bayes factor $B_{21}(z)$ against the null hypothesis $\mu = 0$ for different values of $z = x/\sigma$ and τ

z	0	0.68	1.28	1.96
$\tau^2 = \sigma^2$	0.707	0.794	1.065	1.847
$\tau^2 = 10\sigma^2$	0.302	0.372	0.635	1.728

```
> BaFa=function(z,rat){
#rat denotes the ratio tau^2/sigma^2
sqrt(1/(1+rat))*exp(z^2/(2*(1+1/rat)))}
> BaFa(mean(shift),1)
[1] 0.7071767
> BaFa(mean(shift),10)
[1] 0.3015650
```

which supports the constraint $\mu = 0$ for those two values of τ, since the Bayes factor is less than 1. (For this dataset, the Bayes factor is always less than one, see Exercise 2.13.)

2.3.3 The Ban on Improper Priors

We introduced noninformative priors in Sect. 2.2.4 as a way to handle situations when the prior information was not sufficient to build proper priors. We also saw that, for **normaldata**, a noninformative prior was able to exhibit conflicts between the prior information (based on the Michelson–Morley experiment) and the data (resulting from Illingworth's experiment). Unfortunately, the use of noninformative priors is very much restricted in model choice settings because the fact that they usually are improper leads to the impossibility of comparing the resulting marginal likelihoods.

Looking at the expression of the Bayes factor,

$$B_{21}^{\pi}(\mathscr{D}_n) = \frac{\int_{\Theta_2} \ell_2(\theta_2|\mathscr{D}_n)\pi_2(\theta_2)\,\mathrm{d}\theta_2}{\int_{\Theta_1} \ell_1(\theta_1|\mathscr{D}_n)\pi_1(\theta_1)\,\mathrm{d}\theta_1},$$

it is clear that, when either π_1 or π_2 are improper, it is impossible to normalize the improper measures in a unique manner. Therefore, the Bayes factor becomes completely arbitrary since it can be multiplied by one or two arbitrary constants.

For instance, when comparing $x \sim \mathcal{N}(\mu, 1)$ (model M_1) with $x \sim \mathcal{N}(0, 1)$ (model M_2), the improper Jeffreys prior on model M_1 is $\pi_1(\mu) = 1$. The Bayes factor corresponding to this choice is

$$B_{12}^{\pi}(x) = \frac{e^{-x^2/2}}{\int_{-\infty}^{+\infty} e^{-(x-\theta)^2/2}\,\mathrm{d}\theta} = \frac{e^{-x^2/2}}{\sqrt{2\pi}}.$$

If, instead, we use the prior $\pi_1(\mu) = 100$, the Bayes factor becomes

$$B_{12}^{\pi}(x) = \frac{e^{-x^2/2}}{100 \int_{-\infty}^{+\infty} e^{-(x-\theta)^2/2}\,\mathrm{d}\theta} = \frac{e^{-x^2/2}}{100\sqrt{2\pi}}$$

and is thus one-hundredth of the previous value! Since there is no mathematical way to discriminate between $\pi_1(\mu) = 1$ and $\pi_1(\mu) = 100$, the answer clearly is non-sensical.

Note that, if we are instead comparing model M_1 where $\mu \leq 0$ and model M_2 where $\mu > 0$, then the posterior probability of model M_1 under the flat prior is

$$\mathbb{P}^{\pi}(\mu \leq 0|x) = \frac{1}{\sqrt{2\pi}} \int_{-\infty}^{0} e^{-(x-\theta)^2/2}\,\mathrm{d}\theta = \Phi(-x)\,,$$

which is uniquely defined.

The difficulty in using an improper prior also relates to what is called the *Jeffreys–Lindley paradox*, a phenomenon that shows that limiting arguments are not valid in testing settings. In contrast with estimation settings, the non-informative prior no longer corresponds to the limit of conjugate inferences. For instance, for the comparison of the normal $x \sim \mathcal{N}(\mu, \sigma^2)$ (model M_1) and of the normal $x \sim \mathcal{N}(\mu, \sigma^2)$ (model M_2) models when σ^2 is known, using a conjugate prior $\mu \sim \mathcal{N}(0, \tau^2)$, the Bayes factor

$$B_{21}^{\pi}(x) = \sqrt{\frac{\sigma^2}{\sigma^2 + \tau^2}} \exp\left[\frac{\tau^2 x^2}{2\sigma^2(\sigma^2 + \tau^2)}\right]$$

converges to 0 when τ goes to $+\infty$, for *every* value of x, again a non-sensical procedure.

Since improper priors are an essential part of the Bayesian approach, there are many proposals found in the literature to overcome this ban. Most of those proposals rely on a device that transforms the improper prior into a proper probability distribution by exploiting a fraction of the data $\mathscr{D}_n$ and then restricts itself to the remaining part of the data to run the test as in a standard situation. The variety of available solutions is due to the many possibilities of removing the dependence on the choice of the portion of the data used in the first step. The resulting procedures are called *pseudo-Bayes factors*, although some may actually correspond to true Bayes factors. See Robert (2007, Chap. 5) for more details, although we do not advocate using those procedures.

There is a major exception to this ban on improper priors that we can exploit. If both models under comparison have parameters that have similar enough meanings to share the same prior distribution, as for instance a measurement error σ^2, then the normalization issue vanishes. Note that we are not assuming that parameters are *common* to both models and thus that we do not contradict the earlier warning about different parameters to different models. An illustration is provided by the above remark on the comparison of $\mu < 0$ with $\mu > 0$. This partial opening in the use of improper priors represents an opportunity but it does not apply to parameters of interest, i.e. to parameters on which restrictions are assessed.

Example 2.1. When comparing two id normal samples, $(x_1, \ldots, x_n)$ and $(y_1, \ldots, y_n)$, with respective distributions $\mathscr{N}(\mu_x, \sigma^2)$ and $\mathscr{N}(\mu_y, \sigma^2)$, we can examine whether or not the two means are identical, i.e. $\mu_x = \mu_y$ (corresponding to model M_1). To take advantage of the structure of this model, we can assume that σ^2 is a measurement error with a similar meaning under both models and thus that the same prior $\pi_\sigma(\sigma^2)$ can be used under both models. This means that the Bayes factor

$$B_{21}^{\pi}(\mathscr{D}_n) = \frac{\int \ell_2(\mu_x, \mu_y, \sigma|\mathscr{D}_n)\pi(\mu_x, \mu_y)\pi_\sigma(\sigma^2)\,\mathrm{d}\sigma^2\,\mathrm{d}\mu_x\,\mathrm{d}\mu_y}{\int \ell_1(\mu, \sigma|\mathscr{D}_n)\pi_\mu(\mu)\pi_\sigma(\sigma^2)\,\mathrm{d}\sigma^2\,\mathrm{d}\mu}$$

does not depend on the normalizing constant used for $\pi_\sigma(\sigma^2)$ and thus that we can still use an improper prior such as $\pi_\sigma(\sigma^2) = 1/\sigma^2$ in that case. Furthermore, we can rewrite μ_x and μ_y as $\mu_x = \mu - \xi$ and $\mu_y = \mu + \xi$, respectively, and use a prior of the form $\pi(\mu, \xi) = \pi_\mu(\mu)\pi_\xi(\xi)$ on the new parameterization so that, again, the same prior π_μ can be used under both models. The same cancellation of the normalizing constant occurs for π_μ, which means a Jeffreys prior $\pi_\mu(\mu) = 1$ can be used. However, we need a proper and well-defined prior on ξ, for instance $\xi \sim \mathscr{N}(0, \tau^2)$, which leads to

$$\begin{aligned} B_{21}^{\pi}(\mathscr{D}_n) &= \frac{\displaystyle\int e^{-n\left[(\mu-\xi-\bar{x})^2+(\mu+\xi-\bar{y})^2+s_{xy}^2\right]/2\sigma^2}\,\sigma^{-2n-2}e^{-\xi^2/2\tau^2}/\tau\sqrt{2\pi}\,\mathrm{d}\sigma^2\,\mathrm{d}\mu\,\mathrm{d}\xi}{\displaystyle\int e^{-n\left[(\mu-\bar{x})^2+(\mu-\bar{y})^2+s_{xy}^2\right]/2\sigma^2}\,\sigma^{-2n-2}\,\mathrm{d}\sigma^2\,\mathrm{d}\mu} \\ &= \frac{\displaystyle\int \left[(\mu-\xi-\bar{x})^2+(\mu+\xi-\bar{y})^2+s_{xy}^2\right]^{-n} e^{-\xi^2/2\tau^2}/\tau\sqrt{2\pi}\,\mathrm{d}\mu\,\mathrm{d}\xi}{\displaystyle\int \left[(\mu-\bar{x})^2+(\mu-\bar{y})^2+s_{xy}^2\right]^{-n}\mathrm{d}\mu}, \end{aligned}$$

where s_{xy}^2 denotes the average

$$s_{xy}^2 = \frac{1}{n}\sum_{i=1}^{n}(x_i - \bar{x})^2 + \frac{1}{n}\sum_{i=1}^{n}(y_i - \bar{y})^2 .$$

While the denominator can be completely integrated out, the numerator cannot. A numerical approximation to B_{21}^{π} is thus necessary. (This issue is addressed in Sect. 2.4.) ◀

We conclude this section by a full processing of the assessment of $\mu = 0$ for the single sample normal problem. Comparing models $M_1 : \mathscr{N}(0, \sigma^2)$ under the prior $\pi_1(\sigma^2) = 1/\sigma^2$ and $M_2 : \mathscr{N}(\mu, \sigma^2)$ under the prior made of $\pi_2(\sigma^2) = 1/\sigma^2$ and $\pi_2(\mu|\sigma^2)$ equal to the normal $\mathscr{N}(0, \sigma^2)$ density, the Bayes factor is

$$B_{21}^{\pi}(\mathscr{D}_n) = \frac{\displaystyle\int e^{-[n(\bar{x}-\mu)^2+s^2]/2\sigma^2}\, e^{-\mu^2/2\sigma^2}\, \sigma^{-n-1-2} \frac{\mathrm{d}\mu \mathrm{d}\sigma^2}{\sqrt{2\pi}}}{\displaystyle\int e^{-[n\bar{x}^2+s^2]/2\sigma^2}\, \sigma^{-n-2}\, \mathrm{d}\sigma^2}$$

$$= \frac{\displaystyle\int e^{-(n+1)[\mu-n\bar{x}/(n+1)]^2}\, e^{-[n\bar{x}^2/(n+1)+s^2]/2\sigma^2}\, \sigma^{-n-3} \frac{\mathrm{d}\mu \mathrm{d}\sigma^2}{\sqrt{2\pi}}}{\left[\dfrac{n\bar{x}^2+s^2}{2}\right]^{-n/2} \Big/ \Gamma(n/2)}$$

$$= \frac{\displaystyle\int (n+1)^{-1/2}\, e^{-[n\bar{x}^2/(n+1)+s^2]/2\sigma^2}\, \sigma^{-n-2}\, \mathrm{d}\sigma^2}{\left[\dfrac{n\bar{x}^2+s^2}{2}\right]^{-n/2} \Big/ \Gamma(n/2)}$$

$$= \frac{(n+1)^{-1/2} \left[\dfrac{n\bar{x}^2/(n+1)+s^2}{2}\right]^{-n/2} \Big/ \Gamma(n/2)}{\left[\dfrac{n\bar{x}^2+s^2}{2}\right]^{-n/2} \Big/ \Gamma(n/2)}$$

$$= (n+1)^{-1/2} \left[\frac{n\bar{x}^2+s^2}{n\bar{x}^2/(n+1)+s^2}\right]^{n/2},$$

taking once again advantage of the normalizing constant of the gamma distribution (see also Exercise 2.8). It therefore increases to infinity with $\bar{x}^2/s^2$, starting from $1/\sqrt{n+1}$ when $\bar{x}=0$.

The value of this Bayes factor for Illingworth's data is given by

```
> ratio=n*mean(shift)^2/((n-1)*var(shift))
> ((1+ratio)/(1+ratio/(n+1)))^(n/2)/sqrt(n+1)
[1] 0.1466004
```

which confirms the assessment that the model with $\mu = 0$ is to be preferred.

2.4 Monte Carlo Methods

While, as seen in Sect. 2.3, the Bayes factor and the posterior probability are the only quantities used in the assessment of models (and hypotheses), the analytical derivation of those objects is not always possible, since they involve integrating the likelihood $\ell(\theta|\mathscr{D}_n)$ both on the sets Θ_1 and Θ_2, under the respective priors π_1 and π_2. Fortunately, there exist special numerical techniques for the computation of Bayes factors, which are, mathematically speaking, simply ratios of integrals. We now detail the techniques used in the approximation of intractable integrals, but refer to Chen et al. (2000) and Robert and Casella (2004, 2009) for book-length presentations.

2.4.1 An Approximation Based on Simulations

The technique that is most commonly used for integral approximations in statistics is called the Monte Carlo method[6] and relies on computer simulations of random variables to produce an approximation technique that converges with the number of simulations. Its justification is thus the *law of large numbers*, that is, if $x_1, \ldots, x_N$ are independent and distributed from g, then the empirical average

$$\hat{\mathfrak{I}}_N = (h(x_1) + \ldots + h(x_N))/N$$

converges (almost surely) to the integral

$$\mathfrak{I} = \int h(x)g(x)\,\mathrm{d}x\,.$$

We will not expand on the foundations of the random number generators in this book, except for an introduction to accept–reject methods in Chap. 5 because of their links with Markov chain Monte Carlo techniques (see, instead, Robert and Casella, 2004). The connections of utmost relevance here are (a) that softwares like R can produce pseudo-random series that are indistinguishable from truly random series with a given distribution, as illustrated in Table 1.1 and (b) that those software packages necessarily cover a limited collection of distributions. Therefore, other methods must be found for simulating distributions outside this collection, while relying on the distributions already available, first and foremost the uniform $\mathscr{U}(0,1)$ distribution.

The implementation of the Monte Carlo method is straightforward, at least on a formal basis, with the following algorithmic representation:

Algorithm 2.1 Basic Monte Carlo Method

For $i = 1, \ldots, N$,
  simulate $x_i \sim g(x)$.
Take

$$\hat{\mathfrak{I}}_N = (h(x_1) + \ldots + h(x_N))/N$$

to approximate $\mathfrak{I}$.

as long as the (computer-generated) pseudo-random generation from g is feasible and the $h(x_i)$ values are computable. When simulation from g is a problem because g is nonstandard and usual techniques such as accept–reject algorithms (see Chap. 5) are difficult to devise, more advanced techniques such as Markov Chain Monte Carlo (MCMC) are required. We will introduce those

[6] This method is named in reference to the central district of Monaco, where the famous Monte-Carlo casino lies.

in both next chapters. When the difficulty is with the intractability of the function h, the solution is often to use an integral representation of h and to expand the random variables x_i in (x_i, y_i), where y_i is an auxiliary variable. The use of such representations will be detailed in Chap. 6.

Example 2.2 (Continuation of Example 2.1). As computed in Example 2.1, the Bayes factor $B_{21}^{\pi}(\mathscr{D}_n)$ can be simplified into

$$B_{21}^{\pi}(\mathscr{D}_n) = \frac{\displaystyle\int \left[(\mu-\xi-\bar{x})^2 + (\mu+\xi-\bar{y})^2 + s_{xy}^2\right]^{-n} e^{-\xi^2/2\tau^2}\,\mathrm{d}\mu\,\mathrm{d}\xi/\tau\sqrt{2\pi}}{\displaystyle\int \left[(\mu-\bar{x})^2 + (\mu-\bar{y})^2 + s_{xy}^2\right]^{-n}\mathrm{d}\mu}$$

$$= \frac{\displaystyle\int \left[(2\xi+\bar{x}-\bar{y})^2 + 2\,s_{xy}^2\right]^{-n+1/2} e^{-\xi^2/2\tau^2}\,\mathrm{d}\xi/\tau\sqrt{2\pi}}{\left[(\bar{x}-\bar{y})^2 + 2\,s_{xy}^2\right]^{-n+1/2}}\,,$$

and we are left with a single integral in the numerator that involves the normal $\mathscr{N}(0,\tau^2)$ density and can thus be represented as an expectation against this distribution. This means that simulating a normal $\mathscr{N}(0,\tau^2)$ sample of ξ_i's $(i = 1,\ldots,N)$ and replacing $B_{21}^{\pi}(\mathscr{D}_n)$ with

$$\hat{B}_{21}^{\pi}(\mathscr{D}_n) = \frac{\frac{1}{N}\sum_{i=1}^{N}\left[(2\xi_i+\bar{x}-\bar{y})^2 + 2\,s_{xy}^2+^2\right]^{-n+1/2}}{\left[(\bar{x}-\bar{y})^2 + 2\,s_{xy}^2\right]^{-n+1/2}}$$

is an asymptotically valid approximation scheme. ◀

In **normaldata**, if we compare the fifth and the sixth sessions, both with $n = 10$ observations, we obtain

```
> illing=as.matrix(normaldata)
> xsam=illing[illing[,1]==5,2]
> xbar=mean(xsam)
[1] -0.041
> ysam=illing[illing[,1]==6,2]
> ybar=mean(ysam)
[1] -0.025
> Ssquar=9*(var(xsam)+var(ysam))/10
[1] 0.101474
```

Picking $\tau = 0.75$ as earlier, we get the following approximation to the Bayes factor

```
> Nsim=10^4
> tau=0.75
> xis=rnorm(Nsim,sd=tau)
> BaFa=mean(((2*xis+xbar-ybar)^2+2*Ssquar)^(-8.5))/
+ ((xbar-ybar)^2+2*Ssquar)^(-8.5)
[1] 0.0763622
```

This value of $\widehat{B^{\pi}_{21}}(\mathscr{D}_n)$ implies that $\xi = 0$, i.e. $\mu_x = \mu_y$ is much more likely for the data at hand than $\mu_x \neq \mu_y$. Note that, if we use $\tau = 0.1$ instead, the approximated Bayes factor is 0.4985 which slightly reduces the argument in favor of $\mu_x = \mu_y$.

Obviously, this *Monte Carlo estimate* of $\mathfrak{I}$ is not exact, but generating a sufficiently large number of random variables can render this approximation error arbitrarily small in a suitable probabilistic sense. It is also possible to assess the size of this error for a given number of simulations. If

$$\int |h(x)|^2 g(x)\, \mathrm{d}x < \infty\,,$$

the central limit theorem shows that $\sqrt{N}\,[\hat{\mathfrak{I}}_N - \mathfrak{I}]$ is also normally distributed, and this can be used to construct asymptotic confidence regions for $\hat{\mathfrak{I}}_N$, estimating the asymptotic variance from the simulation output.

For the approximation of $B^{\pi}_{21}(\mathscr{D}_n)$ proposed above, its variability is illustrated in Fig. 2.4, based on 500 replications of the simulation of $N = 1000$ normal variables used in the approximation and obtained as follows

```
> xis=matrix(rnorm(500*10^3,sd=tau),nrow=500)
> BF=((2*xis+xbar-ybar)^2+2*Ssquar)^(-8.5)/
+ ((xbar-ybar)^2+2*Ssquar)^(-8.5)
> estims=apply(BF,1,mean)
> hist(estims,nclass=84,prob=T,col="wheat2",
+ main="",xlab="Bayes Factor estimates")
> curve(dnorm(x,mean=mean(estims),sd=sd(estims)),
+ col="steelblue2",add=TRUE)
```

As can be seen on this figure, the value of 0.076 reported in the previous Monte Carlo approximation is in the middle of the range of possible values. More in connection with the above point, the shape of the histogram is clearly compatible with the normal approximation, as shown by the fitted normal density.

2.4.2 Importance Sampling

An important feature of Example 2.2 is that, for the Monte Carlo approximation of $B^{\pi}_{21}(\mathscr{D}_n)$, we exhibited a normal density within the integral and hence derived a representation of this integral as an expectation under this normal distribution. This seems like a very restrictive constraint in the approximation of integrals but this is only an apparent restriction in that we will now show that there is no need to simulate directly from the normal density and furthermore that there is no intrinsic density corresponding to a given integral, but rather an infinity of densities!

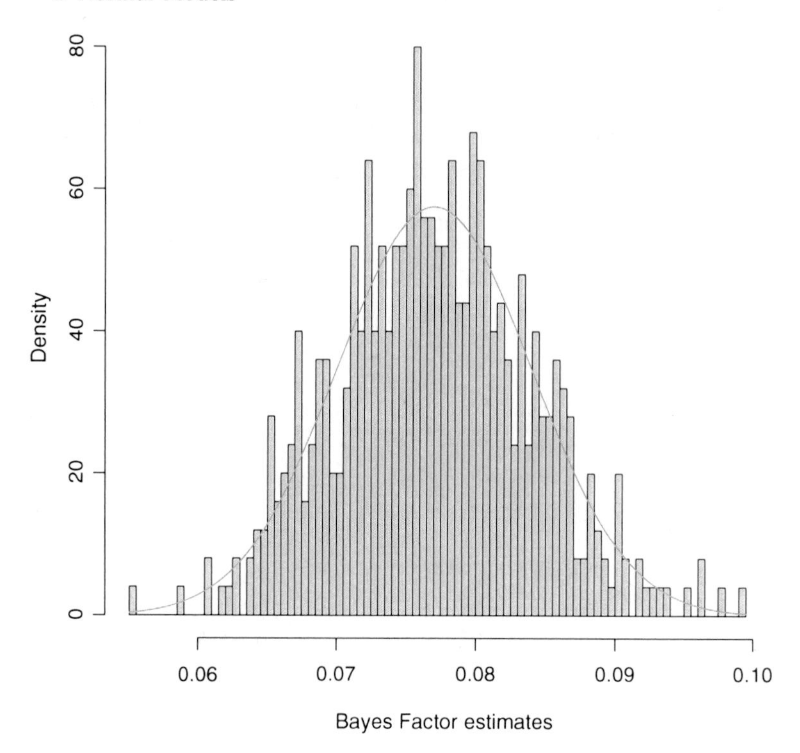

Fig. 2.4. Dataset **normaldata**: Histogram of 500 realizations of the approximation $\widehat{B_{21}(\mathscr{D}_n)}$ based on $N = 1000$ simulations each and normal fit of the sample

Indeed, an arbitrary integral

$$\mathfrak{I} = \int H(x) \, \mathrm{d}x$$

can be represented in infinitely many ways as an expectation, since, for an arbitrary probability density γ, we always have

$$\mathfrak{I} = \int \frac{H(x)}{\gamma(x)} \gamma(x) \, \mathrm{d}x \,, \tag{2.8}$$

under the minimal condition that $\gamma(x) > 0$ when $H(x)$. Therefore, the generation of a sample from γ can provide a converging approximation to $\mathfrak{E}$ and the Monte Carlo method applies in a very wide generality. This method is called *importance sampling* when applied to an expectation under a density g,

$$\mathfrak{I} = \int h(x) g(x) \, \mathrm{d}x \,, H(x) = h(x) g(x)$$

since the values x_i simulated from γ are weighted by the importance weights $g(x_i)/\gamma(x_i)$ in the approximation

$$\hat{\mathfrak{I}}_N = \frac{1}{N} \sum_{i=1}^{N} \frac{g(x_i)}{\gamma(x_i)} h(x_i) .$$

While the representation (2.8) holds for any density γ with a support larger than the support of H, the performance of the empirical average $\hat{\mathfrak{I}}_N$ can deteriorate considerably when the ratio $h(x)g(x)/\gamma(x)$ is not bounded as this raises the possibility for infinite variance in the resulting estimator. When using importance sampling, one must always take heed of a potentially infinite variance of $\hat{\mathfrak{I}}_N$.

An additional incentive in using importance sampling is that this method does not require the density g (or γ) to be known completely. Those densities can be known only up to a normalizing constant, $g(x) \propto \tilde{g}(x)$ and $\gamma(x) \propto \tilde{\gamma}(x)$, since the ratio

$$\sum_{i=1}^{n} h(x_i)\tilde{g}(x_i)/\tilde{\gamma}(x_i) \Big/ \sum_{i=1}^{n} \tilde{g}(x_i)/\tilde{\gamma}(x_i)$$

also converges to $\mathfrak{I}$ when n goes to infinity and when the x_i's are generated from γ.

The equivalent of Algorithm 2.1 for importance sampling is as follows:

Algorithm 2.2 IMPORTANCE SAMPLING METHOD

For $i = 1, \ldots, N$,
  simulate $x_i \sim \gamma(x)$;
  compute $\omega_i = \tilde{g}(x_i)/\gamma(x_i)$.
Take

$$\hat{\mathfrak{I}}_N = \sum_{i=1}^{N} \omega_i \, h(x_i) \Big/ \sum_{i=1}^{N} \omega_i$$

to approximate $\mathfrak{I}$.

This algorithm is straightforward to implement. Since it offers a degree of freedom in the selection of γ, simulation from a manageable distribution can be imposed, keeping in mind the constraint that γ should have flatter tails than g. Unfortunately, as the dimension of x increases, differences between the target density g and the importance density γ have a larger and larger impact.

Example 2.3. Consider almost the same setting as in Exercise 2.11: $\mathscr{D}_n = (x_1, \ldots, x_n)$ is an iid sample from $\mathscr{C}(\theta, 1)$ and the prior on θ is a flat prior. We can use a normal importance function from a $\mathscr{N}(\mu, \sigma^2)$ distribution to produce a sample $\theta_1, \ldots, \theta_N$ that approximates the Bayes estimator of θ, i.e. its posterior mean, by

$$\hat{\delta}^{\pi}(\mathscr{D}_n) = \frac{\sum_{t=1}^{N} \theta_t \exp\left\{(\theta_t - \mu)^2/2\right\} \prod_{i=1}^{n}[1 + (x_i - \theta_t)^2]^{-1}}{\sum_{t=1}^{N} \exp\left\{(\theta_t - \mu)^2/2\right\} \prod_{i=1}^{n}[1 + (x_i - \theta_t)^2]^{-1}} .$$

But this is a very poor estimation (see Exercise 2.17 for an analytic explanation) and it degrades considerably when μ increases. If we run an R simulation experiment producing a sample of estimates when μ increases, as follows,

```
> Nobs=10
> obs=rcauchy(Nobs)
> Nsim=250
> Nmc=500
> sampl=matrix(rnorm(Nsim*Nmc),nrow=1000) # normal samples
> raga=riga=matrix(0,nrow=50,ncol=2) # ranges
> mu=0
> for (j in 1:50){
+   prod=1/dnorm(sampl-mu) # importance sampling
+   for (i in 1:Nobs)
+     prod=prod*dt(obs[i]-sampl,1)
+   esti=apply(sampl*prod,2,sum)/apply(prod,2,sum)
+   raga[j,]=range(esti)
+   riga[j,]=c(quantile(esti,.025),quantile(esti,.975))
+   sampl=sampl+0.1
+   mu=mu+0.1
+   }
> mus=seq(0,4.9,by=0.1)
> plot(mus,0*mus,col="white",xlab=expression(mu),
+ ylab=expression(hat(theta)),ylim=range(raga))
> polygon(c(mus,rev(mus)),c(raga[,1],rev(raga[,2])),col="grey50")
> polygon(c(mus,rev(mus)),c(riga[,1],rev(riga[,2])),col="pink3")
```

as shown by Fig. 2.5, not only does the range of the approximation increase, but it ends up missing the true value $\theta = 0$ when μ is far enough from 0. ◀

2.4.3 Approximation of Bayes Factors

Bayes factors being ratios of integrals, they can be approximated by regular importance sampling tools. However, given their specificity as ratios of marginal likelihoods, hence of normalizing constants of the posterior distributions, there exist more specialized techniques, including a fairly generic method called *bridge sampling*, developed by Gelman and Meng (1998).

When comparing two models with sampling densities $f_1(\mathscr{D}_n|\theta_1)$ (model M_1) and $f_2(\mathscr{D}_n|\theta_2)$ (model M_2), assume that both models share the same parameter space Θ. This is for instance the case when comparing the fit of two densities with the same number of parameters (modulo a potential reparameterization of one of the models). In this setting, if the corresponding prior

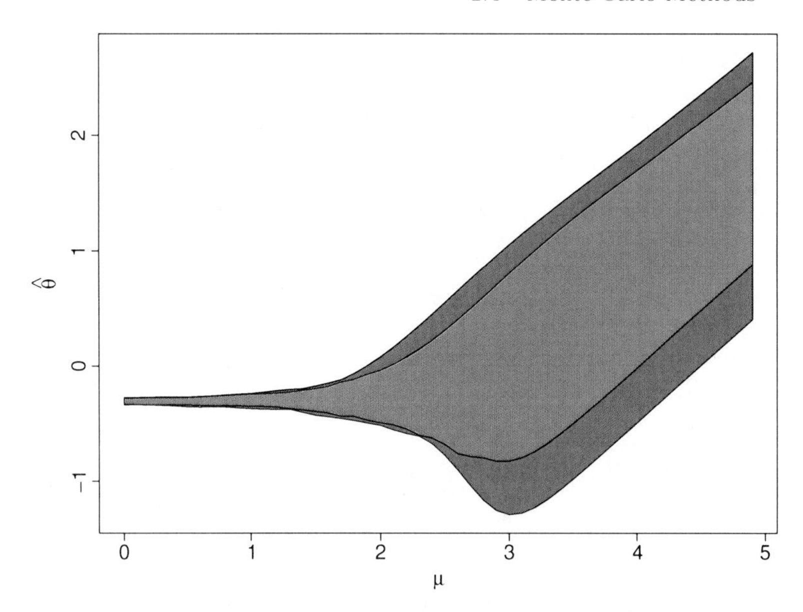

Fig. 2.5. Representation of the whole range (*grey*) and of the 95 % range (*pink*) of variation of the importance sampling approximation to the Bayes estimate for $n = 10$ observations from the $\mathscr{C}(0,1)$ distribution and $N = 250$ simulations of θ from a $\mathscr{N}(\mu, 1)$ distribution as a function of μ. This range is computed using 500 replications of the importance sampling estimates

densities are $\pi_1(\theta)$ and $\pi_2(\theta)$, we only know the unnormalized posterior densities $\tilde{\pi}_1(\theta|\mathscr{D}_n) = f_1(\mathscr{D}_n|\theta)\pi_1(\theta)$ and $\tilde{\pi}_2(\theta|\mathscr{D}_n) = f_2(\mathscr{D}_n|\theta)\pi_2(\theta)$. In this general setting, for any positive function α such that the integrals below exist, the Bayes factor for comparing the two models satisfies

$$\begin{aligned} B^{\pi}_{12}(\mathscr{D}_n) &= \frac{m_1(x)}{m_2(x)} \\ &= \frac{m_1(x)}{m_2(x)} \frac{\displaystyle\int \tilde{\pi}_1(\theta|\mathscr{D}_n)\alpha(\theta)\tilde{\pi}_2(\theta|\mathscr{D}_n)\mathrm{d}\theta}{\displaystyle\int \tilde{\pi}_2(\theta|\mathscr{D}_n)\alpha(\theta)\tilde{\pi}_1(\theta|\mathscr{D}_n)\mathrm{d}\theta} \\ &= \frac{\displaystyle\int \tilde{\pi}_1(\theta|\mathscr{D}_n)\alpha(\theta)\pi_2(\theta|\mathscr{D}_n)\mathrm{d}\theta}{\displaystyle\int \tilde{\pi}_2(\theta|\mathscr{D}_n)\alpha(\theta)\pi_1(\theta|\mathscr{D}_n)\mathrm{d}\theta} . \end{aligned} \tag{2.9}$$

Therefore, the *bridge sampling* approximation

$$\sum_{i=1}^{N} \tilde{\pi}_1(\theta_{2i}|\mathscr{D}_n)\alpha(\theta_{2i}) \bigg/ \sum_{i=1}^{N} \tilde{\pi}_2(\theta_{1i}|\mathscr{D}_n)\alpha(\theta_{1i}) \tag{2.10}$$

is a convergent approximation of the Bayes factor $B^{\pi}_{12}(\mathscr{D}_n)$ when $\theta_{ji} \sim \pi_j(\theta|\mathscr{D}_n)$ $(j = 1,2,\ i = 1,\ldots,N)$. One of the appealing features of the method is that it only requires simulations from the posterior distributions under both models of interest. Another interesting feature is that α is completely arbitrary, which means it can be chosen in the best possible way. Using asymptotic variance arguments, Gelman and Meng (1998) proved that the best choice is

$$\alpha^{\mathrm{O}}(\theta) \propto \frac{1}{\pi_1(\theta|\mathscr{D}_n) + \pi_2(\theta|\mathscr{D}_n)}\,,$$

which bridges both posteriors. This means that the optimal weight of θ_{2i} in (2.10) is

$$\frac{\tilde{\pi}_1(\theta_{2i}|\mathscr{D}_n)}{\pi_1(\theta_{2i}|\mathscr{D}_n) + \pi_2(\theta_{2i}|\mathscr{D}_n)} = \frac{\tilde{\pi}_1(\theta_{2i}|\mathscr{D}_n)}{\tilde{\pi}_1(\theta_{2i}|\mathscr{D}_n) + B^{\pi}_{12}(\mathscr{D}_n)\tilde{\pi}_2(\theta_{2i}|\mathscr{D}_n)}\,,$$

with an appropriate change of indices for the θ_{1i}'s. There is however a caveat with this find in that it cannot be attained because the optimum depends on the very quantity we are trying to approximate! However, the Bayes factor $B^{\pi}_{12}(\mathscr{D}_n)$ can first be approximated on a crude basis and the corresponding construction of α^{O} iterated till the Bayes factor approximation (2.10) stabilizes.

We will now illustrate this derivation in the case of the normal model, with an application to **normaldata**. (We showed in Sect. 2.3.3 that the Bayes factor was available in closed form so this implementation of the bridge sampler is purely for illustrative purposes.) A further implementation is discussed in Chap. 4, Sect. 4.3.2, in connection with the probit model.

When assessing whether or $\mu = 0$ is appropriate for the single sample normal model, the above approximation does not apply directly because there is an extra parameter in the unconstrained model. There are however two easy tricks out of this difficulty. The first one, repeatedly found in the literature, is to add an arbitrary density to make dimensions match. In the normal example, this means introducing an arbitrary (normalized) density $\pi_1^*(\mu|\sigma^2)$ in the constrained model (denoted M_1) and extending the Bayes factor representation (2.9) to

$$B^{\pi}_{12}(\mathscr{D}_n) = \frac{\displaystyle\int \pi_1^*(\mu|\sigma^2)\tilde{\pi}_1(\sigma^2|\mathscr{D}_n)\alpha(\theta)\pi_2(\theta|\mathscr{D}_n)\mathrm{d}\theta}{\displaystyle\int \tilde{\pi}_2(\theta|\mathscr{D}_n)\alpha(\theta)\pi_1(\sigma^2|\mathscr{D}_n)\mathrm{d}\sigma^2\pi_1^*(\mu|\sigma^2)\mathrm{d}\mu}\,.$$

which holds independently of $\pi_1^*(\mu|\sigma^2)$ for the same reason as in (2.9). The choice of the substitute $\pi_1^*(\mu|\sigma^2)$ equal to an approximation of $\pi_2(\mu|\mathscr{D}_n,\sigma^2)$ is suggested by Chen et al. (2000). For instance, we can use as $\pi_1^*(\mu|\sigma^2)$ a normal distribution $\mathscr{N}(\hat{\mu},\hat{\sigma}^2)$ where $\hat{\mu}$ and $\hat{\sigma}^2$ are computed based on a simulation from $\pi_2(\mu,\sigma|\mathscr{D}_n)$.

The exact value of this Bayes factor $B_{12}^{\pi}(\mathscr{D}_n)$ for Illingworth's data is given by

```
> ((1+ratio)/(1+ratio/(n+1)))^(-n/2)*sqrt(n+1)
[1] 6.821262
```

while the bridge sampling solution is obtained as

```
> n=64
> xbar=mean(shift)
> sqar=(n-1)*var(shift)
> Nmc=10^7
> # Simulation from model M2:
> sigma2=1/rgamma(Nmc,shape=n/2,rate=(n*xbar^2/(n+1)+sqar)/2)
> mu2=rnorm(Nmc,n*xbar/(n+1),sd=sqrt(sigma2/(n+1)))
> # Simulation from model M1:
> sigma1=1/rgamma(Nmc,shape=n/2,rate=(n*xbar^2+sqar)/2)
> muhat=mean(mu2)
> tauat=sd(mu2)
> mu1=rnorm(Nmc,mean=muhat,sd=tauat)
> #tilde functions
> tildepi1=function(sigma,mu){
+   exp(-.5*((n*xbar^2+sqar)/sigma+(n+2)*log(sigma))+
+   dnorm(mu,muhat,tauat,log=T))
+   }
> tildepi2=function(sigma,mu){
+   exp(-.5*((n*(xbar-mu)^2+sqar+mu^2)/sigma+(n+3)*log(sigma)+
+   log(2*pi)))}
> #Bayes Factor loop
> K=diff=1
> rationum=tildepi2(sigma1,mu1)/tildepi1(sigma1,mu1)
> ratioden=tildepi1(sigma2,mu2)/tildepi2(sigma2,mu2)
> while (diff>0.01*K){
+   BF=mean(1/(1+K*rationum))/mean(1/(K+ratioden))
+   diff=abs(K-BF)
+   K=BF}
```

and returns the value

```
> BF
[1] 6.820955
```

which is definitely close to the true value!

The second possible trick to overcome the dimension difficulty while using the bridge sampling strategy is to introduce artificial posterior distributions in each of the parameters spaces and to process each marginal likelihood as an integral ratio in itself. For instance, if $\eta_1(\theta_1)$ is an arbitrary normalized density on θ_1, and α is an arbitrary function, we have

$$m_1(\mathscr{D}_n) = \int \tilde{\pi}_1(\theta_1|\mathscr{D}_n)\,\mathrm{d}\theta_1 = \frac{\displaystyle\int \tilde{\pi}_1(\theta_1|\mathscr{D}_n)\alpha(\theta_1)\eta_1(\theta_1)\,\mathrm{d}\theta_1}{\displaystyle\int \eta_1(\theta_1)\alpha(\theta_1)\pi_1(\theta_1|\mathscr{D}_n)\,\mathrm{d}\theta_1}$$

by application of (2.9). Therefore, the optimal choice of α leads to the approximation

$$\hat{m}_1(\mathscr{D}_n) = \frac{\sum_{i=1}^N \tilde{\pi}_1(\theta_{1i}^\eta|\mathscr{D}_n)/\left\{m_1(\mathscr{D}_n)\tilde{\pi}_1(\theta_{1i}^\eta|\mathscr{D}_n)+\eta(\theta_{1i}^\eta)\right\}}{\sum_{i=1}^N \eta(\theta_{1i})/\left\{m_1(\mathscr{D}_n)\tilde{\pi}_1(\theta_{1i}|\mathscr{D}_n)+\eta(\theta_{1i})\right\}}$$

when $\theta_{1i} \sim \pi_1(\theta_1|\mathscr{D}_n)$ and $\theta_{1i}^\eta \sim \eta(\theta_1)$. The choice of the density η is obviously fundamental and it should be close to the true posterior $\pi_1(\theta_1|\mathscr{D}_n)$ to guarantee good convergence approximation. Using a normal approximation to the posterior distribution of θ or a non-parametric approximation based on a sample from $\pi_1(\theta_1|\mathscr{D}_n)$, or yet again an average of MCMC proposals (see Chap. 4) are reasonable choices.

The R implementation of this approach can be done as follows

```
> sigma1=1/rgamma(Nmc,shape=n/2,rate=(n*xbar^2+sqar)/2)
> sihat=mean(log(sigma1))
> tahat=sd(log(sigma1))
> sigma1b=exp(rnorm(Nmc,sihat,tahat))
> #tilde function
> tildepi1=function(sigma){
  exp(-.5*((n*xbar^2+sqar)/sigma+(n+2)*log(sigma)))}
> K=diff=1
> rnum=dnorm(log(sigma1b),sihat,tahat)/
+             (sigma1b*tildepi1(sigma1b))
> rden=sigma1*tildepi1(sigma1)/dnorm(log(sigma1),sihat,tahat)
> while (diff>0.01*K){
>   BF=mean(1/(1+K*rnum))/mean(1/(K+rden))
>   diff=abs(K-BF)
>   K=BF}
> m1=BF
```

when using a normal distribution on $\log(\sigma^2)$ as an approximation to $\pi_1(\theta_1|\mathscr{D}_n)$. When considering the unconstrained model, a bivariate normal density can be used, as in

```
> sigma2=1/rgamma(Nmc,shape=n/2,rate=(n*xbar^2/(n+1)+sqar)/2)
> mu2=rnorm(Nmc,n*xbar/(n+1),sd=sqrt(sigma2/(n+1)))
> temean=c(mean(mu2),mean(log(sigma2)))
```

```
> tevar=cov.wt(cbind(mu2,log(sigma2)))$cov
> te2b=rmnorm(Nmc,mean=temean,tevar)
> mu2b=te2b[,1]
> sigma2b=exp(te2b[,2])
```

leading to

```
> m1/m2
[1] 6.824417
```

The performances of both extensions are obviously highly dependent on the choice of the completion factors, η_1 and η_2 on the one hand and π_1^* on the other hand. The performances of the first solution, which bridges both models via π_1^*, are bound to deteriorate as the dimension gap between those models increases. The impact of the dimension of the models is less keenly felt for the other solution, as the approximation remains local.

As a simple illustration of the performances of both methods, we produce here a comparison between the completions based on a single pseudo-conditional and on two local approximations to the posteriors, by running repeated approximations for **normaldata** and tracing the resulting boxplot as a measure of the variability of those methods. As shown in Fig. 2.6, the variability is quite comparable for both solutions in this specific case.

Note that there exist many other approaches to the approximative computation of marginal likelihoods and of Bayes factors that we cannot cover here. We want however to point out the dangers of the harmonic mean approximation. This approach proceeds from the interesting identity

$$\mathbb{E}^{\pi_1}\left[\left.\frac{\varphi_1(\theta_1)}{\pi_1(\theta_1)\ell_1(\theta_1|\mathscr{D}_n)}\right|\mathscr{D}_n\right] = \int \frac{\varphi_1(\theta_1)}{\pi_1(\theta_1)\ell_1(\theta_1|\mathscr{D}_n)}\,\frac{\pi_1(\theta_1)\ell_1(\theta_1|\mathscr{D}_n)}{m_1(\mathscr{D}_n)}\,\mathrm{d}\theta_1$$
$$= \frac{1}{m_1(\mathscr{D}_n)},$$

which holds, no matter what the density $\varphi_1(\theta_1)$ is—provided $\varphi_1(\theta_1)=0$ when $\pi_1(\theta_1)\ell_1(\theta_1|\mathscr{D}_n)=0$—. The most common implementation in approximations of the marginal likelihood uses $\varphi_1(\theta_1)=\pi_1(\theta_1)$, leading to the approximation

$$\hat{m}_1(\mathscr{D}_n) = 1\bigg/ N^{-1}\sum_{j=1}^{N}\frac{1}{\ell_1(\theta_{1j}|\mathscr{D}_n)}\,.$$

While very tempting, since it allows for a direct processing of simulations from the posterior distribution, this approximation is unfortunately most often associated with an infinite variance (Exercise 2.19) and, thus, *should not be used*. On the opposite, using φ_1's with supports constrained to the 25 % HPD

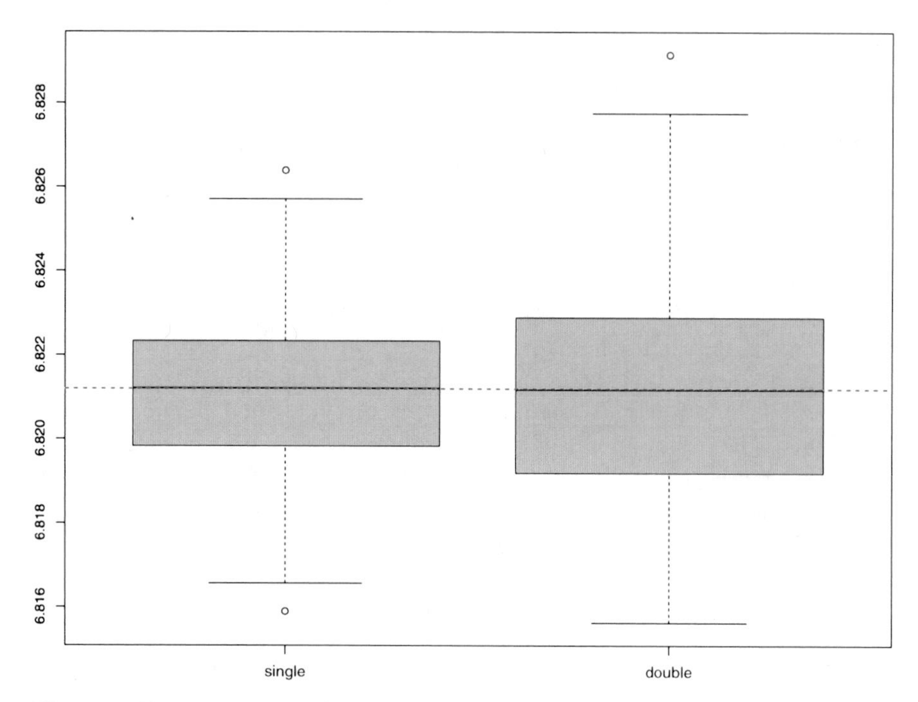

Fig. 2.6. Dataset **normaldata**: Boxplot of the variability of the approximations to the Bayes factor assessing whether or not $\mu = 0$, based on a single and on a double completions. Each approximation is based on 10^5 simulations and the boxplots are based on 250 approximations. The *dotted line* corresponds to the true value of $B_{12}^{\pi}(\mathscr{D}_n)$

regions—approximated by the convex hull of the 10 % or of the 25 % highest simulations—is both completely appropriate and implementable (Marin and Robert, 2010).

2.5 Outlier Detection

The above description of inference in normal models is only an introduction both to Bayesian inference and to normal structures. Needless to say, there exists a much wider range of possible applications. For instance, we will meet the normal model again in Chap. 4 as the original case of the (generalized) linear model. Before that, we conclude this chapter with a simple extension of interest, the detection of outliers.

Since normal modeling is often an approximation to the "real thing," there may be doubts about its adequacy. As already mentioned above, we will deal later with the problem of checking that the normal distribution is appropriate for the whole dataset. Here, we consider the somehow simpler problem of separately assessing whether or not each point in the dataset is compatible with normality. There are many different ways of dealing with this problem. We choose here to use the *predictive distribution*: If an observation x_i is unlikely under the predictive distribution based on the *other observations*, then we can argue against its distribution being equal to the distribution of the other observations.

If x_{n+1} is a future observation from the same distribution $f(\cdot|\theta)$ as the sample $\mathscr{D}_n$, its *predictive distribution* given the current sample is defined as

$$f^\pi(x_{n+1}|\mathscr{D}_n) = \int f(x_{n+1}|\theta, \mathscr{D}_n)\pi(\theta|\mathscr{D}_n)\,\mathrm{d}\theta = \int f(x_{n+1}|\theta)\pi(\theta|\mathscr{D}_n)\,\mathrm{d}\theta\,.$$

This definition is coherent with the Bayesian approach, which considers x_{n+1} as an extra unknown and then integrates out θ if x_{n+1} is the "parameter" of interest.

For the normal $\mathscr{N}(\mu, \sigma^2)$ setup, using a conjugate prior on (μ, σ^2) of the form

$$(\sigma^2)^{-\lambda_\sigma - 3/2} \exp - \left\{\lambda_\mu(\mu - \xi)^2 + \alpha\right\}/2\sigma^2\,,$$

the corresponding posterior distribution on (μ, σ^2) given $\mathscr{D}_n$ is

$$\mathscr{N}\left(\frac{\lambda_\mu\xi + n\overline{x}_n}{\lambda_\mu + n}, \frac{\sigma^2}{\lambda_\mu + n}\right) \times \mathscr{I}\mathscr{G}\left(\lambda_\sigma + n/2, \left[\alpha + s^2 + \frac{n\lambda_\mu}{\lambda_\mu + n}(\overline{x} - \xi)^2\right]/2\right),$$

denoted by

$$\mathscr{N}\left(\xi(\mathscr{D}_n), \sigma^2/\lambda_\mu(\mathscr{D}_n)\right) \times \mathscr{I}\mathscr{G}\left(\lambda_\sigma(\mathscr{D}_n)/2, \alpha(\mathscr{D}_n)/2\right),$$

and the predictive on x_{n+1} is derived as

$$\begin{aligned}
f^\pi(x_{n+1}|\mathscr{D}_n) &\propto \int (\sigma^2)^{-\lambda_\sigma(\mathscr{D}_n)/2-1-1} \exp -(x_{n+1} - \mu)^2/2\sigma^2 \\
&\qquad \times \exp - \left\{\lambda_\mu(\mathscr{D}_n)(\mu - \xi(\mathscr{D}_n))^2 + \alpha(\mathscr{D}_n)\right\}/2\sigma^2\,\mathrm{d}(\mu, \sigma^2) \\
&\propto \int (\sigma^2)^{-\lambda_\sigma(\mathscr{D}_n)/2-3/2} \exp - \left\{(\lambda_\mu(\mathscr{D}_n) + 1)(x_{n+1} - \xi(\mathscr{D}_n))^2\right. \\
&\qquad \left./\lambda_\mu(\mathscr{D}_n) + \alpha(\mathscr{D}_n)\right\}/2\sigma^2\,\mathrm{d}\sigma^2 \\
&\propto \left[\alpha(\mathscr{D}_n) + \frac{\lambda_\mu(\mathscr{D}_n) + 1}{\lambda_\mu(\mathscr{D}_n)}(x_{n+1} - \xi(\mathscr{D}_n))^2\right]^{-(\lambda_\sigma(\mathscr{D}_n)+1)/2}.
\end{aligned}$$

Therefore, the predictive of x_{n+1} given the sample $\mathscr{D}_n$ is a Student t distribution with mean $\xi(\mathscr{D}_n)$ and $\lambda_\sigma(\mathscr{D}_n)$ degrees of freedom. In the special case of the noninformative prior, corresponding to the limiting values $\lambda_\mu = \lambda_\sigma = \alpha = 0$, the predictive is

$$f^\pi(x_{n+1}|\mathscr{D}_n) \propto \left[s^2 + \frac{n+1}{n}1(x_{n+1} - \overline{x}_n)^2\right]^{-(n+1)/2}. \tag{2.11}$$

It is therefore a Student's t distribution with n degrees of freedom, a mean equal to $\overline{x}_n$ and a scale factor equal to $(n-1)s^2/n$, which is equivalent to a variance equal to $(n-1)s^2/n^2$ (to compare with the maximum likelihood estimator $\hat{\sigma}_n^2 = s^2/n$).

In the outlier problem, we process each observation $x_i \in \mathscr{D}_n$ as if it was a "future" observation. Namely, we consider $f_i^\pi(x|\mathscr{D}_n^i)$ as being the predictive distribution based on $\mathscr{D}_n^i = (x_1, \ldots, x_{i-1}, x_{i+1}, \ldots, x_n)$. Considering $f_i^\pi(x_i|\mathscr{D}_n^i)$ or the corresponding cdf $F_i^\pi(x_i|\mathscr{D}_n^i)$ (in dimension one) gives an indication of the level of compatibility of the observation x_i with the sample. To quantify this level, we can, for instance, approximate the distribution of $F_i^\pi(x_i|\mathscr{D}_n^i)$ as being uniform over $[0,1]$ since $F_i^\pi(\cdot|\mathscr{D}_n^i)$ does converge to the true cdf of the model. Simultaneously checking all $F_i^\pi(x_i|\mathscr{D}_n^i)$ over i may signal outliers.

$\lightning$ The detection of outliers must pay attention to the *Bonferroni fallacy*, which is that extreme values do occur in large enough samples. This means that, as n increases, we will see smaller and smaller values of $F_i^\pi(x_i|\mathscr{D}_n^i)$ occurring, and this even when the whole sample is from the same distribution. The significance level must therefore be chosen in accordance with this observation, for instance using a bound a on $F_i^\pi(x_i|\mathscr{D}_n^i)$ such that

$$1 - (1-a)^n = 1 - \alpha\,,$$

where α is the nominal level chosen for outlier detection.

Considering **normaldata**, we can compute the predictive cdf for each of the 64 observations, considering the 63 remaining ones as data.

```
> n=length(shift)
> outl=rep(0,n)
> for (i in 1:n){
+     lomean=-mean(shift[-i])
+     losd=sd(shift[-i])*sqrt((n-2)/n)
+     outl[i]=pt((shift[i]-lomean)/losd,df=n-1)
+     }
```

Figure 2.7 provides the qq-plot of the $F_i^\pi(x_i|\mathscr{D}_n^i)$'s against the uniform quantiles and compares it with a qq-plot based on a dataset truly simulated from the uniform $\mathscr{U}(0,1)$.

```
> plot(c(0,1),c(0,1),lwd=2,ylab="Predictive",xlab="Uniform",
+ type="l")
> points((1:n)/(n+1),sort(outl),pch=19,col="steelblue3")
> points((1:n)/(n+1),sort(runif(n)),pch=19,col="tomato")
```

There is no clear departure from uniformity when looking at this graph, except of course for the multiple values found in **normaldata**.

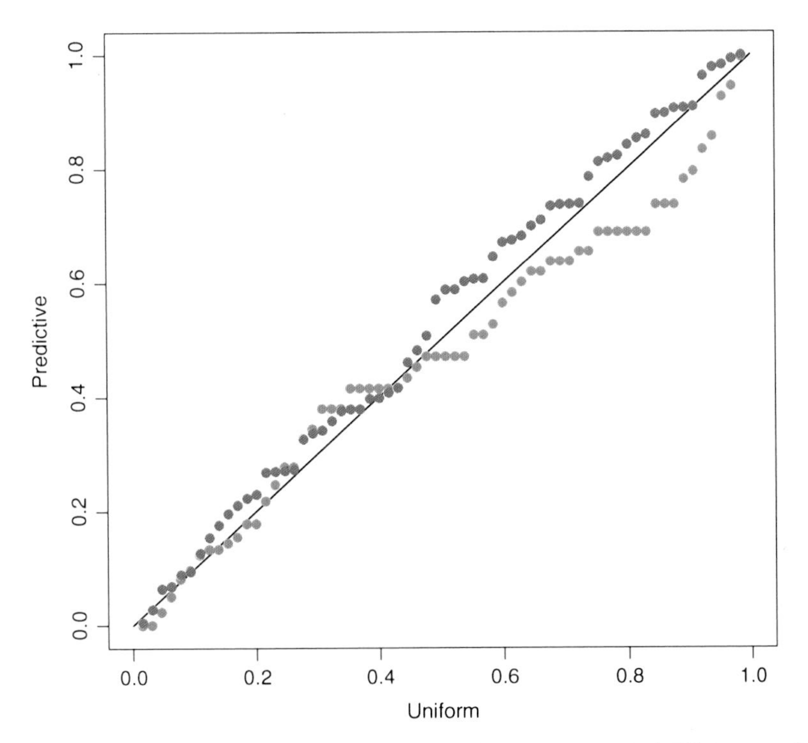

Fig. 2.7. Dataset **normaldata**: qq-plot of the sample of the $F_i^\pi(x_i|\mathscr{D}_n^i)$ for a uniform $\mathscr{U}(0,1)$ distribution (*blue dots*) and comparison with a qq-plot for a uniform $\mathscr{U}(0,1)$ sample (*red dots*)

2.6 Exercises

2.1 Show that, if

$$\mu|\sigma^2 \sim \mathscr{N}(\xi, \sigma^2/\lambda_\mu), \qquad \sigma^2 \sim \mathscr{IG}(\lambda_\sigma/2, \alpha/2),$$

then

$$\mu \sim \mathscr{T}(\lambda_\sigma, \xi, \alpha/\lambda_\mu\lambda_\sigma)$$

a t distribution with λ_σ degrees of freedom, location parameter ξ and scale parameter $\alpha/\lambda_\mu\lambda_\sigma$.

2.2 Show that, if $\sigma^2 \sim \mathscr{IG}(\alpha, \beta)$, then $\mathbb{E}[\sigma^2] = \beta/(\alpha - 1)$. Derive from the density of $\mathscr{IG}(\alpha, \beta)$ that the mode is located in $\beta/(\alpha + 1)$.

2.3 Show that minimizing (in $\hat{\theta}(\mathscr{D}_n)$) the posterior expectation $\mathbb{E}^\pi[||\theta - \hat{\theta}||^2|\mathscr{D}_n]$ produces the posterior expectation as the solution in $\hat{\theta}$.

2.4 Show that the Fisher information on $\theta = (\mu, \sigma^2)$ for the normal $\mathscr{N}(\mu, \sigma^2)$ distribution is given by

$$I^F(\theta) = \mathbb{E}_\theta \left[\begin{pmatrix} 1/\sigma^2 & 2(x-\mu)/2\sigma^4 \\ 2(x-\mu)/2\sigma^4 & (\mu-x)^2/\sigma^6 - 1/2\sigma^4 \end{pmatrix} \right] = \begin{pmatrix} 1/\sigma^2 & 0 \\ 0 & 1/2\sigma^4 \end{pmatrix}$$

and deduce that Jeffreys' prior is $\pi^J(\theta) \propto 1/\sigma^3$.

2.5 Derive each line of Table 2.1 by an application of Bayes' formula, $\pi(\theta|x) \propto \pi(\theta)f(x|\theta)$, and the identification of the standard distributions.

2.6 A Weibull distribution $\mathscr{W}(\alpha, \beta, \gamma)$ is defined as the power transform of a gamma $\mathscr{G}(\alpha, \beta)$ distribution: If $x \sim \mathscr{W}(\alpha, \beta, \gamma)$, then $x^\gamma \sim \mathscr{G}(\alpha, \beta)$. Show that, when γ is known, $\mathscr{W}(\alpha, \beta, \gamma)$ allows for a conjugate family, but that it does not an exponential family when γ is unknown.

2.7 Show that, when the prior on $\theta = (\mu, \sigma^2)$ is $\mathscr{N}(\xi, \sigma^2/\lambda_\mu) \times \mathscr{IG}(\lambda_\sigma, \alpha)$, the marginal prior on μ is a Student t distribution $\mathcal{T}(2\lambda_\sigma, \xi, \alpha/\lambda_\mu\lambda_\sigma)$ (see Example 2.18 for the definition of a Student t density). Give the corresponding marginal prior on σ^2. For an iid sample $\mathscr{D}_n = (x_1, \ldots, x_n)$ from $\mathscr{N}(\mu, \sigma^2)$, derive the parameters of the posterior distribution of (μ, σ^2).

2.8 Show that the normalizing constant for a Student $\mathscr{T}(\nu, \mu, \sigma^2)$ distribution is

$$\frac{\Gamma((\nu+1)/2)/\Gamma(\nu/2)}{\sigma\sqrt{\nu\pi}} .$$

Deduce that the density of the Student t distribution $\mathscr{T}(\nu, \theta, \sigma^2)$ is

$$f_\nu(x) = \frac{\Gamma((\nu+1)/2)}{\sigma\sqrt{\nu\pi}\,\Gamma(\nu/2)} \left(1 + \frac{(x-\theta)^2}{\nu\sigma^2}\right)^{-(\nu+1)/2} .$$

2.9 Show that, for location and scale models, the specific noninformative priors are special cases of Jeffreys' generic prior, i.e., that $\pi^J(\theta) = 1$ and $\pi^J(\theta) = 1/\theta$, respectively.

2.10 Show that, when $\pi(\theta)$ is a probability density, (2.5) necessarily holds for all datasets $\mathscr{D}_n$.

2.11 Consider a dataset $\mathscr{D}_n$ from the Cauchy distribution, $\mathscr{C}(\mu, 1)$.

1. Show that the likelihood function is

$$\ell(\mu|\mathscr{D}_n) = \prod_{i=1}^n f_\mu(x_i) = \frac{1}{\pi^n \prod_{i=1}^n (1 + (x_i - \mu)^2)} .$$

2. Examine whether or not there is a conjugate prior for this problem. (The answer is *no*.)
3. Introducing a normal prior on μ, say $\mathscr{N}(0, 10)$, show that the posterior distribution is proportional to

$$\tilde{\pi}(\mu|\mathscr{D}_n) = \frac{\exp(-\mu^2/20)}{\prod_{i=1}^n (1 + (x_i - \mu)^2)} .$$

4. Propose a numerical solution for solving $\tilde{\pi}(\mu|\mathscr{D}_n) = k$. (*Hint:* A simple trapezoidal integration can be used: based on a discretization size Δ, computing $\tilde{\pi}(\mu|\mathscr{D}_n)$ on a regular grid of width Δ and summing up.)

2.12 Show that the limit of the posterior probability $\mathbb{P}^\pi(\mu < 0|x)$ of (2.7) when τ goes to ∞ is $\Phi(-x/\sigma)$. Show that, when ξ varies in $\mathbb{R}$, the posterior probability can take any value between 0 and 1.

2.13 Define a function `BaRaJ` of the ratio `rat` when `z=mean(shift)/.75` in the function `BaFa`. Deduce from a plot of the function `BaRaJ` that the Bayes factor is always less than one when `rat` varies. (*Note:* It is possible to establish analytically that the Bayes factor is maximal and equal to 1 for $\tau = 0$.)

2.14 In the application part of Example 2.1 to **normaldata**, plot the approximated Bayes factor as a function of τ. (*Hint:* Simulate a single normal $\mathscr{N}(0,1)$ sample and recycle it for all values of τ.)

2.15 In the setup of Example 2.1, show that, when $\xi \sim \mathscr{N}(0,\sigma^2)$, the Bayes factor can be expressed in closed form using the normalizing constant of the t distribution (see Exercise 2.8)

2.16 Discuss what happens to the importance sampling approximation when the support of g is larger than the support of γ.

2.17 Show that, when γ is the normal $\mathscr{N}(0,\nu/(\nu-2))$ density and f_ν is the density of the t distribution with ν degrees of freedom, the ratio

$$\frac{f_\nu^2(x)}{\gamma(x)} \propto \frac{e^{x^2(\nu-2)/2\nu}}{[1+x^2/\nu]^{(\nu+1)}}$$

does not have a finite integral. What does this imply about the variance of the importance weights?

Deduce that the importance weights of Example 2.3 have infinite variance.

2.18 If f_ν denotes the density of the Student t distribution $\mathscr{T}(\nu,0,1)$ (see Exercise 2.8), consider the integral

$$\mathfrak{I} = \int \sqrt{\left|\frac{x}{1-x}\right|}\, f_\nu(x)\,\mathrm{d}x\,.$$

1. Show that $\mathfrak{I}$ is finite but that

$$\int \frac{|x|}{|1-x|} f_\nu(x)\,\mathrm{d}x = \infty\,.$$

2. Discuss the respective merits of the following importance functions γ
   - the density of the Student $\mathscr{T}(\nu,0,1)$ distribution,
   - the density of the Cauchy $\mathscr{C}(0,1)$ distribution,
   - the density of the normal $\mathscr{N}(0,\nu/(\nu-2))$ distribution.

   In particular, show via an R simulation experiment that these different choices all lead to unreliable estimates of $\mathfrak{I}$ and deduce that the three corresponding estimators have infinite variance.
3. Discuss the alternative choice of a gamma distribution folded at 1, that is, the distribution of x symmetric around 1 and such that

$$|x-1| \sim \mathcal{G}a(\alpha,1)\,.$$

Show that

$$h(x)\frac{f^2(x)}{\gamma(x)} \propto \sqrt{x}\, f_\nu^2(x)\, |1-x|^{1-\alpha-1} \exp|1-x|$$

is integrable around $x = 1$ when $\alpha < 1$ but not at infinity. Run a simulation experiment to evaluate the performances of this new proposal.

2.19 Evaluate the harmonic mean approximation

$$\hat{m}_1(\mathscr{D}_n) = 1\Big/ N^{-1} \sum_{j=1}^{N} \frac{1}{\ell_1(\theta_{1j}|\mathscr{D}_n)}\,.$$

when applied to the $\mathscr{N}(0,\sigma^2)$ model, **normaldata**, and an $\mathscr{IG}(1,1)$ prior on σ^2.

3

Regression and Variable Selection

You see, I always keep my sums.
—**Ian Rankin**, ***Strip Jack***.—

Roadmap

Linear regression is one of the most widely used tools in statistics for analyzing the (linear) influence of some variables or some factors on others and thus to uncover explanatory and predictive patterns. This chapter details the Bayesian analysis of the linear (or regression) model both in terms of prior specification (Zellner's G-prior) and in terms of variable selection, the next chapter appearing as a sequel for nonlinear dependence structures. The reader should be warned that, given that these models are the only conditional models where explicit computation can be conducted, this chapter contains a fair amount of matrix calculus. The photograph at the top of this page is a picture of processionary caterpillars, in connection (for once!) with the benchmark dataset used in this chapter.

J.-M. Marin and C.P. Robert, *Bayesian Essentials with R*, Springer Texts in Statistics, DOI 10.1007/978-1-4614-8687-9_3,

3.1 Linear Models

A large proportion of statistical analyses deal with the representation of dependences among several observed quantities. For instance, which social factors influence unemployment duration and the probability of finding a new job? Which economic indicators are best related to recession occurrences? Which physiological levels are most strongly correlated with aneurysm strokes? From a statistical point of view, the ultimate goal of these analyses is thus to find a proper representation of the conditional distribution, $f(y|\theta, \mathbf{x})$, of an observable variable y given a vector of observables $\mathbf{x}$, based on a sample of $\mathbf{x}$ and y. While the overall estimation of the conditional density f is usually beyond our ability, the estimation of θ and possibly of restricted features of f is possible within the Bayesian framework, as shown in this chapter.

The variable of primary interest, y, is called the *response* or the *outcome* variable; we assume here that this variable is continuous, but we will completely relax this assumption in the next chapter. The variables $\mathbf{x} = (x_1, \ldots, x_p)$ are called *explanatory variables* and may be discrete, continuous, or both. One sometimes picks a single variable x_j to be of primary interest. We then call it the *treatment* variable, labeling the other components of x as *control* variables, meaning that we want to address the (linear) influence of x_j on y once the linear influence of all the other variables has been taken into account (as in medical studies). The distribution of y given $\mathbf{x}$ is typically studied in the context of a set of *units* or experimental *subjects*, $i = 1, \ldots, n$, such as patients in a hospital ward, on which both y_i and $x_{i1}, \ldots, x_{ip}$ are measured. The dataset is then made up of the reunion of the vector of outcomes

$$\mathbf{y} = (y_1, \ldots, y_n)$$

and the $n \times p$ matrix of explanatory variables

$$\mathbf{X} = [\mathbf{x}_1 \quad \ldots \quad \mathbf{x}_p] = \begin{bmatrix} x_{11} & x_{12} & \ldots & x_{1p} \\ x_{21} & x_{22} & \ldots & x_{2p} \\ x_{31} & x_{32} & \ldots & x_{3p} \\ \vdots & \vdots & \vdots & \vdots \\ x_{n1} & x_{n2} & \ldots & x_{np} \end{bmatrix} .$$

The **caterpillar** dataset exploited in this chapter was extracted from a 1973 study on pine processionary[1] caterpillars: it assesses the influence of some forest settlement characteristics on the development of caterpillar colonies. This dataset was first published and studied in Tomassone et al. (1993). The response variable is the logarithmic transform of the average number of nests of caterpillars per tree (as the one in the picture at the beginning of this chapter) in an area of 500 m^2 (which corresponds to the last column in **caterpillar**). There are $p = 8$ potential explanatory variables defined on $n = 33$ areas, as follows

[1] These caterpillars derive their name from their habit of moving over the ground in incredibly long head-to-tail monk-like processions when leaving their nest to create a new colony.

x_1 is the altitude (in meters),
x_2 is the slope (in degrees),
x_3 is the number of pine trees in the area,
x_4 is the height (in meters) of the tree sampled at the center of the area,
x_5 is the orientation of the area (from 1 if southbound to 2 otherwise),
x_6 is the height (in meters) of the dominant tree,
x_7 is the number of vegetation strata,
x_8 is the mix settlement index (from 1 if not mixed to 2 if mixed).

The goal of the regression analysis is to decide which explanatory variables have a strong influence on the number of nests and how these influences overlap with one another. As shown by Fig. 3.1, some of these variables clearly have a restricting influence on the number of nests, as for instance with x_5, x_7 and x_8. We use the following R code to produce Fig. 3.1 (the way we created the objects y and X will be described later).

```
> par(mfrow=c(2,4),mar=c(4.2,2,2,1.2))
> for (j in 1:8) plot(X[,j],y,xlab=vnames[j],pch=19,
+ col="sienna4",xaxt="n",yaxt="n")
```

While many models and thus many dependence structures can be proposed for dependent datasets like **caterpillar**, in this chapter we only focus on the Gaussian linear regression model, namely the case when $\mathbb{E}[y|x,\theta]$ is linear in x and the noise is normal.

The *ordinary normal linear regression* model is such that, using a matrix representation,

$$\mathbf{y}|\alpha,\boldsymbol{\beta},\sigma^2 \sim \mathscr{N}_n\left(\alpha\mathbf{1}_n + \mathbf{X}\boldsymbol{\beta}, \sigma^2\mathbf{I}_n\right),$$

where $\mathscr{N}_n$ denotes the normal distribution in dimension n, and thus the y_i's are independent normal random variables with

$$\mathbb{E}[y_i|\alpha,\boldsymbol{\beta},\sigma^2] = \alpha + \beta_1 x_{i1} + \ldots + \beta_p x_{ip}\,, \quad \mathbb{V}[y_i|\alpha,\boldsymbol{\beta},\sigma^2] = \sigma^2\,.$$

Given that the models studied in this chapter are all conditional on the regressors, we omit the conditioning on $\mathbf{X}$ to simplify the notations.

For **caterpillar**, where $n = 33$ and $p = 8$, we thus assume that the expected lognumber y_i of caterpillar nests per tree over an area is modeled as a linear combination of an intercept and eight predictor variables ($i = 1, \ldots, n$),

$$\mathbb{E}[y_i|\alpha,\boldsymbol{\beta},\sigma^2] = \alpha + \sum_{j=1}^{8} \beta_j x_{ij}\,,$$

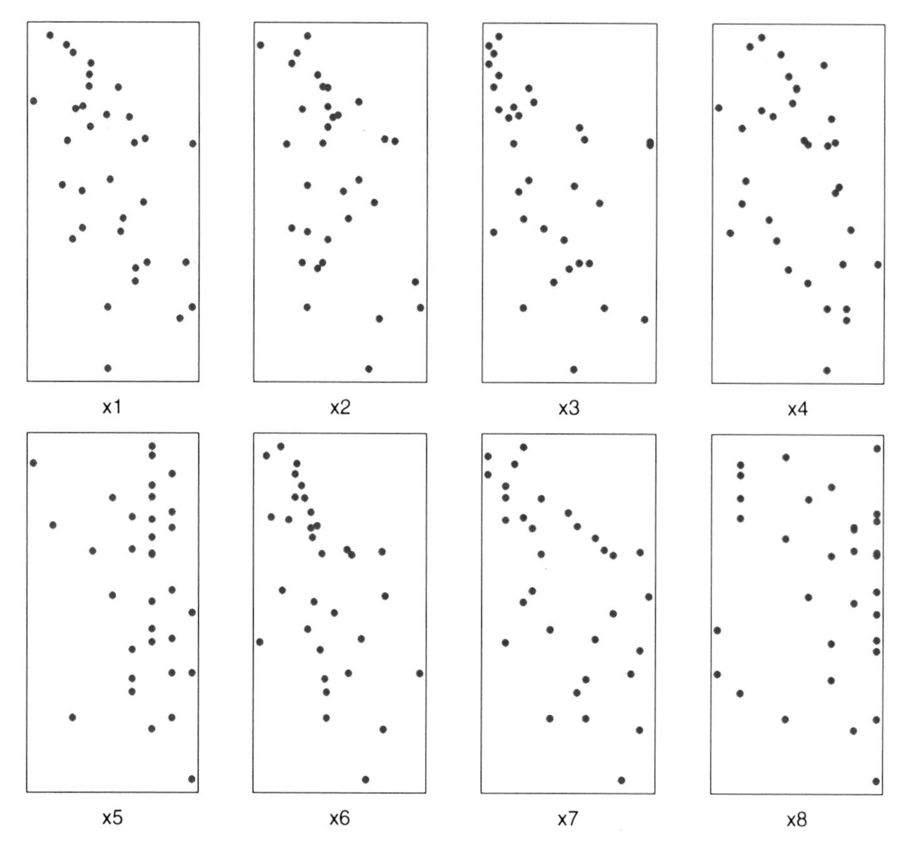

Fig. 3.1. Dataset **caterpillar**: Plot of the pairs $(\mathbf{x}_j, \mathbf{y})$ $(1 \leq j \leq 8)$

while the variation around this expectation is supposed to be normally distributed. Note that it is also customary to assume that the y_i's are conditionally independent.

The **caterpillar** dataset is called by the command `data(caterpillar)` and is made of the following rows:

```
1200 22 1 4 1.1 5.9 1.4 1.4 2.37
1342 28 8 4.4 1.5 6.4 1.7 1.7 1.47
....
1229 21 11 5.8 1.8 10 2.3 2 0.21
1310 36 17 5.2 1.9 10.3 2.6 2 0.03
```

The first eight columns correspond to the explanatory variables and the last column is the response variable, i.e. the lognumber of caterpillar nests. The following R code is an example for starting with this **caterpillar** dataset:

```
> y=log(caterpillar$y)
> X=as.matrix(caterpillar[,1:8])
```

There is a difference between using finite-valued regressors like x_7 in caterpillar and using *categorical* variables (or *factors*), which also take a finite number of values but whose range has no numerical meaning. For instance, if x denotes the socio-professional category of an employee, this variable may range from 1 to 9 for a rough grid of socio-professional activities, or it may range from 1 to 89 on a finer grid, and the numerical values are not comparable. It thus makes little sense to involve x directly in the regression, and the usual approach is to replace the single regressor x (taking values in $\{1, \dots, m\}$, say) with m indicator (or *dummy*) variables $x_1 = \mathbb{I}_1(x)$, ..., $x_m = \mathbb{I}_m(x)$. In essence, a different constant (or *intercept*) β_j is used in the regression for each class of categorical variable: it is invoked in the linear regression under the form

$$\dots + \beta_1 \mathbb{I}_1(x) + \dots + \beta_m \mathbb{I}_m(x) + \dots .$$

Note that there is an identifiability issue related with this model since the sum of the indicators is always equal to one. In a Bayesian perspective, identifiability can be achieved via the prior distribution. However, we can also impose an identifiability constraint on the parameters, for instance by omitting one class (such as $\beta_1 = 0$). We pursue this direction further in Sects. 4.5.1 and 6.2.

3.2 Classical Least Squares Estimator

Before fully launching into the description of the Bayesian approach to the linear model, we recall the basics of the classical processing of this model (in particular, to relate the Bayesian perspective to the results provided by standard software such as R lm output). For instance, the parameter $\boldsymbol{\beta}$ can obviously be estimated via maximum likelihood estimation. In order to avoid non-identifiability and uniqueness problems, we assume that $[\mathbf{1}_n \quad \mathbf{X}]$ is of full rank, that is, rank $[\mathbf{1}_n \quad \mathbf{X}] = p+1$. This also means that there is no redundant structure among the explanatory variables.[2] We suppose in addition that $p + 1 < n$ in order to obtain well-defined estimates for all parameters. Notice that, since the inferential process is conditioned on the design matrix $\mathbf{X}$, we choose to standardize the data, namely to center and to scale the columns of $\mathbf{X}$ so that the estimated values of $\boldsymbol{\beta}$ are truly comparable. For this purpose, we use the R function `scale`:

```
> X=scale(X)
```

[2]Hence, the exclusion of one of the classes for categorical variables.

The likelihood $\ell(\alpha, \boldsymbol{\beta}, \sigma^2|\mathbf{y})$ of the *standard normal linear model* is provided by the following matrix representation:

$$\frac{1}{(2\pi\sigma^2)^{n/2}} \exp\left\{-\frac{1}{2\sigma^2}\left(\mathbf{y}-\alpha\mathbf{1}_n-\mathbf{X}\boldsymbol{\beta}\right)^{\mathrm{T}}\left(\mathbf{y}-\alpha\mathbf{1}_n-\mathbf{X}\boldsymbol{\beta}\right)\right\}. \tag{3.1}$$

The maximum likelihood estimators of α and $\boldsymbol{\beta}$ are then the solution of the (least squares) minimization problem

$$\begin{aligned}
&\min_{\alpha,\boldsymbol{\beta}} \left(\mathbf{y}-\alpha\mathbf{1}_n-\mathbf{X}\boldsymbol{\beta}\right)^{\mathrm{T}}\left(\mathbf{y}-\alpha\mathbf{1}_n-\mathbf{X}\boldsymbol{\beta}\right)\\
&\qquad = \min_{\alpha,\boldsymbol{\beta}} \sum_{i=1}^{n}\left(y_i-\alpha-\beta_1 x_{i1}-\ldots-\beta_p x_{ip}\right)^2 ,
\end{aligned}$$

If we denote by $\bar{\mathbf{y}} = \frac{1}{n}\sum_{i=1}^{n} y_i$ the empirical mean of the y_i's and recall that, $\mathbf{1}_n^{\mathrm{T}}\mathbf{X} = \mathbf{0}_n^{\mathrm{T}}$ because of the standardization step, we have a Pythagorean decomposition of the above norm as

$$\begin{aligned}
&\left(\mathbf{y}-\alpha\mathbf{1}_n-\mathbf{X}\boldsymbol{\beta}\right)^{\mathrm{T}}\left(\mathbf{y}-\alpha\mathbf{1}_n-\mathbf{X}\boldsymbol{\beta}\right)\\
&= \left(\mathbf{y}-\bar{\mathbf{y}}\mathbf{1}_n-\mathbf{X}\boldsymbol{\beta}+(\bar{\mathbf{y}}-\alpha)\mathbf{1}_n\right)^{\mathrm{T}}\left(\mathbf{y}-\bar{\mathbf{y}}\mathbf{1}_n-\mathbf{X}\boldsymbol{\beta}+(\bar{\mathbf{y}}-\alpha)\mathbf{1}_n\right)\\
&= \left(\mathbf{y}-\bar{\mathbf{y}}\mathbf{1}_n-\mathbf{X}\boldsymbol{\beta}\right)^{\mathrm{T}}\left(\mathbf{y}-\bar{\mathbf{y}}\mathbf{1}_n-\mathbf{X}\boldsymbol{\beta}\right)+2(\bar{\mathbf{y}}-\alpha)\mathbf{1}_n^{\mathrm{T}}\left(\mathbf{y}-\bar{\mathbf{y}}\mathbf{1}_n-\mathbf{X}\boldsymbol{\beta}\right)+n(\bar{\mathbf{y}}-\alpha)^2\\
&= \left(\mathbf{y}-\bar{\mathbf{y}}\mathbf{1}_n-\mathbf{X}\boldsymbol{\beta}\right)^{\mathrm{T}}\left(\mathbf{y}-\bar{\mathbf{y}}\mathbf{1}_n-\mathbf{X}\boldsymbol{\beta}\right)+n(\bar{\mathbf{y}}-\alpha)^2 .
\end{aligned}$$

Indeed, $\mathbf{1}_n^{\mathrm{T}}\left(\mathbf{y}-\bar{\mathbf{y}}\mathbf{1}_n-\mathbf{X}\boldsymbol{\beta}\right) = (n\bar{\mathbf{y}}-n\bar{\mathbf{y}}) = 0$. Therefore, the likelihood $\ell(\alpha, \boldsymbol{\beta}, \sigma^2|\mathbf{y})$ is given by

$$\frac{1}{(2\pi\sigma^2)^{n/2}} \exp\left(-\frac{1}{2\sigma^2}\left(\mathbf{y}-\bar{\mathbf{y}}\mathbf{1}_n-\mathbf{X}\boldsymbol{\beta}\right)^{\mathrm{T}}\left(\mathbf{y}-\bar{\mathbf{y}}\mathbf{1}_n-\mathbf{X}\boldsymbol{\beta}\right)\right) \exp\left\{-\frac{n}{2\sigma^2}(\bar{\mathbf{y}}-\alpha)^2\right\}.$$

We get from the above decomposition that

$$\hat{\alpha} = \bar{\mathbf{y}}, \quad \hat{\boldsymbol{\beta}} = (\mathbf{X}^{\mathsf{T}}\mathbf{X})^{-1}\mathbf{X}^{\mathsf{T}}(\mathbf{y}-\bar{y}).$$

In geometrical terms, $(\hat{\alpha}, \hat{\boldsymbol{\beta}})$ is the orthogonal projection of $\mathbf{y}$ on the linear subspace spanned by the columns of $[\mathbf{1}_n \quad \mathbf{X}]$. It is quite simple to check that $(\hat{\alpha}, \hat{\boldsymbol{\beta}})$ is an unbiased estimator of (α, β). In fact, the Gauss–Markov theorem (see, e.g., Christensen, 2002) states that $(\hat{\alpha}, \hat{\beta})$ is the *best* linear unbiased estimator of (α, β). This means that, for all $a \in \mathbb{R}^{p+1}$, and with the abuse of notation that, here, $(\hat{\alpha}, \hat{\beta})$ represents a column vector,

$$\mathbb{V}(a^{\mathsf{T}}(\hat{\alpha}, \hat{\beta})|\alpha, \boldsymbol{\beta}, \sigma^2) \leq \mathbb{V}(a^{\mathsf{T}}(\tilde{\alpha}, \tilde{\beta})|\alpha, \boldsymbol{\beta}, \sigma^2)$$

for any unbiased linear estimator $(\tilde{\alpha}, \tilde{\beta})$ of (α, β). (Note that the property of unbiasedness is not particularly appealing when considered on its own.)

An unbiased estimator of σ^2 is

$$\hat{\sigma}^2 = \frac{1}{n-p-1}(\mathbf{y} - \hat{\alpha}\mathbf{1}_n - \mathbf{X}\hat{\beta})^\mathsf{T}(\mathbf{y} - \hat{\alpha}\mathbf{1}_n - \mathbf{X}\hat{\beta}) = \frac{s^2}{n-p-1},$$

and $\hat{\sigma}^2(\mathbf{X}^\mathsf{T}\mathbf{X})^{-1}$ approximates the covariance matrix of $\hat{\boldsymbol{\beta}}$. Note that the MLE of σ^2 is not $\hat{\sigma}^2$ but $\tilde{\sigma}^2 = s^2/n$.

The standard *t-statistics* are defined as $(j = 1, \ldots, p)$

$$T_j = \frac{\hat{\beta}_j - \beta_j}{\sqrt{\hat{\sigma}^2 \omega_{jj}}} \sim \mathscr{T}(n-p-1, 0, 1),$$

where ω_{jj} denotes the (j,j)-th element of the matrix $(\mathbf{X}^\mathsf{T}\mathbf{X})^{-1}$. These t-statistic are used in classical tests, for instance for testing $H_0 : \beta_j = 0$ versus $H_1 : \beta_j \neq 0$, the former being accepted at level γ if

$$|\hat{\beta}_j|/\hat{\sigma}\sqrt{\omega_{jj}} < F_{n-p-1}^{-1}(1-\gamma/2)$$

the $(1-\gamma/2)$th quantile of the Student's t $\mathscr{T}(n-p-1, 0, 1)$ distribution (with location parameter 0 and scale parameter 1). The frequentist argument in using this bound (see Casella and Berger, 2001) is that the so-called *p-value* is smaller than γ,

$$p_j = P_{H_0}(|T_j| > |t_j|) < \gamma.$$

Note that these statistics T_j can also be used when constructing marginal frequentist confidence intervals on the β_j's like

$$\left\{\beta_j;\ \left|\beta_j - \hat{\beta}_j\right| \leq \hat{\sigma}\sqrt{\omega_{jj}}\, F_{n-p-1}^{-1}(1-\gamma/2)\right\} = \left\{\beta_j;\ |T_j| \leq \hat{\sigma}\sqrt{\omega_{jj}}\, F_{n-p-1}^{-1}(1-\gamma/2)\right\}.$$

↯ From a Bayesian perspective, we far from advocate the use of p-values in Bayesian settings or elsewhere since they suffer many defects (exposed for instance in Robert, 2007, Chap. 5), one being that they are often wrongly interpreted as probabilities of the null hypotheses.

For **caterpillar**, the unbiased estimate of σ^2 is equal to 0.7781 and the maximum likelihood estimates of α and of the components β_j produced by the R command

```
> summary(lm(y~X))
```

are given in Fig. 3.2, along with the least squares estimates of their respective standard deviations and p-values. According to the classical paradigm, the coefficients β_1, β_2 and β_7 are the only ones considered to be *significant*.

We stress here that conditioning on $\mathbf{X}$ is valid only when $\mathbf{X}$ is *exogenous*, that is, only when we can write the joint distribution of $(\mathbf{y}, \mathbf{X})$ as

$$f(\mathbf{y}, \mathbf{X}|\alpha, \boldsymbol{\beta}, \sigma^2, \delta) = f(\mathbf{y}|\alpha, \boldsymbol{\beta}, \sigma^2, \mathbf{X}) f(\mathbf{X}|\delta),$$

where $(\alpha, \boldsymbol{\beta}, \sigma^2)$ and δ are fixed parameters. We can thus ignore $f(\mathbf{X}|\delta)$ if the parameter δ is only a nuisance parameter since this part is independent[3] of $(\alpha, \boldsymbol{\beta}, \sigma^2)$. The practical advantage of using a regression model as above is that it is much easier to specify a realistic conditional distribution of one variable given p others rather than a joint distribution on all $p+1$ variables. Note that if $\mathbf{X}$ is not *exogenous*, for instance when $\mathbf{X}$ involves past values of $\mathbf{y}$ (see Chap. 7), the joint distribution must be used instead.

```
Residuals:
    Min      1Q  Median      3Q     Max
-1.4710 -0.4474 -0.1769  0.6121  1.5602

lm(formula = y ~ X)

Residuals:
    Min      1Q  Median      3Q     Max
-1.4710 -0.4474 -0.1769  0.6121  1.5602

Coefficients:
            Estimate Std. Error t value Pr(>|t|)
(Intercept) -0.81328    0.15356  -5.296 1.97e-05 ***
Xx1         -0.52722    0.21186  -2.489   0.0202 *
Xx2         -0.39286    0.16974  -2.315   0.0295 *
Xx3          0.65133    0.38670   1.684   0.1051
Xx4         -0.29048    0.31551  -0.921   0.3664
Xx5         -0.21645    0.16865  -1.283   0.2116
Xx6          0.29361    0.53562   0.548   0.5886
Xx7         -1.09027    0.47020  -2.319   0.0292 *
Xx8         -0.02312    0.17225  -0.134   0.8944
---
Signif. codes:  0 ?***? 0.001 ?**? 0.01 ?*? 0.05 ?.? 0.1 ? ? 1

Residual standard error: 0.8821 on 24 degrees of freedom
Multiple R-squared: 0.6234,Adjusted R-squared: 0.4979
```

Fig. 3.2. Dataset **caterpillar**: R output providing the least squares estimates of the regression coefficients along with their standard significance analysis

[3]From a Bayesian point of view, note that we would also need to impose prior independence between $(\alpha, \boldsymbol{\beta}, \sigma^2)$ and δ to achieve this separation.

3.3 The Jeffreys Prior Analysis

Considering only the case of a complete lack of prior information on the parameters of the linear model, we first describe a noninformative solution based on the Jeffreys prior. It is rather easy to show that the Jeffreys prior in this case is

$$\pi^J(\alpha, \boldsymbol{\beta}, \sigma^2) \propto \sigma^{-2},$$

which is equivalent to a flat prior on $(\alpha, \boldsymbol{\beta}, \log \sigma^2)$. We recall that

$$\begin{aligned}
\ell(\alpha, \boldsymbol{\beta}, \sigma^2|\mathbf{y}) &= \frac{1}{(2\pi\sigma^2)^{n/2}} \exp\left\{-\frac{1}{2\sigma^2}(\mathbf{y} - \bar{\mathbf{y}}\mathbf{1}_n - \mathbf{X}\boldsymbol{\beta})^{\mathrm{T}}(\mathbf{y} - \bar{\mathbf{y}}\mathbf{1}_n - \mathbf{X}\boldsymbol{\beta})\right\} \times \\
&\quad \exp\left\{-\frac{n}{2\sigma^2}(\bar{\mathbf{y}} - \alpha)^2\right\} \\
&= \frac{1}{(2\pi\sigma^2)^{n/2}} \exp\left\{-\frac{1}{2\sigma^2}\left(\mathbf{y} - \hat{\alpha}\mathbf{1}_n - \mathbf{X}\hat{\boldsymbol{\beta}}\right)^{\mathrm{T}}\left(\mathbf{y} - \hat{\alpha}\mathbf{1}_n - \mathbf{X}\hat{\boldsymbol{\beta}}\right)\right\} \times \\
&\quad \exp\left\{-\frac{n}{2\sigma^2}(\hat{\alpha} - \alpha)^2 - \frac{1}{2\sigma^2}(\boldsymbol{\beta} - \hat{\boldsymbol{\beta}})^{\mathsf{T}}\mathbf{X}^{\mathsf{T}}\mathbf{X}(\boldsymbol{\beta} - \hat{\boldsymbol{\beta}})\right\}.
\end{aligned}$$

The corresponding posterior distribution is therefore

$$\begin{aligned}
\pi^J(\alpha, \boldsymbol{\beta}, \sigma^2|\mathbf{y}) &\propto \left(\sigma^{-2}\right)^{-n/2} \exp\left\{-\frac{1}{2\sigma^2}(\mathbf{y} - \hat{\alpha}\mathbf{1}_n - \mathbf{X}\hat{\boldsymbol{\beta}})^{\mathsf{T}}(\mathbf{y} - \hat{\alpha}\mathbf{1}_n - \mathbf{X}\hat{\boldsymbol{\beta}})\right\} \times \\
&\quad \sigma^{-2} \exp\left\{-\frac{n}{2\sigma^2}(\hat{\alpha} - \alpha)^2 - \frac{1}{2\sigma^2}(\boldsymbol{\beta} - \hat{\boldsymbol{\beta}})^{\mathsf{T}}\mathbf{X}^{\mathsf{T}}\mathbf{X}(\boldsymbol{\beta} - \hat{\boldsymbol{\beta}})\right\} \\
&\propto \left(\sigma^{-2}\right)^{-p/2} \exp\left\{-\frac{1}{2\sigma^2}(\boldsymbol{\beta} - \hat{\boldsymbol{\beta}})^{\mathsf{T}}\mathbf{X}^{\mathsf{T}}\mathbf{X}(\boldsymbol{\beta} - \hat{\boldsymbol{\beta}})\right\} \times \\
&\quad \left(\sigma^{-2}\right)^{-1/2} \exp\left\{-\frac{n}{2\sigma^2}(\hat{\alpha} - \alpha)^2\right\} \\
&\quad \left(\sigma^{-2}\right)^{-(n-p-1)/2-1} \exp\left\{-\frac{1}{2\sigma^2}s^2\right\}.
\end{aligned}$$

From this expression, we deduce the following (conditional and marginal) posterior distributions

$$\begin{aligned}
\alpha|\sigma^2, \mathbf{y} &\sim \mathscr{N}\left(\hat{\alpha}, \sigma^2/n\right), \\
\boldsymbol{\beta}|\sigma^2, \mathbf{y} &\sim \mathscr{N}_p\left(\hat{\boldsymbol{\beta}}, \sigma^2(\mathbf{X}^{\mathsf{T}}\mathbf{X})^{-1}\right), \\
\sigma^2|\mathbf{y} &\sim \mathscr{IG}((n-p-1)/2, s^2/2).
\end{aligned}$$

↯ As in every analysis involving an improper prior, one needs to check that the corresponding posterior distribution is proper. In this case, $\pi(\alpha, \boldsymbol{\beta}, \sigma^2|\mathbf{y})$ is proper when both $n > p+1$ and rank $[\mathbf{1}_n \quad \mathbf{X}] = p+1$. The former constraint requires that there be at least as many data points as there are parameters in the model, and, as already explained above, the latter is obviously necessary for identifiability reasons.

The corresponding Bayesian estimates of α, $\boldsymbol{\beta}$ and σ^2 are thus given by

$$\mathbb{E}^\pi[\alpha|\mathbf{y}] = \hat{\alpha}\,, \quad \mathbb{E}^\pi[\boldsymbol{\beta}|\mathbf{y}] = \hat{\boldsymbol{\beta}} \quad \text{and} \quad \mathbb{E}^\pi[\sigma^2|\mathbf{y}] = \frac{s^2}{n-p-3}\,,$$

respectively. Unsurprisingly, the Jeffreys prior estimate of α is the empirical mean. Further, the posterior expectation of $\boldsymbol{\beta}$ is the maximum likelihood estimate. Note also that the Jeffreys prior estimate of σ^2 is larger (and thus more pessimistic) than both the maximum likelihood estimate s^2/n and the classical unbiased estimate $s^2/(n-p-1)$.

The marginal posterior distribution of β_j associated with the above joint distribution is

$$\mathscr{T}(n-p-1, \hat{\beta}_j, \omega_{jj}s^2/(n-p-1))\,,$$

(recall that $\omega_{jj} = (\mathbf{X}^\mathsf{T}\mathbf{X})^{-1}_{(j,j)}$). Hence, the similarity with a frequentist analysis of this model is very strong since the classical $(1-\gamma)$ confidence interval and the Bayesian HPD interval on β_j coincide, even though they have different interpretations. They are both equal to

$$\left\{\beta_j; |\beta_j - \hat{\beta}_j| \le F^{-1}_{n-p-1}(1-\gamma/2)\sqrt{\omega_{jj}s^2/(n-p-1)}\right\}.$$

For **caterpillar**, the Bayes estimate of σ^2 is equal to 0.8489. Figure 3.3 provides the corresponding (marginal) 95 % HPD intervals for each component of β. (It is obtained by the `plotCI` function, part of the `gplots` package.) Note that while some of these credible intervals include the value $\beta_j = 0$ (represented by the dashed line), they do not necessarily validate acceptance of the null hypothesis $H_0 : \beta_j = 0$, which must be tested through a Bayes factor, as described below. This distinction is a major difference from the classical approach, where confidence intervals are dual sets of acceptance regions.

3.4 Zellner's G-Prior Analysis

From this section onwards,[4] we concentrate on a different noninformative approach which was proposed by Arnold Zellner[5] to handle linear regression from a Bayesian perspective. This approach is a middle-ground perspective where some prior information may be available on $\boldsymbol{\beta}$ and it is called *Zellner's G-prior*, the "*G*" being the symbol used by Zellner in the prior variance.

[4]In order to keep this coverage of G-priors simple and self-contained, we made several choices in the presentation that the most mature readers will possibly find arbitrary, but this cannot be avoided if we want to keep the chapter at a reasonable length.

[5]Arnold Zellner was a famous Bayesian econometrician, who wrote two reference books on Bayesian econometrics (Zellner, 1971, 1984)

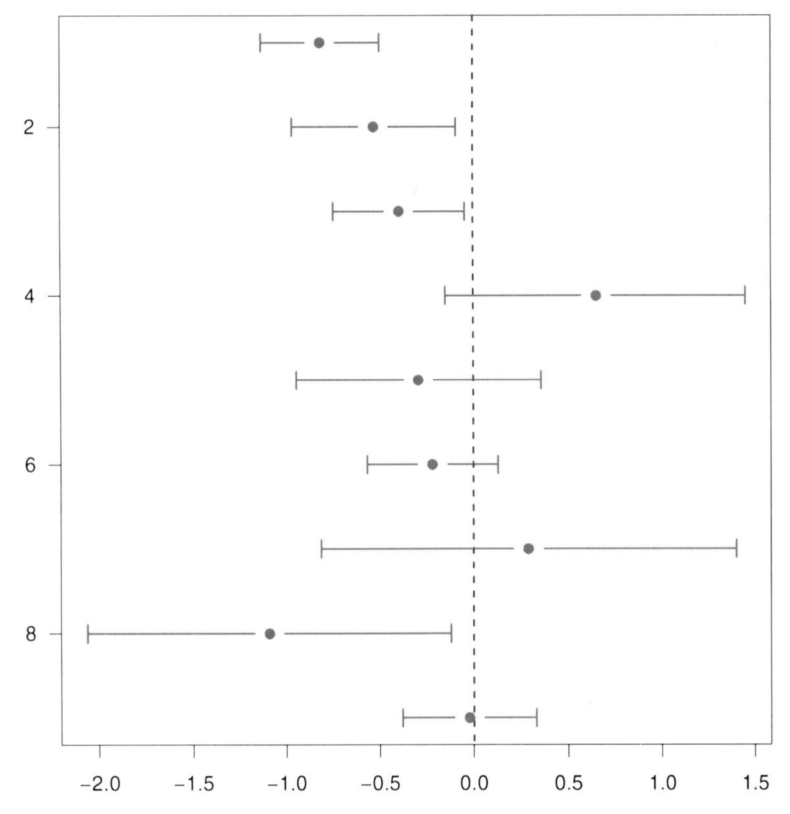

Fig. 3.3. Dataset **caterpillar**: Range of the credible 95 % HPD intervals for α (*top row*) and each component of β when using the Jeffreys prior

3.4.1 A Semi-noninformative Solution

When considering the likelihood (3.1) its shape is both Gaussian and Inverse Gamma, indeed, $\boldsymbol{\beta}$ given σ^2 appears in a Gaussian-like expression, while σ^2 involves an Inverse Gamma expression. This structure leads to a natural conjugate prior family, of the form

$$(\alpha, \boldsymbol{\beta})|\sigma^2 \sim \mathscr{N}_{p+1}((\tilde{\alpha}, \tilde{\boldsymbol{\beta}}), \sigma^2 M^{-1}),$$

conditional on σ^2, where M is a $(p+1, p+1)$ positive definite symmetric matrix, and for σ^2,

$$\sigma^2 \sim \mathscr{I}\mathscr{G}(a, b), \qquad a, b > 0.$$

(The conjugacy can be easily checked by the reader.) Even in the presence of genuine information on the parameters, the hyperparameters M, a and b are very difficult to specify. Moreover, the posterior distributions, notably the posterior variances are sensitive to the specification of these hyper-parameters.

Therefore, given that a natural conjugate prior for the linear regression model has severe limitations, a more elaborate strategy is called for. The idea at the core of Zellner's G-prior modeling is to allow the experimenter to introduce (possibly weak) information about the location parameter of the regression but to bypass the most difficult aspects of the prior specification, namely the derivation of the prior correlation structure. This structure is fixed in Zellner's proposal since the prior corresponds to

$$\boldsymbol{\beta}|\alpha,\sigma^2 \sim \mathscr{N}_p\left(\tilde{\boldsymbol{\beta}}, g\sigma^2(\mathbf{X}^{\mathrm{T}}\mathbf{X})^{-1}\right), \tag{3.2}$$

and a noninformative prior distribution is imposed on the pair (α,σ^2),

$$\pi\left(\alpha,\sigma^2\right) \propto \sigma^{-2}. \tag{3.3}$$

Zellner's G-prior is thus decomposed as a (conditional) Gaussian prior for $\boldsymbol{\beta}$ and an improper (Jeffreys) prior for (α,σ^2). This modelling somehow appears as a data-dependent prior through its dependence on $\mathbf{X}$, but this is not a genuine issue[6] since the *whole* model is conditional on $\mathbf{X}$. The experimenter thus restricts prior determination to the choices of $\tilde{\boldsymbol{\beta}}$ and of the constant g. As we will see once the posterior distribution is constructed, the factor g can be interpreted as being inversely proportional to the amount of information available in the prior relative to the sample. For instance, setting $g = n$ gives the prior the same weight as one observation of the sample. We will use this as our default value.

↯ Genuine data-dependent priors are not acceptable in a Bayesian analysis because they use the data *twice* and fail to enjoy the basic convergence properties of the Bayes estimators. (See Carlin and Louis, 1996, for a comparative study of the corresponding so-called *empirical Bayes* estimators.)

Note that, in the initial proposition of Zellner (1984), the parameter α is not modelled by a flat prior distribution. It was instead considered to be a component of the vector $\boldsymbol{\beta}$. (This was also the approach adopted in Marin and Robert 2007.) However, endowing α with a flat prior ensures the location-scale invariance of the analysis, which means that changes in location or scale on $\mathbf{y}$ (like a switch from Celsius to Fahrenheit degrees for temperatures) do not impact on the resulting inference.

We are now engaging into some algebra that will expose the properties of the G-posterior. First, we assume $p > 0$, meaning that there is at least one explanatory variable in the model. We define the matrix $\mathbf{P} = \mathbf{X}\left\{\mathbf{X}^{\mathrm{T}}\mathbf{X}\right\}^{-1}\mathbf{X}^{\mathrm{T}}$. The prior $\pi\left(\alpha,\boldsymbol{\beta},\sigma^2\right)$ can then be decomposed as

[6]This choice is more problematic when conditioning on $\mathbf{X}$ is no longer possible, as for instance when $\mathbf{X}$ contains lagged dependent variables (Chap. 7) or endogenous variables.

$$\pi\left(\alpha,\boldsymbol{\beta},\sigma^2\right) \propto (\sigma^2)^{-p/2}\exp\left[-\frac{1}{2g\sigma^2}\left\{\boldsymbol{\beta}^{\mathrm{T}}\mathbf{X}^{\mathrm{T}}\mathbf{X}\boldsymbol{\beta}-2\boldsymbol{\beta}^{\mathrm{T}}\mathbf{X}^{\mathrm{T}}\mathbf{P}\mathbf{X}\tilde{\boldsymbol{\beta}}\right\}\right]\times$$
$$\sigma^{-2}\exp\left(-\frac{1}{2g\sigma^2}\tilde{\boldsymbol{\beta}}^{\mathrm{T}}\mathbf{X}^{\mathrm{T}}\mathbf{P}\mathbf{X}\tilde{\boldsymbol{\beta}}\right),$$

since $\mathbf{X}^{\mathrm{T}}\mathbf{P}\mathbf{X}=\mathbf{X}^{\mathrm{T}}\mathbf{X}$. Therefore,

$$\pi\left(\alpha,\boldsymbol{\beta},\sigma^2|\mathbf{y}\right) \propto (\sigma^2)^{-n/2-p/2-1}$$
$$\exp\left\{-\frac{1}{2\sigma^2}\left(\mathbf{y}-\bar{\mathbf{y}}\mathbf{1}_n-\mathbf{X}\boldsymbol{\beta}\right)^{\mathrm{T}}\left(\mathbf{y}-\bar{\mathbf{y}}\mathbf{1}_n-\mathbf{X}\boldsymbol{\beta}\right)\right\}\times$$
$$\exp\left\{-\frac{n}{2\sigma^2}(\bar{\mathbf{y}}-\alpha)^2\right\}\times\exp\left\{-\frac{1}{2g\sigma^2}\tilde{\boldsymbol{\beta}}^{\mathrm{T}}\mathbf{X}^{\mathrm{T}}\mathbf{P}\mathbf{X}\tilde{\boldsymbol{\beta}}\right\}\times$$
$$\exp\left\{-\frac{1}{2g\sigma^2}\left[\boldsymbol{\beta}\mathrm{T}\mathbf{X}^{\mathrm{T}}\mathbf{X}\boldsymbol{\beta}-2\boldsymbol{\beta}^{\mathrm{T}}\mathbf{X}^{\mathrm{T}}\mathbf{P}\mathbf{X}\tilde{\boldsymbol{\beta}}\right]\right\}.$$

Since $\mathbf{1}_n^{\mathrm{T}}\mathbf{X}=\mathbf{0}_p$, we deduce that

$$\pi\left(\alpha,\boldsymbol{\beta},\sigma^2|\mathbf{y}\right) \propto (\sigma^2)^{-n/2-p/2-1}\exp\left\{-\frac{1}{2\sigma^2}\left[\boldsymbol{\beta}^{\mathrm{T}}\mathbf{X}^{\mathrm{T}}\mathbf{X}\boldsymbol{\beta}-2\mathbf{y}^{\mathrm{T}}\mathbf{X}\boldsymbol{\beta}\right]\right\}\times$$
$$\exp\left\{-\frac{1}{2\sigma^2}\left(\mathbf{y}-\bar{\mathbf{y}}\mathbf{1}_n\right)^{\mathrm{T}}\left(\mathbf{y}-\bar{\mathbf{y}}\mathbf{1}_n\right)\right\}\times$$
$$\exp\left\{-\frac{n}{2\sigma^2}(\bar{\mathbf{y}}-\alpha)^2\right\}\times\exp\left\{-\frac{1}{2g\sigma^2}\tilde{\boldsymbol{\beta}}^{\mathrm{T}}\mathbf{X}^{\mathrm{T}}\mathbf{P}\mathbf{X}\tilde{\boldsymbol{\beta}}\right\}\times$$
$$\exp\left\{-\frac{1}{2g\sigma^2}\left[\boldsymbol{\beta}^{\mathrm{T}}\mathbf{X}^{\mathrm{T}}\mathbf{X}\boldsymbol{\beta}-2\boldsymbol{\beta}^{\mathrm{T}}\mathbf{X}^{\mathrm{T}}\mathbf{P}\mathbf{X}\tilde{\boldsymbol{\beta}}\right]\right\}.$$

Since $\mathbf{P}\mathbf{X}=\mathbf{X}$, we deduce that, conditionally on $\mathbf{y}$, $\mathbf{X}$ and σ^2, the parameters α and $\boldsymbol{\beta}$ are independent and such that

$$\alpha|\sigma^2,\mathbf{y}\sim\mathscr{N}_1\left(\bar{\mathbf{y}},\sigma^2/n\right),$$

$$\boldsymbol{\beta}|\mathbf{y},\sigma^2\sim\mathscr{N}_p\left(\frac{g}{g+1}\left(\hat{\boldsymbol{\beta}}+\mathbf{X}\tilde{\boldsymbol{\beta}}/g\right),\frac{\sigma^2 g}{g+1}\left\{\mathbf{X}^{\mathrm{T}}\mathbf{X}\right\}^{-1}\right),$$

where $\hat{\boldsymbol{\beta}}=\left\{\mathbf{X}^{\mathrm{T}}\mathbf{X}\right\}^{-1}\mathbf{X}^{\mathrm{T}}\mathbf{y}$ is the maximum likelihood (and least squares) estimator of $\boldsymbol{\beta}$. The posterior independence between α and $\boldsymbol{\beta}$ is due to the fact that $\mathbf{X}$ is centered and that α and $\boldsymbol{\beta}$ are a priori independent.

Moreover, the posterior distribution of σ^2 is given by

$$\sigma^2|\mathbf{y}\sim I\mathscr{G}\left[(n-1)/2,s^2+(\tilde{\boldsymbol{\beta}}-\hat{\boldsymbol{\beta}})^{\mathrm{T}}\mathbf{X}^{\mathrm{T}}\mathbf{X}(\tilde{\boldsymbol{\beta}}-\hat{\boldsymbol{\beta}})/(g+1)\right]$$

where $I\mathscr{G}(a,b)$ is an inverse Gamma distribution with mean $b/(a-1)$ and where $s^2=(\mathbf{y}-\bar{\mathbf{y}}\mathbf{1}_n-\mathbf{X}\hat{\boldsymbol{\beta}})^{\mathrm{T}}(\mathbf{y}-\bar{\mathbf{y}}\mathbf{1}_n-\mathbf{X}\hat{\boldsymbol{\beta}})$ corresponds to the (classical) residual sum of squares.

↯ The previous derivation assumes that $p > 0$. In the special case $p = 0$, which will later be used as a null model in hypothesis testing, similar arguments lead to

$$\alpha|\mathbf{y},\sigma^2 \sim \mathscr{N}\left(\bar{\mathbf{y}},\sigma^2/n\right),$$

$$\sigma^2|\mathbf{y} \sim \mathscr{IG}\left[(n-1)/2,(\mathbf{y}-\bar{\mathbf{y}}\mathbf{1}_n)^{\mathrm{T}}(\mathbf{y}-\bar{\mathbf{y}}\mathbf{1}_n)/2\right].$$

(There is no β when $p = 0$, as this corresponds to the constant mean model.)

Recalling the double expectation formulas

$$\mathbb{E}\left[\mathbb{E}\left[X|Y\right]\right] = \mathbb{E}\left[X\right] \quad \text{and} \quad \mathbb{V}(X) = \mathbb{V}[\mathbb{E}(X|Y)] + \mathbb{E}[\mathbb{V}(X|Y)]$$

for $\mathbb{V}(X|Y) = \mathbb{E}[(X - \mathbb{E}(X|Y))^2|Y]$, we can derive from the previous derivations that

$$\mathbb{E}^\pi\left[\alpha|\mathbf{y}\right] = \mathbb{E}^\pi\left[\mathbb{E}^\pi\left(\alpha|\sigma^2,\mathbf{y}\right)|\mathbf{y}\right] = \mathbb{E}^\pi\left[\bar{\mathbf{y}}|\mathbf{y}\right] = \bar{\mathbf{y}}$$

and that

$$\mathbb{V}^\pi(\alpha|\mathbf{y}) = \mathbb{V}(\bar{\mathbf{y}}|\mathbf{y}) + \mathbb{E}\left[\frac{\sigma^2}{n}\middle|\mathbf{y}\right] = \kappa/n(n-3),$$

where

$$\begin{aligned}\kappa &= (\mathbf{y}-\bar{\mathbf{y}}\mathbf{1}_n)^{\mathrm{T}}(\mathbf{y}-\bar{\mathbf{y}}\mathbf{1}_n) + \frac{1}{g+1}\left\{-g\mathbf{y}^{\mathrm{T}}\mathbf{P}\mathbf{y} + \tilde{\beta}^{\mathrm{T}}\mathbf{X}^{\mathrm{T}}\mathbf{P}\mathbf{X}\tilde{\beta} - 2\mathbf{y}^{\mathrm{T}}\mathbf{P}\mathbf{X}\tilde{\beta}\right\} \\ &= s^2 + (\tilde{\boldsymbol{\beta}}-\hat{\boldsymbol{\beta}})^{\mathrm{T}}\mathbf{X}^{\mathrm{T}}\mathbf{X}(\tilde{\boldsymbol{\beta}}-\hat{\boldsymbol{\beta}})/(g+1).\end{aligned}$$

With a bit of extra algebra, we can recover the whole distribution of α from

$$\pi(\alpha,\sigma^2|\mathbf{y}) \propto (\sigma^{-2})^{(n-1)/2+1+1/2}\exp\left\{-\frac{n}{2\sigma^2}(\alpha-\bar{\mathbf{y}})^2\right\}\exp\left\{-\frac{\kappa}{2\sigma^2}\kappa\right\},$$

namely

$$\pi(\alpha|\mathbf{y}) \propto \left[1 + \frac{n(\alpha-\bar{\mathbf{y}})^2}{\kappa}\right]^{-n/2}.$$

This means that the marginal posterior distribution of α—the distribution of α given only $\mathbf{y}$ and $\mathbf{X}$—is a Student's t distribution with $n-1$ degrees of freedom, a location parameter equal to $\bar{\mathbf{y}}$ and a scale parameter equal to $\kappa/n(n-1)$.

If we now turn to the parameter $\boldsymbol{\beta}$, by the same double expectation formula, we derive that

$$\begin{aligned}\mathbb{E}^\pi\left[\boldsymbol{\beta}|\mathbf{y}\right] &= \mathbb{E}^\pi\left[\mathbb{E}^\pi\left(\boldsymbol{\beta}|\sigma^2,\mathbf{y}\right)|\mathbf{y}\right] \\ &= \mathbb{E}^\pi\left[\frac{g}{g+1}(\hat{\boldsymbol{\beta}}+\tilde{\boldsymbol{\beta}}/g)|\mathbf{y}\right] \\ &= \frac{g}{g+1}(\hat{\boldsymbol{\beta}}+\tilde{\boldsymbol{\beta}}/g).\end{aligned}$$

This result gives its meaning to the above point relating g with the amount of information contained in the dataset. For instance, when $g = 1$, the prior information has the same weight as this amount. In this case, the Bayesian estimate of $\boldsymbol{\beta}$ is the average between the least square estimator and the prior expectation. The larger g is, the weaker the prior information and the closer the Bayesian estimator is to the least squares estimator. For instance, when g goes to infinity, the posterior mean converges to $\hat{\boldsymbol{\beta}}$.

Based on similar derivations, we can compute the posterior variance of $\boldsymbol{\beta}$. Indeed,

$$\begin{aligned}\mathbb{V}^{\pi}(\boldsymbol{\beta}|\mathbf{y}) &= \mathbb{V}\left[\frac{g}{g+1}(\hat{\boldsymbol{\beta}}+\tilde{\boldsymbol{\beta}}/g)|\mathbf{y}\right] + \mathbb{E}\left[\frac{g\sigma^2}{g+1}(\mathbf{X}^{\mathrm{T}}\mathbf{X})^{-1}\right] \\ &= \frac{\kappa g}{(g+1)(n-3)}(\mathbf{X}^{\mathrm{T}}\mathbf{X})^{-1}.\end{aligned}$$

Once more, it is possible to integrate out σ^2 in

$$\begin{aligned}\pi(\boldsymbol{\beta},\sigma^2|\mathbf{y}) \propto (\sigma^2)^{-p/2}\exp\left(-\frac{g+1}{2g\sigma^2}\{\boldsymbol{\beta}-\mathbb{E}^{\pi}[\boldsymbol{\beta}|\mathbf{y}]\}^{\mathrm{T}}\mathbf{X}^{\mathrm{T}}\mathbf{X}\{\boldsymbol{\beta}-\mathbb{E}^{\pi}[\boldsymbol{\beta}|\mathbf{y}]\}\right) \\ \times(\sigma^2)^{-(n-1)/2-1}\exp\left(-\frac{1}{2\sigma^2}\kappa\right),\end{aligned}$$

leading to

$$\pi(\boldsymbol{\beta}|\mathbf{y}) \propto \left[1+\frac{g+1}{g\kappa}\{\boldsymbol{\beta}-\mathbb{E}^{\pi}[\boldsymbol{\beta}|\mathbf{y}]\}^{\mathrm{T}}\mathbf{X}^{\mathrm{T}}\mathbf{X}\{\boldsymbol{\beta}-\mathbb{E}^{\pi}[\boldsymbol{\beta}|\mathbf{y}]\}\right].$$

Therefore, the marginal posterior distribution of $\boldsymbol{\beta}$ is also a multivariate Student's t distribution with $n-1$ degrees of freedom, location parameter equal to $\frac{g}{g+1}(\hat{\boldsymbol{\beta}}+\tilde{\boldsymbol{\beta}}/g)$ and scale parameter equal to $\frac{g\kappa}{(g+1)(n-1)}(\mathbf{X}^{\mathrm{T}}\mathbf{X})^{-1}$.

The standard Bayes estimator of σ^2 for this model is the posterior expectation

$$\mathbb{E}^{\pi}\left[\sigma^2|\mathbf{y}\right] = \frac{\kappa}{n-3} = \frac{s^2+(\tilde{\boldsymbol{\beta}}-\hat{\boldsymbol{\beta}})^{\mathrm{T}}\mathbf{X}^{\mathrm{T}}\mathbf{X}(\tilde{\boldsymbol{\beta}}-\hat{\boldsymbol{\beta}})/(g+1)}{n-3}.$$

↯ In the special case $p = 0$, by using similar arguments, we get

$$\mathbb{E}^{\pi}\left[\sigma^2|\mathbf{y}\right] = \frac{(\mathbf{y}-\bar{\mathbf{y}}\mathbf{1}_n)^{\mathrm{T}}(\mathbf{y}-\bar{\mathbf{y}}\mathbf{1}_n)}{n-3} = \frac{s^2}{n-3},$$

which is the same expectation as with the Jeffreys prior.

HPD regions on subvectors of the parameter $\boldsymbol{\beta}$ can be derived in a straightforward manner from this marginal posterior distribution of $\boldsymbol{\beta}$. For a single parameter, we have for instance

$$\beta_j|\mathbf{y} \sim \mathscr{T}\left(n-1, \frac{g}{g+1}\left(\frac{\tilde{\beta}_j}{g}+\hat{\beta}_j\right), \frac{g\kappa}{(n-1)(g+1)}\omega_{jj}\right),$$

where ω_{jj} is the (j,j)-th element of the matrix $(\mathbf{X}^\mathsf{T}\mathbf{X})^{-1}$. If we set

$$\zeta = (\tilde{\boldsymbol{\beta}} + g\hat{\boldsymbol{\beta}})/(g+1)$$

the transform

$$\beta_j - \zeta_j \Big/ \sqrt{\frac{g\kappa}{(n-1)(g+1)}\omega_{jj}}$$

is (marginally) distributed as a standard t distribution with $n-1$ degrees of freedom. A $(1-\gamma)$ HPD interval on β_j has therefore

$$\zeta_j \pm \sqrt{\frac{g\kappa}{(n-1)(g+1)}\omega_{jj}}F_{n-1}^{-1}(1-\gamma/2)$$

as bounds, where F_{n-1}^{-1} denotes the quantile function of the $\mathscr{T}(n-1,0,1)$ distribution.

3.4.2 The BayesReg R Function

We have created in `bayess` an R function called `BayesReg` to implement Zellner's G-prior analysis within R. The purpose is dual: first, this R function shows how easily automated this approach can be. Second, it also illustrates how it is possible to get exactly the same type of output as the standard R function `summary(lm(y~X))`.

The following R code is extracted from this function `BayesReg` and used to calculate the Bayes estimates. As an aside, notice that we use the function `stop` in order to end the calculations if the matrix $\mathbf{X}^\mathsf{T}\mathbf{X}$ is not invertible.

```
if (det(t(X)%*%X)<=1e-7)
stop("Design matrix has too low a rank!",call.=FALSE)
```

We also stress the use of `scale` below to standardize the explanatory variables.

```
X=as.matrix(X)
n=length(y)
p=dim(X)[2]
X=scale(X)
U=solve(t(X)%*%X)%*%t(X)
# MLE
alphaml=mean(y)
betaml=U%*%y
s2=t(y-alphaml-X%*%betaml)%*%(y-alphaml-X%*%betaml)
kappa=as.numeric(s2+t(betatilde-betaml)%*%t(X)%*%X%*%
    (betatilde-betaml)/(g+1))
```

```
malphabayes=alphaml
mbetabayes=g/(g+1)*(betaml+betatilde/g)
msigma2bayes=kappa/(n-3)
valphabayes=kappa/(n*(n-3))
vbetabayes=diag(kappa*g/((g+1)*(n-3))*solve(t(X)%*%X))
vsigma2bayes=2*kappa^2/((n-3)*(n-4))
postmean=c(malphabayes,mbetabayes)
postsd=sqrt(c(valphabayes,vbetabayes))
# evidence of the model
intlike=(g+1)^(-p/2)*kappa^(-(n-1)/2)
```

We will see further aspects of BayesReg in the following sections.

3.4.3 Bayes Factors and Model Comparison

One important inferential issue pertaining to linear models is to test whether or not a specific explanatory variable is truly explanatory or, in other words, to decide which explanatory variables should be kept within the model. This leads to tests on the nullity of some elements of the parameter $\boldsymbol{\beta}$. Following the general testing methodology presented in Chap. 2, these tests can be conducted using Bayes factors. In the case of linear models and under Zellner's G-priors, those Bayes factors are actually available in closed form.

When considering the marginal likelihood (or evidence) at the core of the Bayes factors, we have, if $p \neq 0$,

$$f(\mathbf{y}) = \int \left(\int \int f(\mathbf{y}|\alpha, \boldsymbol{\beta}, \sigma^2)\pi(\boldsymbol{\beta}|\alpha, \sigma^2)\pi(\sigma^2, \alpha)\mathrm{d}\alpha\mathrm{d}\boldsymbol{\beta} \right) \mathrm{d}\sigma^2 ,$$

with

$$\begin{aligned} f(\mathbf{y}|\alpha, \boldsymbol{\beta}, \sigma^2)\pi(\boldsymbol{\beta}|\alpha, \sigma^2) &= \frac{\left|\mathbf{X}^{\mathrm{T}}\mathbf{X}\right|^{1/2}}{(2\pi\sigma^2)^{(n+p)/2} g^{p/2}} \exp\left\{ -\frac{n}{2\sigma^2}(\alpha - \bar{\mathbf{y}})^2 \right\} \times \\ &\quad \exp\left\{ -\frac{1}{2\sigma^2}(\mathbf{y} - \bar{\mathbf{y}}\mathbf{1}_n - \mathbf{X}\boldsymbol{\beta})^{\mathrm{T}}(\mathbf{y} - \bar{\mathbf{y}}\mathbf{1}_n - \mathbf{X}\boldsymbol{\beta}) \right\} \times \\ &\quad \exp\left\{ -\frac{1}{2g\sigma^2}(\boldsymbol{\beta} - \tilde{\boldsymbol{\beta}})^{\mathrm{T}}\mathbf{X}^{\mathrm{T}}\mathbf{X}(\boldsymbol{\beta} - \tilde{\boldsymbol{\beta}}) \right\} , \end{aligned}$$

and $\pi(\alpha, \sigma^2) = \delta\sigma^{-2}$ (where δ is an arbitrary constant). Thus

$$\begin{aligned} f(\mathbf{y}) &= \delta n^{-1/2}(g+1)^{-p/2}(2\pi)^{-(n-1)/2} \int (\sigma^2)^{-(n-1)/2-1} \exp\left(-\frac{1}{2\sigma^2}\kappa \right) \mathrm{d}\sigma^2 \\ &= \frac{\delta \Gamma((n-1)/2)}{\pi^{(n-1)/2} n^{1/2}}(g+1)^{-p/2} \left[s^2 + (\tilde{\boldsymbol{\beta}} - \hat{\boldsymbol{\beta}})^{\mathrm{T}}\mathbf{X}^{\mathrm{T}}\mathbf{X}(\tilde{\boldsymbol{\beta}} - \hat{\boldsymbol{\beta}})/(g+1) \right]^{-(n-1)/2} , \\ &= \frac{\delta \Gamma((n-1)/2)}{\pi^{(n-1)/2} n^{1/2}}(g+1)^{-p/2}\kappa^{-(n-1)/2} . \end{aligned} \tag{3.4}$$

↯ If $p = 0$, a similar expression emerges:

$$f(\mathbf{y}) = \int \left(\int f(\mathbf{y}|\alpha, \sigma^2)\pi(\alpha, \sigma^2)\mathrm{d}\alpha \right) \mathrm{d}\sigma^2 ,$$

with

$$\begin{aligned} f(\mathbf{y}|\alpha, \sigma^2)\pi(\alpha, \sigma^2) &= \frac{\delta(\sigma^2)^{-1}}{(2\pi\sigma^2)^{n/2}} \exp\left\{ -\frac{1}{2\sigma^2}(\mathbf{y} - \bar{\mathbf{y}}\mathbf{1}_n)^{\mathsf{T}}(\mathbf{y} - \bar{\mathbf{y}}\mathbf{1}_n) \right\} \\ &= \frac{\delta(\sigma^2)^{-n/2-1}}{(2\pi)^{n/2}} \exp\left(-\frac{1}{2\sigma^2}(\mathbf{y} - \bar{\mathbf{y}}\mathbf{1}_n)^{\mathsf{T}}(\mathbf{y} - \bar{\mathbf{y}}\mathbf{1}_n) \right\} \times \\ &\quad \exp\left\{ -\frac{n}{2\sigma^2}(\alpha - \bar{\mathbf{y}})^2 \right\} . \end{aligned}$$

The integration in both α and σ^2 can then be conducted in closed form and we obtain

$$f(\mathbf{y}) = \frac{\delta\Gamma((n-1)/2)}{\pi^{(n-1)/2}n^{1/2}} \left[(\mathbf{y} - \bar{\mathbf{y}}\mathbf{1}_n)^{\mathsf{T}}(\mathbf{y} - \bar{\mathbf{y}}\mathbf{1}_n) \right]^{-(n-1)/2}$$

as the evidence associated with this "null" model. The evidence corresponds to `intlike0` in the `BayesReg` code.

As pointed out in Chap. 2, the computation of Bayes factors is plagued by the inability to include generic improper prior distributions. In order to bypass this difficulty, we will assume that all the linear models under comparison do include the parameter α, which means that each regression model includes an intercept. This assumption allows us to take the *same* improper prior (and hence the *same* arbitrary constant δ) on (α, σ^2) for all of those models. Otherwise, the Bayes factors simply cannot be correctly defined.

When we compare two sets of regressors, we have to handle two regression matrices, $\mathbf{X}^1$ and $\mathbf{X}^2$, with respective dimensions (n, p_1) and (n, p_2), extracted from the original matrix $\mathbf{X}$ by removing some columns. From a Bayesian perspective, using Zellner's G-prior modelling in both cases, we are thus comparing model $\mathfrak{M}_1$

$$\begin{aligned} \mathbf{y}|\alpha, \boldsymbol{\beta}^1, \sigma^2 &\sim \mathscr{N}_n \left(\alpha\mathbf{1}_n + \mathbf{X}^1\boldsymbol{\beta}^1, \sigma^2 \mathbf{I}_n \right) , \\ \boldsymbol{\beta}^1|\alpha, \sigma^2 &\sim \mathscr{N}_{p_1} \left(\tilde{\boldsymbol{\beta}}^1, g_1\sigma^2((\mathbf{X}^1)^{\mathrm{T}}\mathbf{X}^1)^{-1} \right), \quad p_1 \neq 0 \\ \pi\left(\alpha, \sigma^2\right) &\propto \sigma^{-2} , \end{aligned}$$

with model $\mathfrak{M}_2$:

$$\begin{aligned} \mathbf{y}|\alpha, \boldsymbol{\beta}^2, \sigma^2 &\sim \mathscr{N}_n \left(\alpha\mathbf{1}_n + \mathbf{X}^2\boldsymbol{\beta}^2, \sigma^2 \mathbf{I}_n \right) , \\ \boldsymbol{\beta}^2|\alpha, \sigma^2 &\sim \mathscr{N}_{p_2} \left(\tilde{\boldsymbol{\beta}}^2, g_2\sigma^2((\mathbf{X}^2)^{\mathrm{T}}\mathbf{X}^2)^{-1} \right), \quad p_2 \neq 0 \\ \pi\left(\alpha, \sigma^2\right) &\propto \sigma^{-2} . \end{aligned}$$

Using the above derivations, the Bayes factor between model $\mathfrak{M}_1$ and model $\mathfrak{M}_2$ is then given by

$$B_{12}(\mathbf{y}) = \frac{(g_1+1)^{-p_1/2}\left[s_1^2 + (\tilde{\boldsymbol{\beta}}^1 - \hat{\boldsymbol{\beta}}^1)^{\mathrm{T}}(\mathbf{X}^1)^{\mathrm{T}}\mathbf{X}^1(\tilde{\boldsymbol{\beta}}^1 - \hat{\boldsymbol{\beta}}^1)/(g_1+1)\right]^{-(n-1)/2}}{(g_2+1)^{-p_2/2}\left[s_2^2 + (\tilde{\boldsymbol{\beta}}^2 - \hat{\boldsymbol{\beta}}^2)^{\mathrm{T}}(\mathbf{X}^2)^{\mathrm{T}}\mathbf{X}^2(\tilde{\boldsymbol{\beta}}^2 - \hat{\boldsymbol{\beta}}^2)/(g_2+1)\right]^{-(n-1)/2}}\,.$$

For **caterpillar**, if we have to test the null hypothesis $H_0 : \beta_8 = \beta_9 = 0$, using $\tilde{\beta}^1 = 0_8$, $\tilde{\beta}^2 = 0_6$, and an arbitrary[7] $g_1 = g_2 = 100$, in Zellner's G-priors, we obtain $B_{12}^{\pi} = 0.0165$ when model $\mathfrak{M}_2$ corresponds to H_0. Using Jeffreys' scale of evidence (provided in Chap. 2), this implies that $\log_{12}(B_{12}^{\pi}) = -1.78$, hence that the posterior distribution appears to strongly favor H_0.

More generally, using $\tilde{\boldsymbol{\beta}} = 0_8$ and $g = 100$, we can produce a Bayesian regression output, programmed in R, which mimics a standard software regression output like `lm`: besides the estimation of the β_j's via their posterior expectation, we include the Bayes factors B_{12}^j, in the log scale $\log_{10}\left(B_{12}^j\right)$, corresponding to testing the null hypotheses $H_0 : \beta_j = 0$. (The stars are related to Jeffreys' scale of evidence.)

The R code corresponding to this "standard" output is also part of the R function `BayesReg`:

```
bayesfactor=rep(0,p)
p0=p-1 # remove one variate
X0=X[,-j]
U0=solve(t(X0)%*%X0)%*%t(X0)
betatilde0=U0%*%X%*%betatilde
betaml0=U0%*%y
s20=t(y-alphaml-X0%*%betaml0)%*%(y-alphaml-X0%*%betaml0)
kappa0=as.numeric(s20+t(betatilde0-betaml0)%*%t(X0)%*%
X0%*%(betatilde0-betaml0)/(g+1))
intlike0=(g+1)^(-p0/2)*kappa0^(-(n-1)/2)
bayesfactor[j]=intlike/intlike0
```

where `intlike` is the marginal likelihood for the full model. (The way this computation is repeated and used to mimic the output of the `lm` function can be found by reading the function `BayesReg`.)

For the **caterpillar** dataset, $\tilde{\boldsymbol{\beta}} = 0_8$ and $g = n = 33$, the G-prior estimate of σ^2 is equal to 0.653, while the posterior means and standard variations of the β_j's are given below. We can immediately spot that the (most) significant explanatory variables are the same ones as those selected by `lm`, x_1, x_2, and x_7. Note, however, that this output does not rigorously validate the selection of the submodel with the covariates x_1, x_2, and x_7, as it does not produce the Bayes factor associated with this (sub)model and the full model.

---

[7]Arbitrary means here that this choice is no more justified than any other. We will see later that $g_j = n$ is the recommended or default value for non-informative settings.

```
> res1=BayesReg(y,X)

            PostMean PostStError Log10bf EvidAgaH0
Intercept  -0.8133      0.1407
x1         -0.5039      0.1883  0.7224       (**)
x2         -0.3755      0.1508  0.5392       (**)
x3          0.6225      0.3436 -0.0443
x4         -0.2776      0.2804 -0.5422
x5         -0.2069      0.1499 -0.3378
x6          0.2806      0.4760 -0.6857
x7         -1.0420      0.4178  0.5435       (**)
x8         -0.0221      0.1531 -0.7609

Posterior Mean of Sigma2: 0.6528
Posterior StError of Sigma2: 0.939
```

3.4.4 Prediction

The prediction of $m \geq 1$ future observations from units for which the explanatory variables $\tilde{\mathbf{X}}$—but not the outcome variable $\tilde{\mathbf{y}}$—have been observed or set is also based on the posterior distribution. Logically enough, were α, $\boldsymbol{\beta}$ and σ^2 known quantities, the m-vector $\tilde{\mathbf{y}}$ would then have a Gaussian distribution with mean $\alpha\mathbf{1}_m + \tilde{\mathbf{X}}\boldsymbol{\beta}$ and variance $\sigma^2\mathbf{I}_m$. The *predictive distribution* on $\tilde{\mathbf{y}}$ is defined as the marginal in $\mathbf{y}$ of the joint posterior distribution on $(\tilde{\mathbf{y}}, \alpha, \boldsymbol{\beta}, \sigma^2)$.

Conditional on σ^2, the vector $\tilde{\mathbf{y}}$ of future observations has a Gaussian distribution and we can derive its expectation—used as our Bayesian estimator—by averaging over α and $\boldsymbol{\beta}$,

$$\begin{aligned}
\mathbb{E}^\pi[\tilde{\mathbf{y}}|\sigma^2, \mathbf{y}] &= \mathbb{E}^\pi[\mathbb{E}^\pi(\tilde{\mathbf{y}}|\alpha, \boldsymbol{\beta}, \sigma^2, \mathbf{y})|\sigma^2, \mathbf{y}] \\
&= \mathbb{E}^\pi[\alpha\mathbf{1}_m + \tilde{\mathbf{X}}\boldsymbol{\beta}|\sigma^2, \mathbf{y}] \\
&= \hat{\alpha}\mathbf{1}_m + \tilde{\mathbf{X}}\frac{\tilde{\boldsymbol{\beta}} + g\hat{\boldsymbol{\beta}}}{g+1},
\end{aligned}$$

which is independent from σ^2. This representation is quite intuitive, being the product of the matrix of explanatory variables $\tilde{\mathbf{X}}$ by the Bayesian estimator of $\boldsymbol{\beta}$. Similarly, we can compute

$$\begin{aligned}
\mathbb{V}^\pi(\tilde{\mathbf{y}}|\sigma^2, \mathbf{y}) &= \mathbb{E}^\pi[\mathbb{V}^\pi(\tilde{\mathbf{y}}|\alpha, \boldsymbol{\beta}, \sigma^2, \mathbf{y})|\sigma^2, \mathbf{y}] \\
&\quad + \mathbb{V}^\pi(\mathbb{E}^\pi(\tilde{\mathbf{y}}|\alpha, \boldsymbol{\beta}, \sigma^2)|\sigma^2, \mathbf{y}) \\
&= \mathbb{E}^\pi[\sigma^2 I_m|\sigma^2, \mathbf{y}] + \mathbb{V}^\pi(\alpha\mathbf{1}_m + \tilde{X}\boldsymbol{\beta}|\sigma^2, \mathbf{y}) \\
&= \sigma^2\left(I_m + \frac{g}{g+1}\tilde{\mathbf{X}}(\mathbf{X}^\mathsf{T}\mathbf{X})^{-1}\tilde{\mathbf{X}}^\mathsf{T}\right).
\end{aligned}$$

Due to this factorization, and the fact that the conditional expectation does not depend on σ^2, we thus obtain

$$\mathbb{V}^\pi(\tilde{\mathbf{y}}|\mathbf{y}) = \hat{\sigma}^2 \left(I_m + \frac{g}{g+1} \tilde{\mathbf{X}}(\mathbf{X}^\mathsf{T}\mathbf{X})^{-1}\tilde{\mathbf{X}}^\mathsf{T} \right) .$$

This decomposition of the variance makes perfect sense: Conditionally on σ^2, the posterior predictive variance has two terms, the first term being $\sigma^2 I_m$, which corresponds to the sampling variation, and the second one being $\sigma^2 \frac{g}{g+1} \tilde{\mathbf{X}}(\mathbf{X}^\mathsf{T}\mathbf{X})^{-1}\tilde{\mathbf{X}}^\mathsf{T}$, which corresponds to the uncertainty about $\boldsymbol{\beta}$.

HPD credible regions and tests can then be conducted based on this conditional predictive distribution

$$\tilde{\mathbf{y}}|\sigma^2, \mathbf{y}, \sigma^2 \sim \mathscr{N}\left(\mathbb{E}^\pi[\tilde{\mathbf{y}}], \mathbb{V}^\pi(\tilde{\mathbf{y}}|\mathbf{y}, \sigma^2)\right) .$$

Integrating σ^2 out to produce the marginal distribution of $\tilde{\mathbf{y}}$ leads to a multivariate Student's t distribution

$$\tilde{\mathbf{y}}|\mathbf{y} \sim \mathscr{T}_m \Bigg(n, \hat{\alpha}\mathbf{1}_m + g\tilde{\boldsymbol{\beta}}/(g+1), \frac{s^2 + \hat{\boldsymbol{\beta}}^\mathsf{T}\mathbf{X}^\mathsf{T}\mathbf{X}\hat{\boldsymbol{\beta}}}{n} \left\{ \mathbf{I}_m + \tilde{\mathbf{X}}(\mathbf{X}^\mathsf{T}\mathbf{X})^{-1}\tilde{\mathbf{X}}^\mathsf{T} \right\} \Bigg) .$$

(following a straightforward but lengthy derivation that is very similar to the one we conducted at the end of Chap. 2, see (2.11)).

3.5 Markov Chain Monte Carlo Methods

Given the complexity of most models encountered in Bayesian modeling, standard simulation methods are not a sufficiently versatile solution. We now present the rudiments of a technique that emerged in the late 1980s as the core of Bayesian computing and that has since then revolutionized the field.

This technique is based on *Markov chains*, but we will not make many incursions into the theory of Markov chains (see Meyn and Tweedie, 1993), focusing rather on the practical implementation of these algorithms and trusting that the underlying theory is sound enough to validate them (Robert and Casella, 2004). At this point, it is sufficient to recall that a Markov chain $(\mathbf{x}_t)_{t\in\mathbb{N}}$ is a sequence of dependent random vectors whose dependence on the past values $\mathbf{x}_0, \ldots, \mathbf{x}_{t-1}$ stops at the value immediately before, $\mathbf{x}_{t-1}$, and that is entirely defined by its *kernel*—that is, the conditional distribution of $\mathbf{x}_t$ given $\mathbf{x}_{t-1}$.

The central idea behind these new methods, called *Markov chain Monte Carlo* (MCMC) algorithms, is that, to simulate from a distribution π (for instance, the posterior distribution), it is actually sufficient to produce a Markov chain $(\mathbf{x}_t)_{t\in\mathbb{N}}$ whose *stationary distribution* is π: when $\mathbf{x}_t$ is marginally distributed according to π, then $\mathbf{x}_{t+1}$ is also marginally distributed according to

π. If an algorithm that generates such a chain can be constructed, the ergodic theorem guarantees that, in almost all settings, the average

$$\frac{1}{T}\sum_{t=1}^{T} g(\mathbf{x}_t)$$

converges to $\mathbb{E}[g(\mathbf{x})]$, no matter what the starting value.[8]

More informally, this property means that, for large enough t, $\mathbf{x}_t$ is approximately distributed from π and can thus be used like the output from a more standard simulation algorithm (even though one must take care of the correlation between the $\mathbf{x}_t$'s created by the Markovian structure). For integral approximation purposes, the difference from regular Monte Carlo approximations is that the variance structure of the estimator is more complex because of the Markovian dependence. These methods being central to the cases studied from this stage onward, we hope that the reader will become sufficiently proficient with them by the end of the book! In this chapter, we detail a particular type of MCMC algorithm, the Gibbs sampler, that is currently sufficient for our needs. The next chapter will introduce a more universal type of algorithm.

3.5.1 Conditionals

A first remark that motivates the use of the Gibbs sampler[9] is that, within structures such as

$$\pi(x_1) = \int \pi_1(x_1|x_2)\tilde{\pi}(x_2)\,\mathrm{d}x_2\,, \tag{3.5}$$

to simulate from the joint distribution

$$\pi(x_1, x_1) = \pi_1(x_1|x_2)\tilde{\pi}(x_2) \tag{3.6}$$

automatically produces (marginal) simulation from $\pi(x_1)$. Therefore, in settings where (3.5) holds, it is not necessary to simulate from $\pi(x_1)$ when one can jointly simulate (x_1, x_2) from (3.6).

For example, consider $(x_1, x_2) \in \mathbb{N} \times [0, 1]$ distributed from the joint density

$$\pi(x_1, x_2) \propto \binom{n}{x_1} x_2^{x_1+\alpha-1}(1-x_2)^{n-x_1+\beta-1}\,.$$

This is a joint distribution where

$$x_1|x_2 \sim \mathscr{B}(n, x_2) \quad \text{and} \quad x_2|\alpha,\beta \sim \mathscr{B}e(\alpha,\beta)\,.$$

---

[8]In probabilistic terms, if the Markov chains produced by these algorithms are *irreducible*, then these chains are both *positive recurrent* with stationary distribution π and *ergodic*, that is, asymptotically independent of the starting value $\mathbf{x}_0$.

[9]In the literature, both the denominations Gibbs *sampler* and Gibbs *sampling* can be found. In this book, we will use Gibbs sampling for the simulation technique and Gibbs sampler for the simulation algorithm.

Therefore, although

$$\pi(x_1) = \binom{n}{x_1} \frac{B(\alpha + x_1, \beta + n - x_1)}{B(\alpha, \beta)}$$

is available in closed form as the *beta-binomial distribution*, it is unnecessary to work with this marginal when one can simulate an iid sample $(x_1^{(i)}, x_2^{(i)})$ $(t = 1, \ldots, N)$ as

$$x_2^{(t)} \sim \mathscr{B}e(\alpha, \beta) \text{ and } x_1^{(t)} \sim \mathscr{B}(n, x_2^{(t)}) .$$

Integrals such as $\mathbb{E}[x_1/(x_1+1)]$ can then be approximated by

$$\frac{1}{N} \sum_{i=1}^{N} \frac{x_1^{(t)}}{x_1^{(t)} + 1} ,$$

using a regular Monte Carlo approach.

Unfortunately, even when one works with a representation such as (3.6) that is naturally associated with the original model, it is often the case that the mixing density $\tilde{\pi}(x_2)$ itself is neither available in closed form nor amenable to simulation. However, both *conditional posterior distributions*,

$$\pi_1(x_1|x_2) \quad \text{and} \quad \pi_2(x_2|x_1) ,$$

can often be simulated, and the following method takes full advantage of this feature.

3.5.2 Two-Stage Gibbs Sampler

The availability of both conditionals of (3.6) in terms of simulation can be exploited to build a transition kernel and a corresponding Markov chain, somewhat analogous to the derivation of the maximum of a multivariate function via an iterative device that successively maximizes the function in each of its arguments until a fixed point is reached.

The corresponding Markov kernel is built by simulating successively from each conditional distribution, with the conditioning variable being updated on the run. It is called the *two-stage Gibbs sampler* or sometimes the *data augmentation* algorithm, although both terms are rather misleading.[10]

[10]Gibbs sampling got its name from *Gibbs fields*, used in image analysis, when Geman and Geman (1984) proposed an early version of this algorithm, while data augmentation refers to Tanner's (1996) special use of this algorithm in missing-data settings, as seen in Chap. 6.

Algorithm 3.3 TWO-STAGE GIBBS SAMPLER

Initialization: Start with an arbitrary value $x_2^{(0)}$.
Iteration t: Given $x_2^{(t-1)}$, generate
1. $x_1^{(t)}$ according to $\pi_1(x_1|x_2^{(t-1)})$,
2. $x_2^{(t)}$ according to $\pi_2(x_2|x_1^{(t)})$.

Note that, in the second step of the algorithm, $x_2^{(t)}$ is generated conditional on $x_1 = x_1^{(t)}$, not $x_1^{(t-1)}$. The validation of this algorithm is that, for both generations, π is a stationary distribution. Therefore, the limiting distribution of the chain $(x_1^{(t)}, x_2^{(t)})_t$ is π if the chain is *irreducible*; that is, if it can reach any region in the support of π in a finite number of steps. (Note that there is a difference between the *stationary* distribution and the *limiting* distribution only in cases when the chain is not ergodic, as shown in Exercise 3.9.)

The practical implementation of Gibbs sampling involves solving two types of difficulties: the first type corresponds to deriving an efficient decomposition of the joint distribution in easily-simulated conditionals and the second one to deciding when to stop the algorithm. Evaluating the efficiency of the decomposition includes assessing the ease of simulating from both conditionals and the level of correlation between the $\mathbf{x}^{(t)}$'s, as well as the *mixing* behavior of the chain, that is, its ability to explore the support of π sufficiently fast. While deciding whether or not a given conditional can be simulated is easy enough, it is not always possible to find a manageable conditional, and more robust alternatives such as the *Metropolis–Hastings algorithm* will be described in the following chapters (see Sect. 4.2).

Choosing a stopping rule also relates to the mixing performances of the algorithm, as well as to its ability to approximate posterior expectations under π. Many indicators have been proposed in the literature (see Robert and Casella, 2004, Chap. 12) to signify convergence, or lack thereof, although none of these is foolproof. In the easiest cases, the lack of convergence is blatant and can be spotted on the raw plot of the sequence of the $\mathbf{x}^{(t)}$'s, while, in other cases, the Gibbs sampler explores very satisfactorily one mode of the posterior distribution but fails altogether to visit the *other* modes of the posterior: we will encounter such cases in Chap. 6 with mixtures of distributions. Throughout this chapter and the following ones, we give hints on how to implement these recommendations in practice.

Consider the posterior distribution derived in Exercise 2.11, for $n = 2$ observations,

$$\pi(\mu|\mathscr{D}_2) \propto \frac{e^{-\mu^2/20}}{\{1 + (x_1 - \mu)^2)(1 + (x_2 - \mu)^2\}} .$$

Even though this is a univariate distribution, it can still be processed by a Gibbs sampler through a data augmentation step, thus illustrating the idea behind (3.5). In fact, since $(j = 1, 2)$

$$\frac{1}{1+(x_j-\mu)^2} = \int_0^\infty e^{-\omega_j[1+(x_j-\mu)^2]}\, \mathrm{d}\omega_j\,,$$

we can define $\boldsymbol{\omega} = (\omega_1, \omega_2)$ and envision $\pi(\mu|\mathscr{D}_2)$ as the marginal distribution of

$$\pi(\mu, \boldsymbol{\omega}|\mathscr{D}_2) \propto e^{-\mu^2/20} \times \prod_{j=1}^{2} e^{-\omega_j[1+(x_j-\mu)^2]}\,.$$

For this multivariate distribution, a corresponding Gibbs sampler is associated with the following two steps:

1. Generate $\mu^{(t)} \sim \pi(\mu|\boldsymbol{\omega}^{(t-1)}, \mathscr{D}_2)$.
2. Generate $\boldsymbol{\omega}^{(t)} \sim \pi(\boldsymbol{\omega}|\mu^{(t)}, \mathscr{D}_2)$.

The second step is straightforward: the ω_i's are conditionally independent and distributed as $\mathscr{E}xp(1 + (x_i - \mu^{(t)})^2)$. The first step is also well-defined since $\pi(\mu|\boldsymbol{\omega}, \mathscr{D}_2)$ is a normal distribution with mean $\sum_i \omega_i x_i / (\sum_i \omega_i + 1/20)$ and variance $1/(2\sum_i \omega_i + 1/10)$. The corresponding R program then simplifies into two lines

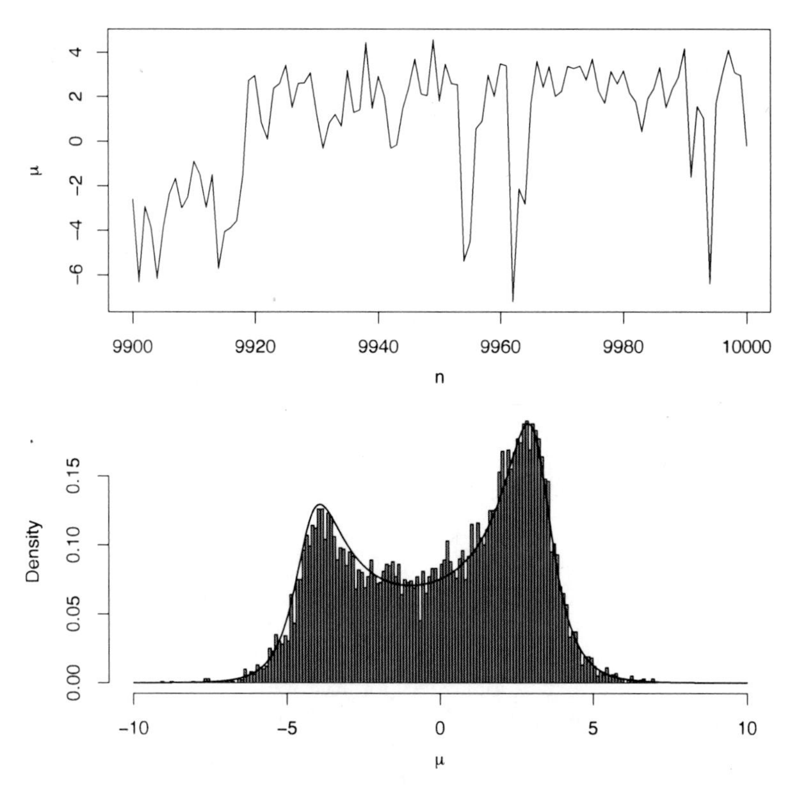

Fig. 3.4. (*Top*) Last 100 iterations of the chain $(\mu^{(t)})$; (*bottom*) histogram of the chain $(\mu^{(t)})$ and comparison with the target density for 10,000 iterations

```
> mu = rnorm(1,sum(x*omega)/sum(omega+.05),
+ sqrt(1/(.1+2*sum(omega))))
> omega = rexp(2,1+(x-mu)^2)
```

and the output of the simulation is represented in Fig. 3.4, with a very satisfying fit between the histogram of the simulated values and the target. A detailed zoom on the last 100 iterations shows how the chain $(\mu^{(t)})$ moves around, alternatively visiting each mode of the target.

↯ When running a Gibbs sampler, the number of iterations should never be fixed in advance: it is usually impossible to predict the performance of a given sampler before producing a corresponding chain. Deciding on the length of an MCMC run is therefore a sequential process where output behaviors are examined after pilot runs and new simulations (or new samplers) are chosen on the basis of these pilot runs.

3.5.3 The General Gibbs Sampler

For a joint distribution $\pi(x_1, \ldots, x_p)$ with full conditionals $\pi_1, \ldots, \pi_p$ where π_j is the distribution of x_j conditional on $(x_1, \ldots, x_{j-1}, x_{j+1}, \ldots, x_p)$, the Gibbs sampler simulates successively from all conditionals, modifying one component of $\mathbf{x}$ at a time. The corresponding algorithmic representation is given in Algorithm 3.4.

Algorithm 3.4 GIBBS SAMPLER

Initialization: Start with an arbitrary value $\mathbf{x}^{(0)} = (x_1^{(0)}, \ldots, x_p^{(0)})$.
Iteration t: Given $(x_1^{(t-1)}, \ldots, x_p^{(t-1)})$, generate
1. $x_1^{(t)}$ according to $\pi_1(x_1|x_2^{(t-1)}, \ldots, x_p^{(t-1)})$,
2. $x_2^{(t)}$ according to $\pi_2(x_2|x_1^{(t)}, x_3^{(t-1)}, \ldots, x_p^{(t-1)})$,

⋮

p. $x_p^{(t)}$ according to $\pi_p(x_p|x_1^{(t)}, \ldots, x_{p-1}^{(t)})$.

Quite logically, the validation of this generalization of Algorithm 3.3 is identical: for each of the p steps of the t-th iteration, the joint distribution $\pi(\mathbf{x})$ is stationary. Under the same restriction on the irreducibility of the chain, it also converges to π for every possible starting value. Note that the order in which the components of $\mathbf{x}$ are simulated can be modified at each iteration, either deterministically or randomly, without putting the validity of the algorithm in jeopardy.

The two-stage Gibbs sampler naturally appears as a special case of Algorithm 3.4 for $p = 2$. It is, however, endowed with higher theoretical properties, as detailed in Robert and Casella (2004, Chap. 9) and Robert and Casella (2009, Chap. 7).

To conclude this section, let us stress that the impact of MCMC on Bayesian statistics has been considerable. Since the 1990s, which saw the emergence of MCMC methods in the statistical community, the occurrence of Bayesian methods in applied statistics has greatly increased, and the frontier between Bayesian and "classical" statistics is now so fuzzy that in some fields, it has completely disappeared. From a Bayesian point of view, the access to far more advanced computational means has induced a radical modification of the way people work with models and prior assumptions. In particular, it has opened the way to process much more complex structures, such as graphical models and latent variable models (see Chap. 6). It has also freed inference by opening for good the possibility of Bayesian model choice (see, e.g., Robert, 2007, Chap. 7). This expansion is much more visible among academics than among applied statisticians, though, given that the use of the MCMC technology requires some "hard" thinking to process every new problem. The availability of specific software such as BUGS has nonetheless given access to MCMC techniques to a wider community, starting with the medical field. New modules in R and other languages like Python are also helping to bridge the gap.

3.6 Variable Selection

3.6.1 Deciding on Explanatory Variables

In an ideal world, when building a regression model, we should include all relevant pieces of information, which in the regression context means including all predictor variables that might possibly help in explaining $\mathbf{y}$. However, there are obvious drawbacks to the advice of increasing the number of explanatory variables. For one thing, in noninformative settings, this eventually clashes with the constraint $p < n$. For another, using a huge number of explanatory variables leaves little information available to obtain precise estimators. In other words, when we increase the explanatory scope of the regression model, we do not necessarily increase its explanatory power because it gets harder and harder to estimate the coefficients.[11] It is thus important to be

[11]This phenomenon is related to the *principle of parsimony*, also called *Occam's razor*, which states that, among two models with similar explanatory powers, the simplest one should always be preferred. It is also connected with the *learning curve effect* found in information theory and neural networks, where the performance of a model increases on the learning dataset but decreases on a testing dataset as its complexity increases.

able to decide which variables—within a large pool of potential explanatory variables—should be kept in a model that balances good explanatory power with good estimation performance.

This is truly a *decision* problem in that all potential models have to be considered in parallel against a criterion that ranks them. This variable-selection problem can be formalized as follows. We consider a dependent random variable y and a set of p potential explanatory variables. At this stage, we assume that every subset of q explanatory variables could make a proper set of explanatory variables for the regression of y. The only restriction we impose is that the intercept (that is, the constant variable) is included in every model. There are thus 2^p models in competition and we are looking for a procedure that selects the "best" model, that is, the "most relevant" explanatory variables. Note that this variable-selection procedure can alternatively be seen as a two-stage estimation setting where we first estimate the indicator of the model (within the collection of models), which also amounts to estimating variable indicators, as detailed below, and we then estimate the parameters corresponding to this very model.

Each of the 2^p models under comparison is in fact associated with a binary indicator vector $\gamma \in \Gamma = \{0,1\}^p$, where $\gamma_j = 1$ means that the variable x_j is included in the model, denoted by $\mathfrak{M}_\gamma$. This notation is quite handy since γ=(1,0,1,0,0,...,1,0) clearly indicates which explanatory variables are in and which are not. We also use the notation

$$q_\gamma = \mathbf{1}_p^\mathsf{T}\gamma$$

for computing the number of variables included in the model $\mathfrak{M}_\gamma$. We define β^γ as a sub-vector of β containing only the components such that x_j is included in the model $\mathfrak{M}_\gamma$ and $\mathbf{X}^\gamma$ as the sub-matrix of $\mathbf{X}$ where only the columns such that x_j is included in the model $\mathfrak{M}_\gamma$ have been left. The model $\mathfrak{M}_\gamma$ is thus defined as

$$\mathbf{y}|\alpha,\beta^\gamma,\sigma^2,\gamma \sim \mathscr{N}_n\left(\alpha\mathbf{1}_n + \beta^\gamma\mathbf{X}^\gamma\beta^\gamma, \sigma^2 I_n\right) .$$

↯ Once again, and apparently in contradiction to our basic tenet that different models should enjoy completely different parameters, we are compelled to denote by σ^2 and α the variance and intercept terms common to *all models*, respectively. Although this is more of a mathematical trick than a true modeling reason, the prior independence of (α,σ^2) and γ allows for the simultaneous use of Bayes factors and an improper prior. Despite the possibly confusing notation, β^γ and β are completely unrelated in that they are parameters of different models.

3.6.2 G-Prior Distributions for Model Choice

Because so many models are in competition and thus considered in the global model all at once, we cannot expect a practitioner to specify one's own prior on every model $\mathfrak{M}_\gamma$ in a completely subjective and autonomous manner. We thus now proceed to derive *all* priors from a single global prior associated with the so-called *full model* that corresponds to $\gamma = (1, \ldots, 1)$. The argument goes as follows:

(1) For the full model, we use Zellner's G-prior as defined in Sect. 3.4,

$$\beta|\sigma^2 \sim \mathscr{N}_p(\tilde{\beta}, g\sigma^2(\mathbf{X}^\mathsf{T}\mathbf{X})^{-1}) \quad \text{and} \quad \pi(\alpha, \sigma^2) \propto \sigma^{-2}.$$

(2) For each (sub-)model $\mathfrak{M}_\gamma$, the prior distribution of β^γ conditional on σ^2 is fixed as

$$\beta^\gamma|\sigma^2, \gamma \sim \mathscr{N}_{q_\gamma}\left(\tilde{\beta}^\gamma, g\sigma^2\left(\mathbf{X}^{\gamma\mathsf{T}}\mathbf{X}^\gamma\right)^{-1}\right),$$

where $\tilde{\beta}^\gamma = \left(\mathbf{X}^{\gamma T}\mathbf{X}^\gamma\right)^{-1}\mathbf{X}^{\gamma T}\tilde{\mathbf{X}}\tilde{\beta}$ and we use the same prior on (α, σ^2).

↯ This distribution is conditional on γ; in particular, this implies that, while the variance notation σ^2 is common to all models, its distribution varies with γ.

Although there are many possible ways of defining the prior on the model index[12] γ, we opt for the uniform prior

$$\pi(\gamma) = 2^{-p}.$$

The posterior distribution of γ (that is, the distribution of γ given $\mathbf{y}$) is central to the variable-selection methodology since it is proportional to the marginal density of $\mathbf{y}$ in $\mathfrak{M}_\gamma$. In addition, for prediction purposes, the prediction distribution can be obtained by averaging over all models, the weights being the model probabilities (this is called *model averaging*).

The posterior distribution of γ is

$$\begin{aligned}\pi(\gamma|\mathbf{y}) &\propto f(\mathbf{y}|\gamma)\pi(\gamma) \propto f(\mathbf{y}|\gamma)\\ &\propto (g+1)^{-(q_\gamma+1)/2}\left[\mathbf{y}^\mathsf{T}\mathbf{y} - \frac{g}{g+1}\mathbf{y}^\mathsf{T}\mathbf{X}^\gamma\left(\mathbf{X}^{\gamma\mathsf{T}}\mathbf{X}^\gamma\right)^{-1}\mathbf{X}^{\gamma\mathsf{T}}\mathbf{y}\right.\\ &\qquad \left. - \frac{1}{g+1}\tilde{\beta}^{\gamma\mathsf{T}}\mathbf{X}^{\gamma\mathsf{T}}\mathbf{X}^\gamma\tilde{\beta}^\gamma\right]^{-(n-1)/2}. \qquad (3.7)\end{aligned}$$

When the number of explanatory variables is less than 15, say, the exact derivation of the posterior probabilities for all submodels can be undertaken.

[12] For instance, one could instead use a uniform prior on the number q_γ of explanatory variables or a more parsimonious prior such as $\pi(\gamma) = 1/q_\gamma$.

Indeed, $2^{15} = 32768$ means that the problem remains tractable. The following R code (part of the function ModChoBayesReg) is used to calculate those posterior probabilities and returns the top most probable models. The integrated likelihood for the null model is computed as intlike0.

```
intlike=rep(intlike0,2^p)
for (j in 2:2^p){
  gam=as.integer(intToBits(i-1)[1:p]==1)
  pgam=sum(gam)
  Xgam=X[,which(gam==1)]
  Ugam=solve(t(Xgam)%*%Xgam)%*%t(Xgam)
  betatildegam=b1=Ugam%*%X%*%betatilde
  betamlgam=b2=Ugam%*%y
  s2gam=t(y-alphaml-Xgam%*%b2)%*%(y-alphaml-Xgam%*%b2)
  kappagam=as.numeric(s2gam+t(b1-b2)%*%t(Xgam)%*%
  Xgam%*%(b1-b2)/(g+1))
  intlike[j]=(g+1)^(-pgam/2)*kappagam^(-(n-1)/2)
  }
intlike=intlike/sum(intlike)
modcho=order(intlike)[2^p:(2^p-9)]
probtop10=intlike[modcho]
```

The above R code uses the generic function intToBits to turn an integer i into the indicator vector gam. The remainder of the code is quite similar to the model choice code when computing the Bayes factors.

For the **caterpillar** data, we set $\tilde{\beta} = 0_8$ and $g = 1$. The models corresponding to the top 10 posterior probabilities are then given by

```
> ModChoBayesReg(y,X,g=1)

Number of variables less than 15
Model posterior probabilities are calculated exactly

   Top10Models  PostProb
1      1 2 3 7    0.0142
2    1 2 3 5 7    0.0138
3        1 2 7    0.0117
4    1 2 3 4 7    0.0112
5  1 2 3 4 5 7    0.0110
6      1 2 5 7    0.0108
7    1 2 3 7 8    0.0104
8    1 2 3 6 7    0.0102
9  1 2 3 5 6 7    0.0100
10 1 2 3 5 7 8    0.0098
```

In a basic $0-1$ decision setup, we would choose the model $\mathfrak{M}_\gamma$ with the highest posterior probability—that is, the model with explanatory variables x_1, x_2, x_3 and x_7—which corresponds to the variables

- altitude,
- slope,
- the number of pine trees in the area, and
- the number of vegetation strata.

The model selected by the procedure thus fails to correspond to the three variables identified in the R output at the end of Sect. 3.4. But interestingly, even under this strong shrinkage prior $g = 1$ (where the prior has the same weight as the data), all top ten models contain the explanatory variables x_1, x_2 and x_7, which have the most stars in this R analysis.

Now, the default or noninformative calibration of the G-prior corresponds to the choice $\tilde{\beta} = 0_p$ and $g = n$, which reduces the prior input to the equivalent of a *single* observation. Pushing g to a smaller value results in a paradoxical behaviour of the procedure which then usually picks the simpler model: this is another illustration of the *Jeffreys-Lindley paradox*, mentioned in Chap. 2.

For $\tilde{\beta} = 0_p$ and $g = n$, the ten most likely models and their posterior probabilities are:

```
> ModChoBayesReg(y,X)

Number of variables less than 15
Models's posterior probabilities are calculated exactly

   Top10Models PostProb
1        1 2 7   0.0767
2          1 7   0.0689
3      1 2 3 7   0.0686
4        1 3 7   0.0376
5        1 2 6   0.0369
6    1 2 3 5 7   0.0326
7      1 2 5 7   0.0294
8          1 6   0.0205
9      1 2 4 7   0.0201
10           7   0.0198
```

For this different prior modelling, we chose the same model as the `lm` classical procedure, rather than when $g = 1$; however, the posterior probabilities of the most likely models are much lower for $g = 1$, which is logical given that the current prior is less informative. Therefore, the top model is not as strongly supported as in the informative case. Once again, we stress that the choice $g = 1$ is rather arbitrary and that it is used here merely for illustrative purposes. The default value we recommend is $g = n$.

3.6.3 A Stochastic Search for the Most Likely Model

When the number p of variables is large, it becomes impossible to compute the posterior probabilities for the whole series of 2^p models. We then need a tailored algorithm that samples from $\pi(\boldsymbol{\gamma}|\mathbf{y})$ and thus selects the most likely models, without computing first all the values of $\pi(\boldsymbol{\gamma}|\mathbf{y})$. This can be done rather naturally by Gibbs sampling, given the availability of the full conditional posterior probabilities of the γ_j's.

Indeed, if $\boldsymbol{\gamma}_{-j}$ $(1 \leq j \leq p)$ is the vector $(\gamma_1, \ldots, \gamma_{j-1}, \gamma_{j+1}, \ldots, \gamma_p)$, the full conditional distribution $\pi(\gamma_j|\mathbf{y}, \boldsymbol{\gamma}_{-j})$ of γ_j is proportional to $\pi(\boldsymbol{\gamma}|\mathbf{y})$ and can be computed in both $\gamma_j = 0$ and $\gamma_j = 1$ at no cost (since these are the only possible values of γ_j).

Algorithm 3.5 GIBBS SAMPLER FOR VARIABLE SELECTION

Initialization: Draw $\boldsymbol{\gamma}^0$ from the uniform distribution on Γ.
Iteration t: Given $(\gamma_1^{(t-1)}, \ldots, \gamma_p^{(t-1)})$, generate
1. $\gamma_1^{(t)}$ according to $\pi(\gamma_1|\mathbf{y}, \gamma_2^{(t-1)}, \ldots, \gamma_p^{(t-1)})$,
2. $\gamma_2^{(t)}$ according to $\pi(\gamma_2|\mathbf{y}, \gamma_1^{(t)}, \gamma_3^{(t-1)}, \ldots, \gamma_p^{(t-1)})$,

$\vdots$

p. $\gamma_p^{(t)}$ according to $\pi(\gamma_p|\mathbf{y}, \gamma_1^{(t)}, \ldots, \gamma_{p-1}^{(t)})$.

After a large number of iterations of this algorithm (that is, when the sampler is supposed to have converged or, more accurately, when the sampler has sufficiently explored the support of the target distribution), its output can be used to approximate the posterior probabilities $\pi(\boldsymbol{\gamma}|\mathbf{y}, X)$ by empirical averages based on the Gibbs output,

$$\hat{\mathbb{P}}^\pi(\boldsymbol{\gamma} = \boldsymbol{\gamma}^*|\mathbf{y}) = \left(\frac{1}{T - T_0 + 1}\right) \sum_{t=T_0}^{T} \mathbb{I}_{\boldsymbol{\gamma}^{(t)} = \boldsymbol{\gamma}^*},$$

where the T_0 first values are eliminated as *burn-in*. (The number T_0 is therefore the number of iterations roughly needed to "reach" convergence.) The Gibbs output can also be used to approximate the inclusion of a given variable, $P^\pi(\gamma_j = 1|\mathbf{y}, X)$, as

$$\hat{\mathbb{P}}^\pi(\gamma_j = 1|\mathbf{y}) = \left(\frac{1}{T - T_0 + 1}\right) \sum_{t=T_0}^{T} \mathbb{I}_{\gamma_j^{(t)} = 1},$$

with the same asymptotic validation.

The following R code (again part of the function `ModChoBayesReg`) describes our implementation of the above variable-selection Gibbs sampler.

The code uses the null model with only the intercept α as a reference, based on the integrated likelihood `intlike0` as above. It then starts at random in the collection of models:

```
gamma=rep(0,niter)
mcur=sample(c(0,1),p,replace=TRUE)
gamma[1]=sum(2^(0:(p-1))*mcur)+1
pcur=sum(mcur)
```

and computes the corresponding integrated likelihood `intlikecur`

```
if (pcur==0) intlikecur=intlike0 else{ #integrated likelihood
 Xcur=X[,which(mcur==1)]
 Ucur=solve(t(Xcur)%*%Xcur)%*%t(Xcur)
 betatildecur=b1=Ucur%*%X%*%betatilde
 betamlcur=b2=Ucur%*%y
 s2cur=t(y-alphaml-Xcur%*%b2)%*%(y-alphaml-Xcur%*%b2)
 kappacur=as.numeric(s2cur+t(b1-b2)%*%t(Xcur)%*%
 Xcur%*%(b1-b2)/(g+1))
 intlikecur=(g+1)^(-pcur/2)*kappacur^(-(n-1)/2)
 }
```

It then proceeds according to Algorithm 3.5, proposing to change one variable indicator γ_j and accepting this move with a Metropolis–Hastings (defined and justified in Chap. 4) probability:

```
if (runif(1)<=(intlikeprop/intlikecur))
```

This modification is more efficient than directly simulating from the conditional as it avoids proposing the same value for γ_j twice.

```
for (t in 1:(niter-1)){ #iteration index
 mprop=mcur
 j=sample(1:p,1)
 mprop[j]=abs(mcur[j]-1)
 pprop=sum(mprop)
 if (pprop==0) intlikeprop=intlike0 else{ #integrated
   likelihood Xprop=X[,which(mprop==1)]
   Uprop=solve(t(Xprop)%*%Xprop)%*%t(Xprop)
   betatildeprop=b1=Uprop%*%X%*%betatilde
   betamlprop=b2=Uprop%*%y
   s2prop=t(y-alphaml-Xprop%*%betamlprop)%*
    %(y-alphaml-Xprop%*%betamlprop)
   kappaprop=as.numeric(s2prop+t(betatildeprop-betamlprop)%*
    %t(Xprop)%*%Xprop%*%
   (betatildeprop-betamlprop)/(g+1))
   intlikeprop=(g+1)^(-pprop/2)*kappaprop^(-(n-1)/2)
   }
 if (runif(1)<=(intlikeprop/intlikecur)){
```

```
    mcur=mprop
    intlikecur=intlikeprop
    }
  gamma[t+1]=sum(2^(0:(p-1))*mcur)+1
  }
gamma=gamma[20001:niter] #20,000 burnin steps
res=as.data.frame(table(as.factor(gamma)))
odo=order(res$Freq)[length(res$Freq):(length(res$Freq)-9)]
modcho=res$Var1[odo]
probtop10=res$Freq[odo]/(niter-20000)
```

In this setting of **caterpillar**, handling only eight (potential) explanatory variables means that it is possible to compute all of the 2^8 probabilities $\pi(\gamma|\mathbf{y})$ and to thus deduce the normalizing constant in (3.7). We can therefore compare these exact values with the approximations produced by the Gibbs sampler. Using $T_0 = 20{,}000$ and $T_0 = 80{,}000$, i.e. a total of 10^5 simulations, we obtain the following results for the top five models:

	Models	PostProb	Gibbs estimates of the PostProb
1	1 2 7	0.0767	0.0740
2	1 7	0.0689	0.0675
3	1 2 3 7	0.0686	0.0668
4	1 3 7	0.0376	0.0376
5	1 2 6	0.0369	0.0370

The comparison is quite comforting for the Gibbs sampler as the differences are truly minor! Rather naturally, as the number of variables grows, the number of simulations needed to provide a good approximation grows as well. Once more, we recommend running the code several times (with different random sequences) to ensure the stability of the approximation.

3.7 Exercises

3.1 Show that the matrix $\mathbf{Z}$ is of full rank if and only if the matrix $\mathbf{Z}^\mathsf{T}\mathbf{Z}$ is invertible (where $\mathbf{Z}^\mathsf{T}$ denotes the transpose of the matrix $\mathbf{Z}$, which can be produced in R using the `t(Z)` command). Apply to $\mathbf{Z} = [\mathbf{1}_n \quad \mathbf{X}]$ and deduce that this cannot happen when $p + 1 > n$.

3.2 Show that solving the minimization program

$$\min_\beta \, (\mathbf{y} - \mathbf{X}\beta)^\mathsf{T}(\mathbf{y} - \mathbf{X}\beta)$$

requires solving the system of equations $(\mathbf{X}^\mathsf{T}\mathbf{X})\beta = \mathbf{X}^\mathsf{T}\mathbf{y}$. Check that this can be done via the R command `solve(t(X)%*%(X),t(X)%*%y)`.

3.3 Show that the variance of the maximum likelihood estimator of β in the regression model is given by $\mathbb{V}(\hat{\boldsymbol{\beta}}|\sigma^2) = \sigma^2(\mathbf{X}^\mathsf{T}\mathbf{X})^{-1}$.

3.4 For the model

$$\mathbf{y}|\boldsymbol{\beta}, \sigma^2 \sim \mathscr{N}_n\left(\mathbf{X}\boldsymbol{\beta}, \sigma^2\mathbf{I}_n\right)$$

a conjugate prior distribution is as follows: the conditional distribution of $\boldsymbol{\beta}$ is given by

$$\boldsymbol{\beta}|\sigma^2 \sim \mathscr{N}_p(\tilde{\boldsymbol{\beta}}, \sigma^2\mathbf{M}^{-1}),$$

where $\mathbf{M}$ is a (p,p) positive definite symmetric matrix, and the marginal prior on σ^2 is an inverse Gamma distribution

$$\sigma^2 \sim \mathscr{IG}(a,b), \qquad a,b > 0\,.$$

Taking advantage of the matrix identities

$$\begin{aligned}\left(\mathbf{M}+\mathbf{X}^\mathsf{T}\mathbf{X}\right)^{-1} &= \mathbf{M}^{-1} - \mathbf{M}^{-1}\left(\mathbf{M}^{-1}+(\mathbf{X}^\mathsf{T}\mathbf{X})^{-1}\right)^{-1}\mathbf{M}^{-1}\\ &= (\mathbf{X}^\mathsf{T}\mathbf{X})^{-1} - (\mathbf{X}^\mathsf{T}\mathbf{X})^{-1}\left(\mathbf{M}^{-1}+(\mathbf{X}^\mathsf{T}\mathbf{X})^{-1}\right)^{-1}(\mathbf{X}^\mathsf{T}\mathbf{X})^{-1}\end{aligned}$$

and

$$\begin{aligned}\mathbf{X}^\mathsf{T}\mathbf{X}(\mathbf{M}+\mathbf{X}^\mathsf{T}\mathbf{X})^{-1}\mathbf{M} &= \left(\mathbf{M}^{-1}(\mathbf{M}+\mathbf{X}^\mathsf{T}\mathbf{X})(\mathbf{X}^\mathsf{T}\mathbf{X})^{-1}\right)^{-1}\\ &= \left(\mathbf{M}^{-1}+(\mathbf{X}^\mathsf{T}\mathbf{X})^{-1}\right)^{-1},\end{aligned}$$

establish that

$$\boldsymbol{\beta}|\mathbf{y},\sigma^2 \sim \mathscr{N}_p\left((\mathbf{M}+\mathbf{X}^\mathsf{T}\mathbf{X})^{-1}\{(\mathbf{X}^\mathsf{T}\mathbf{X})\hat{\boldsymbol{\beta}}+\mathbf{M}\tilde{\boldsymbol{\beta}}\}, \sigma^2(\mathbf{M}+\mathbf{X}^\mathsf{T}\mathbf{X})^{-1}\right) \tag{3.8}$$

where $\hat{\boldsymbol{\beta}} = (\mathbf{X}^\mathsf{T}\mathbf{X})^{-1}\mathbf{X}^\mathsf{T}\mathbf{y}$ and

$$\sigma^2|\mathbf{y} \sim \mathscr{IG}\left(\frac{n}{2}+a, b+\frac{s^2}{2}+\frac{(\tilde{\boldsymbol{\beta}}-\hat{\boldsymbol{\beta}})^\mathsf{T}\left(\mathbf{M}^{-1}+(\mathbf{X}^\mathsf{T}\mathbf{X})^{-1}\right)^{-1}(\tilde{\boldsymbol{\beta}}-\hat{\boldsymbol{\beta}})}{2}\right) \tag{3.9}$$

where $s^2 = (\mathbf{y}-\hat{\boldsymbol{\beta}}\mathbf{X})^\mathsf{T}(\mathbf{y}-\hat{\boldsymbol{\beta}}\mathbf{X})$ are the correct posterior distributions. Give a $(1-\alpha)$ HPD region on $\boldsymbol{\beta}$.

3.5 The regression model of Exercise 3.4 can also be used in a predictive sense: for a given $(m, p+1)$ explanatory matrix $\tilde{\mathbf{X}}$, *i.e.*, when predicting m unobserved variates $\tilde{y}_i$, the corresponding outcome $\tilde{\mathbf{y}}$ can be inferred through the *predictive distribution* $\pi(\tilde{\mathbf{y}}|\sigma^2,\mathbf{y})$. Show that $\pi(\tilde{\mathbf{y}}|\sigma^2,\mathbf{y})$ is a Gaussian density with mean

$$\mathbb{E}^\pi[\tilde{\mathbf{y}}|\sigma^2,\mathbf{y}] = \tilde{\mathbf{X}}(\mathbf{M}+\mathbf{X}^\mathsf{T}\mathbf{X})^{-1}(\mathbf{X}^\mathsf{T}\mathbf{X}\hat{\boldsymbol{\beta}}+\mathbf{M}\tilde{\boldsymbol{\beta}})$$

and covariance matrix

$$\mathbb{V}^{\pi}(\tilde{\mathbf{y}}|\sigma^2, \mathbf{y}) = \sigma^2(\mathbf{I}_m + \tilde{\mathbf{X}}(\mathbf{M} + \mathbf{X}^{\mathsf{T}}\mathbf{X})^{-1}\tilde{\mathbf{X}}^{\mathsf{T}}).$$

Deduce that

$$\begin{aligned}\tilde{\mathbf{y}}|\mathbf{y} \sim \mathscr{T}_m \Big(& n + 2a, \tilde{\mathbf{X}}(\mathbf{M} + \mathbf{X}^{\mathsf{T}}\mathbf{X})^{-1}(\mathbf{X}^{\mathsf{T}}\mathbf{X}\hat{\beta} + \mathbf{M}\tilde{\beta}), \\ & \frac{2b + s^2 + (\tilde{\beta} - \hat{\beta})^{\mathsf{T}} \left(\mathbf{M}^{-1} + (\mathbf{X}^{\mathsf{T}}\mathbf{X})^{-1}\right)^{-1} (\tilde{\beta} - \hat{\beta})}{n + 2a} \\ & \times \left\{\mathbf{I}_m + \tilde{\mathbf{X}}(\mathbf{M} + \mathbf{X}^{\mathsf{T}}\mathbf{X})^{-1}\tilde{\mathbf{X}}^{\mathsf{T}}\right\}\Big).\end{aligned}$$

3.6 Show that the marginal distribution of $\mathbf{y}$ associated with (3.8) and (3.9) is given by

$$\mathbf{y} \sim \mathscr{T}_n\left(2a, \mathbf{X}\tilde{\beta}, \frac{b}{a}(\mathbf{I}_n + \mathbf{X}\mathbf{M}^{-1}\mathbf{X}^{\mathsf{T}})\right).$$

3.7 Show that the matrix $(\mathbf{I}_n + g\mathbf{X}(\mathbf{X}^{\mathsf{T}}\mathbf{X})^{-1}\mathbf{X}^{\mathsf{T}})$ has 1 and $g + 1$ as only eigenvalues. (*Hint:* Show that the eigenvectors associated with $g + 1$ are of the form $\mathbf{X}\beta$ and that the eigenvectors associated with 1 are those orthogonal to $\mathbf{X}$.) Deduce that the determinant of the matrix $(\mathbf{I}_n + g\mathbf{X}(\mathbf{X}^{\mathsf{T}}\mathbf{X})^{-1}\mathbf{X}^{\mathsf{T}})$ is indeed $(g+1)^{p+1}$.

3.8 Under the Jeffreys prior, give the predictive distribution of $\tilde{\mathbf{y}}$, m dimensional vector corresponding to the (m, p) matrix of explanatory variables $\tilde{\mathbf{X}}$.

3.9 If (x_1, x_2) is distributed from the uniform distribution on

$$\left\{(x_1, x_2);\, (x_1 - 1)^2 + (x_2 - 1)^2 \leq 1\right\} \cup \left\{(x_1, x_2);\, (x_1 + 1)^2 + (x_2 + 1)^2 \leq 1\right\},$$

show that the Gibbs sampler does not produce an irreducible chain. For this distribution, find an alternative Gibbs sampler that works. (*Hint:* Consider a rotation of the coordinate axes.)

3.10 If a joint density $g(y_1, y_2)$ corresponds to the conditional distributions $g_1(y_1|y_2)$ and $g_2(y_2|y_1)$, show that it is given by

$$g(y_1, y_2) = \frac{g_2(y_2|y_1)}{\int g_2(v|y_1)/g_1(y_1|v)\, dv}.$$

3.11 Considering the model

$$\eta|\theta \sim \mathcal{B}\text{in}(n, \theta), \quad \theta \sim \mathcal{B}e(a, b),$$

derive the joint distribution of (η, θ) and the corresponding full conditional distributions. Implement a Gibbs sampler associated with those full conditionals and compare the outcome of the Gibbs sampler on θ with the true marginal distribution of θ.

3.12 Take the posterior distribution on (θ, σ^2) associated with the joint model

$$x_i|\theta, \sigma^2 \sim \mathscr{N}(\theta, \sigma^2), \quad i = 1, \ldots, n,$$
$$\theta \sim \mathscr{N}(\theta_0, \tau^2), \quad \sigma^2 \sim \mathscr{IG}(a, b).$$

Show that the full conditional distributions are given by

$$\theta|\mathbf{x}, \sigma^2 \sim \mathscr{N}\left(\frac{\sigma^2}{\sigma^2 + n\tau^2}\,\theta_0 + \frac{n\tau^2}{\sigma^2 + n\tau^2}\,\bar{\mathbf{x}},\ \frac{\sigma^2\tau^2}{\sigma^2 + n\tau^2}\right)$$

and

$$\sigma^2|\mathbf{x}, \theta \sim \mathscr{IG}\left(\frac{n}{2} + a, \frac{1}{2}\sum_i (x_i - \theta)^2 + b\right),$$

where $\bar{x}$ is the empirical average of the observations. Implement the Gibbs sampler associated with these conditionals.

4

Generalized Linear Models

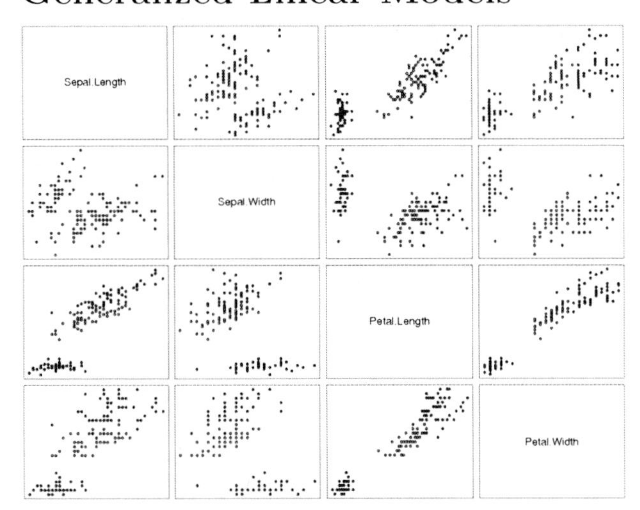

This was the sort of thing that impressed
Rebus: not nature, but ingenuity.
—**Ian Rankin**, ***A Question of Blood***.—

Roadmap

Generalized linear models are extensions of the linear regression model described in the previous chapter. In particular, they avoid the selection of a single transformation of the data that must achieve the possibly conflicting goals of normality and linearity imposed by the linear regression model, which is for instance impossible for binary or count responses. The trick that allows both a feasible processing and an extension of linear regression is first to turn the covariates into a real number by a linear projection and then to transform this value so that it fits the support of the response. We focus here on the Bayesian analysis of probit and logit models for binary data and of log-linear models for contingency tables.

On the methodological side, we present a general MCMC method, the Metropolis–Hastings algorithm, which is used for the simulation of complex distributions where both regular and Gibbs sampling fail. This includes in particular the random walk Metropolis–Hastings algorithm, which acts like a plain vanilla MCMC algorithm.

J.-M. Marin and C.P. Robert, *Bayesian Essentials with R*, Springer Texts in Statistics, DOI 10.1007/978-1-4614-8687-9_4,

4.1 A Generalization of the Linear Model

4.1.1 Motivation

In the previous chapter, we modeled the connection between a response variable y and a vector $\mathbf{x}$ of explanatory variables by a linear dependence relation with normal perturbations. There are many instances where both the linearity and the normality assumptions are not appropriate, especially when the support of y is restricted to $\mathbb{R}_+$ or $\mathbb{N}$. For instance, in dichotomous models, y takes its values in $\{0, 1\}$ as it represents the indicator of occurrence of a particular event (death in a medical study, unemployment in a socioeconomic study, migration in a capture–recapture study, etc.); in this case, a linear conditional expectation $\mathbb{E}[y|\mathbf{x}, \boldsymbol{\beta}] = \mathbf{x}^{\mathsf{T}}\boldsymbol{\beta}$ would be fairly cumbersome to handle, both in terms of the constraints on $\boldsymbol{\beta}$ and the corresponding distribution of the error $\varepsilon = y - \mathbb{E}[y|\mathbf{x}, \boldsymbol{\beta}]$.

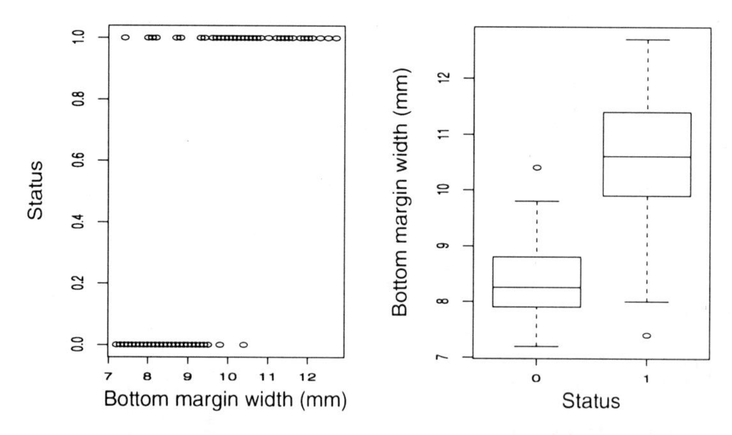

Fig. 4.1. Dataset **bank**: (*left*) Plot of the status indicator versus the bottom margin width; (*right*) boxplots of the bottom margin width for both counterfeit statuses

The **bank** dataset we analyze in the first part of this chapter comes from Flury and Riedwyl (1988) and is made of four measurements on 100 genuine Swiss banknotes and 100 counterfeit ones. The response variable y is thus the status of the banknote, where 0 stands for genuine and 1 stands for counterfeit, while the explanatory factors are the length of the bill x_1, the width of the left edge x_2, the width of the right edge x_3, and the bottom margin width x_4, all expressed in millimeters. We want a probabilistic model that predicts the type of banknote (i.e., that detects counterfeit banknotes) based on the four measurements above. To motivate the introduction of the generalized linear models, we only consider here the dependence of y on the fourth measure, x_4, which again is the bottom margin width of the banknote. To start, the y_i's being binary, the conditional distribution of y given x_4 cannot be normal. Nonetheless, as shown by Fig. 4.1, the variable x_4 clearly has a strong influence on whether the banknote is or is not counterfeit. To model this dependence in a proper manner, we must devise a realistic (if not real!) connection between y and x_4. The fact that y is binary implies a specific form of dependence: Indeed, both its marginal and conditional distributions necessarily are Bernoulli distributions. This means that, for instance, the conditional distribution of y given x_4 is a Bernoulli $\mathscr{B}(p(x_4))$ distribution; that is, for $x_4 = x_{4i}$, there exists $0 \leq p_i = p(x_{4i}) \leq 1$ such that

$$\mathbb{P}(y_i = 1|x_4 = x_{4i}) = p_i \,,$$

which turns out to be also the conditional expectation of y_i, $\mathbb{E}[y_i|x_{4i}]$. If we do impose a linear dependence on the p_i's, namely,

$$p(x_{4i}) = \beta_0 + \beta_1 x_{4i} \,,$$

the maximum likelihood estimates of β_0 and β_1 are then equal to -2.02 and 0.268, leading to the estimated prediction equation

$$\hat{p}_i = -2.02 + 0.268 x_{i4} \,. \tag{4.1}$$

This implies that a banknote with bottom margin width equal to 8 is counterfeit with probability

$$-2.02 + 0.268 \times 8 = 0.120 \,.$$

Thus, this banknote has a relatively small probability of having been counterfeited, which coincides with the intuition drawn from Fig. 4.1. However, if we now consider a banknote with bottom margin width equal to 12, (4.1) implies that this banknote is counterfeited with probability

$$-2.02 + 0.268 \times 12 = 1.192 \,,$$

which is certainly embarrassing for a probability estimate! We could try to modify the result by truncating the probability to (0, 1) and by deciding that this value of x_4 almost certainly indicates a counterfeit, but still there is a fundamental difficulty with this model. The fact that an ordinary linear dependence can predict values outside $(0, 1)$ suggests that the connection between this explanatory variable and the probability of a counterfeit cannot be modeled through a linear function but rather can be achieved using functions of x_{4i} that take their values within the interval $(0, 1)$.

4.1.2 Link Functions

As shown by the previous analysis, while linear models are nice to work with, they also have strong limitations. Therefore, we need a broader class of models to cover various dependence structures. The class selected for this chapter is called the family of *generalized linear models* (GLM), which has been formalized in McCullagh and Nelder (1989). This nomenclature stems from the fact that the dependence of y on $\mathbf{x}$ is partly *linear* in the sense that the conditional distribution of y given x is defined in terms of a linear combination $\mathbf{x}^{\mathsf{T}}\boldsymbol{\beta}$ of the components of $\mathbf{x}$,

$$y|\boldsymbol{\beta} \sim f(y|\mathbf{x}^{\mathsf{T}}\boldsymbol{\beta}) .$$

As in the previous chapter, we use the notation $\mathbf{y} = (y_1, \ldots, y_n)$ for a sample of n responses and

$$\mathbf{X} = [\mathbf{x}_1 \quad \ldots \quad \mathbf{x}_k] = \begin{bmatrix} x_{11} & x_{12} & \ldots & x_{1k} \\ x_{21} & x_{22} & \ldots & x_{2k} \\ x_{31} & x_{32} & \ldots & x_{3k} \\ \vdots & \vdots & \vdots & \vdots \\ x_{n1} & x_{n2} & \ldots & x_{nk} \end{bmatrix}$$

for the $n \times k$ matrix of corresponding explanatory variables, possibly with $x_{11} = \ldots = x_{n1} = 1$. We use y and $\mathbf{x}$ as generic notations for single-response and covariate vectors, respectively. Once again, we will omit the dependence on $\mathbf{x}$ or $\mathbf{X}$ to simplify notations.

A *generalized linear model* is specified by two functions:

1. a conditional density f of y given $\mathbf{x}$ that belongs to an exponential family (Sect. 2.2.3) and that is parameterized by an expectation parameter $\mu = \mu(\mathbf{x}) = \mathbb{E}[y|\mathbf{x}]$ and possibly a dispersion parameter $\varphi > 0$ that does not depend on $\mathbf{x}$; and
2. a *link* function g that relates the mean $\mu = \mu(\mathbf{x})$ of f and the covariate vector, $\mathbf{x}$, as $g(\mu) = (\mathbf{x}^{\mathsf{T}}\boldsymbol{\beta})$, $\boldsymbol{\beta} \in \mathbb{R}^k$.

For identifiability reasons, the link function g is a one-to-one function and we have

$$\mathbb{E}[y|\boldsymbol{\beta}, \varphi] = g^{-1}\left(\mathbf{x}^{\mathsf{T}}\boldsymbol{\beta}\right) .$$

We can thus write the (conditional) likelihood as

$$\ell(\boldsymbol{\beta}, \varphi|\mathbf{y}) = \prod_{i=1}^{n} f\left(y_i|\mathbf{x}^{i\mathsf{T}}\boldsymbol{\beta}, \varphi\right)$$

if we choose to reparameterize f with the transform $g(\mu_i)$ of its mean and if we denote by $\mathbf{x}^i$ the covariate vector for the ith observation.[1]

The ordinary linear regression is obviously a special case of GLM where $g(x) = x$, $\varphi = \sigma^2$ and $y|\boldsymbol{\beta}, \sigma^2 \sim \mathscr{N}\left(\mathbf{x}^{\mathsf{T}}\boldsymbol{\beta}, \sigma^2\right)$. However, outside the linear model, the interpretation of the coefficients β_i is much more delicate because these coefficients do not relate directly to the observables, due to the presence of a link function that cannot be the identity. For instance, in the logistic regression model (defined in the following paragraph), the linear dependence is defined in terms of the *log-odds ratio* $\log\{p_1/(1-p_i)\}$.

The most widely used GLMs are presumably those that analyze binary data, as in **bank**, that is, when $y_i \sim \mathscr{B}(1, p_i)$ (with $\mu_i = p_i = p(\mathbf{x}^{i\mathsf{T}}\boldsymbol{\beta})$). The mean function p thus transforms a real value into a value between 0 and 1, and a possible choice of link function is the *logit transform*,

$$g(p) = \log\{p/(1-p)\}\,,$$

associated with the *logistic regression model*. Because of the limited support of the responses y_i, there is no dispersion parameter in this model and the corresponding likelihood function is

$$\begin{aligned}\ell(\boldsymbol{\beta}|\mathbf{y}) &= \prod_{i=1}^{n} \left(\frac{\exp(\mathbf{x}^{i\mathsf{T}}\boldsymbol{\beta})}{1+\exp(\mathbf{x}^{i\mathsf{T}}\boldsymbol{\beta})}\right)^{y_i} \left(\frac{1}{1+\exp(\mathbf{x}^{i\mathsf{T}}\boldsymbol{\beta})}\right)^{1-y_i} \qquad (4.2)\\ &= \exp\left\{\sum_{i=1}^{n} y_i\, \mathbf{x}^{i\mathsf{T}}\boldsymbol{\beta}\right\} \Big/ \prod_{i=1}^{n}\left[1+\exp(\mathbf{x}^{i\mathsf{T}}\boldsymbol{\beta})\right].\end{aligned}$$

It thus fails to factorize conveniently because of the denominator: there is no manageable conjugate prior for this model, called the *logit model*.

There exists a specific form of link function for each exponential family which is called the *canonical link*. This canonical function is chosen as the function $g^\star$ of the expectation parameter that appears in the exponent of the natural exponential family representation of the probability density, namely

$$g^\star(\mu) = \theta \quad \text{if} \quad f(y|\mu,\varphi) = h(y)\exp\varphi\{T(y)\cdot\theta - \Psi(\theta)\}\,.$$

Since the logistic regression model can be written as

$$f(y_i|p_i) = \exp\left\{y_i \log\left(\frac{p_i}{1-p_i}\right) + \log(1-p_i)\right\}\,,$$

the logit link function is the canonical version for the Bernoulli model. Note that, while it is customary to use the canonical link, there is no compelling reason to do so, besides following custom!

[1] This upper indexing allows for the distinction between x_i, the ith component of the covariate vector, and $\mathbf{x}^i$, the ith vector of covariates in the sample.

For binary response variables, many link functions can be substituted for the logit link function. For instance, the *probit* link function, $g(\mu_i) = \Phi^{-1}(\mu_i)$, where Φ is the standard normal cdf, is often used in econometrics. The corresponding likelihood is

$$\ell(\boldsymbol{\beta}|\mathbf{y}) \propto \prod_{i=1}^{n} \Phi(\mathbf{x}^{i\mathsf{T}}\boldsymbol{\beta})^{y_i} \left[1 - \Phi(\mathbf{x}^{i\mathsf{T}}\boldsymbol{\beta})\right]^{1-y_i} . \tag{4.3}$$

Although this alternative is also quite arbitrary and any other cdf could be used as a link function (such as the logistic cdf associated with (4.2)), the probit link function enjoys a missing-data (Chap. 6) interpretation that clearly boosted its popularity: This model can indeed be interpreted as a degraded linear regression model in the sense that observing $y_i = 1$ corresponds to the case $z_i \geq 0$, where z_i is a latent (that is, unobserved) variable such that $z_i \sim \mathscr{N}\left(\mathbf{x}^{i\mathsf{T}}\boldsymbol{\beta}, 1\right)$. In other words, $y = \mathbb{I}(z_i \geq 0)$ appears as a dichotomized linear regression response. Of course, this perspective is only an *interpretation* of the probit model in the sense that there may be no hidden z_i's at all in the real world! In addition, the probit and logistic regression models have quite similar behaviors, differing mostly in the tails.

Another type of GLM deals with unbounded integer-valued variables. The *Poisson regression model* starts from the assumption that the y_i's are Poisson $\mathscr{P}(\mu_i)$ and it selects a link function connecting $\mathbb{R}^+$ bijectively with $\mathbb{R}$, such as, for instance, the logarithmic function, $g(\mu_i) = \log(\mu_i)$. This model is thus a *count* model in the sense that the responses are integers, for instance the number of deaths due to lung cancer in a county or the number of speeding tickets issued on a particular stretch of highway, and it is quite common in epidemiology. The corresponding likelihood is

$$\ell(\boldsymbol{\beta}|\mathbf{y}) = \prod_{i=1}^{n} \left(\frac{1}{y_i!}\right) \exp\left\{y_i\, \mathbf{x}^{i\mathsf{T}}\boldsymbol{\beta} - \exp(\mathbf{x}^{i\mathsf{T}}\boldsymbol{\beta})\right\} ,$$

where the factorial terms $(1/y_i!)$ are irrelevant for both likelihood and posterior computations. Note that it does not factorize conveniently because of the exponential terms within the exponential.

The three examples above are simply illustrations of the versatility of generalized linear modeling. In this chapter, we discuss only two types of data for which generalized linear modeling is appropriate. We refer the reader to McCullagh and Nelder (1989) and Gelman et al. (2013) for a much more detailed coverage.

4.2 Metropolis–Hastings Algorithms

As partly hinted by the previous examples, posterior inference in GLMs is much harder than in linear models because of less manageable (and non-factorizing) likelihoods, which explains the longevity and versatility of linear model studies over the past centuries! Working with a GLM typically requires specific numerical or simulation tools. We take the opportunity of this requirement to introduce a universal MCMC method called the *Metropolis–Hastings* algorithm. Its range of applicability is incredibly broad (meaning that it is by no means restricted to GLM applications) and its inclusion in the Bayesian toolbox in the early 1990s has led to considerable extensions of the Bayesian field.[2]

4.2.1 Definition

When compared with the Gibbs sampler, Metropolis–Hastings algorithms are generic (or off-the-shelf) MCMC algorithms in the sense that they can be tuned toward a much wider range of possibilities. Those algorithms are also a natural extension of standard simulation algorithms such as accept–reject (see Chap. 5) or sampling importance resampling methods since they are all based on a *proposal* distribution. However, a major difference is that, for the Metropolis–Hastings algorithms, the proposal distribution is *Markov*, with kernel density $q(x, y)$. If the *target* distribution has density π, the Metropolis–Hastings algorithm is as follows:

Algorithm 4.6 GENERIC METROPOLIS–HASTINGS SAMPLER

Initialization: Choose an arbitrary starting value $x^{(0)}$.
Iteration t ($t \geq 1$):

1. Given $x^{(t-1)}$, generate $\tilde{x} \sim q(x^{(t-1)}, x)$.
2. Compute

$$\rho(x^{(t-1)}, \tilde{x}) = \min\left(\frac{\pi(\tilde{x})/q(x^{(t-1)}, \tilde{x})}{\pi(x^{(t-1)})/q(\tilde{x}, x^{(t-1)})}, 1\right).$$

3. With probability $\rho(x^{(t-1)}, \tilde{x})$, accept $\tilde{x}$ and set $x^{(t)} = \tilde{x}$; otherwise reject $\tilde{x}$ and set $x^{(t)} = x^{(t-1)}$.

The distribution q is also called the *instrumental* distribution. As in the accept–reject method (Sect. 5.4), we only need to know either π or q up to a proportionality constant since both constants cancel in the calculation of ρ. Note also the advantage of this approach compared with the Gibbs sampler: it is not necessary to use the conditional distributions of π.

The strong appeal of this algorithm is that it is rather universal in its formulation as well as in its use. Indeed, we only need to simulate from a

[2]This algorithm had been used by particle physicists, including Metropolis, since the late 1940s, but, as is often the case, the connection with statistics was not made until much later!

proposal q that can be chosen quite freely. There is, however, a theoretical constraint, namely that the chain produced by this algorithm must be able to explore the support of $\pi(y)$ in a finite number of steps. As discussed below, there also are many practical difficulties that are such that the algorithm may lose its universal feature and that it may require some specific tuning for each new application.

The theoretical validation of this algorithm is the same as with other MCMC algorithms: The target distribution π is the limiting distribution of the Markov chain produced by Algorithm 4.6. This is due to the choice of the acceptance probability $\rho(x, y)$ since the so-called *detailed balance equation*

$$\pi(x)q(x,y)\rho(x,y) = \pi(y)q(y,x)\rho(y,x)$$

holds and thus implies that π is stationary by integrating out x.

While theoretical guarantees that the algorithm converges are very high, the choice of q remains essential in practice. Poor choices of q may indeed result either in a very high rejection rate, meaning that the Markov chain $(x^{(t)})_t$ hardly moves, or in a myopic exploration of the support of π, that is, in a dependence on the starting value $x^{(0)}$ such that the chain is stuck in a neighborhood region of $x^{(0)}$. A particular choice of proposal q may thus work well for one target density but be extremely poor for another one. While the algorithm is indeed universal, it is impossible to prescribe application-independent strategies for choosing q.

We thus consider below two specific cases of proposals and briefly discuss their pros and cons (see Robert and Casella, 2004, Chap. 7, for a detailed discussion).

4.2.2 The Independence Sampler

The choice of q closest to the accept–reject method (see Algorithm 5.9) is to pick a constant q that is independent of its first argument,

$$q(x,y) = q(y)\,.$$

In that case, ρ simplifies into

$$\rho(x,y) = \min\left(1, \frac{\pi(y)/q(y)}{\pi(x)/q(x)}\right).$$

In the special case in which q is proportional to π, we obtain $\rho(x, y) = 1$ and the algorithm reduces, as expected, to iid sampling from π. The analogy with the accept–reject algorithm is that the maximum of the ratio π/q is replaced with the current value $\pi(x^{(t-1)})/q(x^{(t-1)})$ but the sequence of accepted $x^{(t)}$'s is not iid because of the acceptance step.

The convergence properties of the algorithm depend on the density q. First, q needs to be positive everywhere on the support of π. Second, for good exploration of this support, it appears that the ratio π/q needs to be bounded (see Robert and Casella, 2004, Theorem 7.8). Otherwise, the chain may take too long to reach some regions with low q/π values. This constraint obviously reduces the appeal of using an independence sampler, even though the fact that it does not require an explicit upper bound on π/q may sometimes be a plus.

This type of MH sampler is thus very model-dependent, and it suffers from the same drawbacks as the importance sampling methodology, namely that tuning the "right" proposal becomes much harder as the dimension increases.

4.2.3 The Random Walk Sampler

Since the independence sampler requires too much global information about the target distribution that is difficult to come by in complex or high-dimensional problems, an alternative is to opt for a local gathering of information, clutching to the hope that the accumulated information will provide, in the end, the global picture. Practically, this means exploring the neighborhood of the current value $x^{(t)}$ in search of other points of interest. The simplest exploration device is based on random walk dynamics.

A *random walk* proposal is based on a symmetric transition kernel $q(x, y) = q_{RW}(y - x)$ with $q_{RW}(x) = q_{RW}(-x)$. Symmetry implies that the acceptance probability $\rho(x, y)$ reduces to the simpler form

$$\rho(x, y) = \min\left(1, \pi(y)/\pi(x)\right) .$$

The appeal of this scheme is obvious when looking at the acceptance probability, since it only depends on the target π and since this version accepts all proposed moves that increase the value of π. There is considerable flexibility in the choice of the distribution q_{RW}, at least in terms of scale (i.e., the size of the neighborhood of the current value) and tails. Note that while from a probabilistic point of view random walks usually have no stationary distribution, the algorithm biases the random walk by moving toward modes of π more often than moving away from them.

The ambivalence of MCMC methods like the Metropolis–Hastings algorithm is that they can be applied to virtually any target. This is a terrific plus in that they can tackle new models, but there is also a genuine danger that they simultaneously fail to converge and fail to signal that they have failed to converge! Indeed, these algorithms can produce seemingly reasonable results, with all outer aspects of stability, while they are missing major modes of the target distribution. For instance, particular attention must be paid to models where the number of parameters exceeds by far the size of the dataset.

4.2.4 Output Analysis and Proposal Design

An important problem with the implementation of an MCMC algorithm is to gauge when convergence has been achieved; that is, to assess at what point the distribution of the chain is sufficiently close to its asymptotic distribution for all practical purposes or, more practically, when it has covered the whole support of the target distribution with sufficient regularity. The number of iterations T_0 that is required to achieve this goal is called the *burn-in* period. It is usually sensible to discard simulated values within this burn-in period in the Monte Carlo estimation so that the bias caused by the starting value is reduced. However, and this is particularly true in high dimensions, the empirical assessment of MCMC convergence is extremely delicate, to the point that it is rarely possible to be certain that an algorithm has converged.[3] Nevertheless, some partial convergence diagnostic procedures can be found in the literature (see Robert and Casella, 2004, Chap. 12, and Robert and Casella, 2009, Chap. 8). In particular, the latter describes the R package coda in Sect. 8.2.4.

A first way to assess whether or not a chain is in its stationary regime is to visually compare trace plots of sequences started at different values, as it may expose difficulties related, for instance, to multimodality. In practice, when chains of length T from two starting values have visited substantially different parts of the state space, the burn-in period for at least one of the chains should be greater than T. Note, however, that the problem of obtaining overdispersed starting values can be difficult when little is known about the target density, especially in large dimensions.

Autocorrelation plots of particular components provide in addition good indications of the chain's mixing behavior. If ρ_k ($k \in \mathbb{N}^*$) denotes the kth-order autocorrelation,

$$\rho_k = \text{cov}\left(x^{(t)}, x^{(t+k)}\right),$$

these quantities can be estimated from the observed chain itself,[4] at least for small values of k, and an *effective sample size* factor can be deduced from these estimates,

$$T^{\text{ess}} = T\left(1 + 2\sum_{k=1}^{T_0} \hat{\rho}_k\right)^{-1/2},$$

where $\hat{\rho}_k$ is the empirical autocorrelation function. This quantity represents the sample size of an equivalent iid sample when running T iterations. Conversely, the ratio T/T^{ess} indicates the multiplying factor on the minimum number of iid iterations required to run a simulation. Note, however, that this is only a partial indicator: Chains that remain stuck in one of the modes of the target distribution may well have a high effective ratio.

While we cannot discuss at length the selection of the proposal distribution (see Robert and Casella, 2004, Chap. 7, and Robert and Casella, 2009,

[3]Guaranteed convergence as in accept–reject algorithms is sometimes achievable with MCMC methods using techniques such as *perfect sampling* or *renewal*. But such techniques require a much more advanced study of the target distribution and the transition kernel of the algorithm. These conditions are not met very often in practice (see Robert and Casella 2004, Chap. 13).

[4]In R, this estimation can be conducted using the acf function.

Chap. 6), we stress that this is an important choice that has deep consequences for the convergence properties of the simulated Markov chain and thus for the exploration of the target distribution. As for prior distributions, we advise the simultaneous use of different kernels to assess their performances on the run. When considering a random walk proposal, for instance, a quantity that needs to be calibrated against the target distribution is the scale of this random walk. Indeed, if the variance of the proposal is too small with respect to the target distribution, the exploration of the target support will be small and may fail in more severe cases. Similarly, if the variance is too large, this means that the proposal will most often generate values that are outside the support of the target and that the algorithm will reject a large portion of attempted transitions.

↯ It seems reasonable to tune the proposal distribution in terms of its past performances, for instance by increasing the variance if the acceptance rate is high or decreasing it otherwise (or moving the location parameter toward the mean estimated over the past iterations). This must not be implemented outside a burn-in step, though, because a permanent modification of the proposal distribution amounts to taking into account the whole past of the sequence and thus it cancels both its Markovian nature and its convergence guarantees.

Consider, solely for illustration purposes, the standard normal distribution $\mathscr{N}(0,1)$ as a target. If we use Algorithm 4.6 with a normal random walk, i.e.,

$$\tilde{x}|x^{(t-1)} \sim \mathscr{N}\left(x^{(t-1)}, \sigma^2\right),$$

the performance of the sampler depends on the value σ. An R function that implements the associated Hastings–Metropolis sampler is coded as

```
hm=function(n,x0,sigma2){
  x=rep(x0,n)
  for (i in 2:n){
    y=rnorm(1,x[i-1],sqrt(sigma2))
    if (runif(1)<=exp(-0.5*(y^2-x[i-1]^2))) x[i]=y
    else x[i]=x[i-1]
    }
  x
  }
```

For instance, picking σ^2 equal to either 10^{-4} or 10^3 provides two extreme cases: As shown in Fig. 4.2, the chain has a high acceptance rate but a low exploration ability and a high autocorrelation in the former case, while its acceptance rate is low but its ability to move around the normal range is high in the latter case (with a quickly decreasing autocorrelation). Both cases use the "wrong scale", though, in that the histograms of the simulation outputs are quite far from the target distribution after 10,000 iterations, and this indicates that a much larger number of iterations must be used. A comparison with Fig. 4.3, which corresponds to $\sigma = 1$, clearly makes this point but also illustrates the fact that the large variance still induces large autocorrelations.

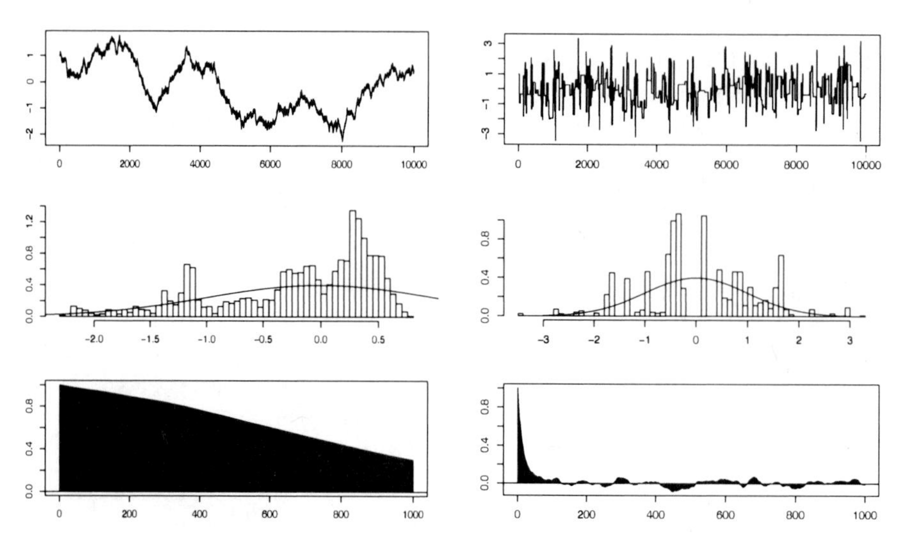

Fig. 4.2. Simulation of a $\mathscr{N}(0,1)$ target with (*left*) a $\mathscr{N}(x,10^{-4})$ and (*right*) a $\mathscr{N}(x,10^{3})$ random walk proposal. *Top*: Sequence of 10,000 iterations; *middle*: histogram of the last 2,000 iterations compared with the target density; *bottom*: empirical autocorrelations using R function plot.acf

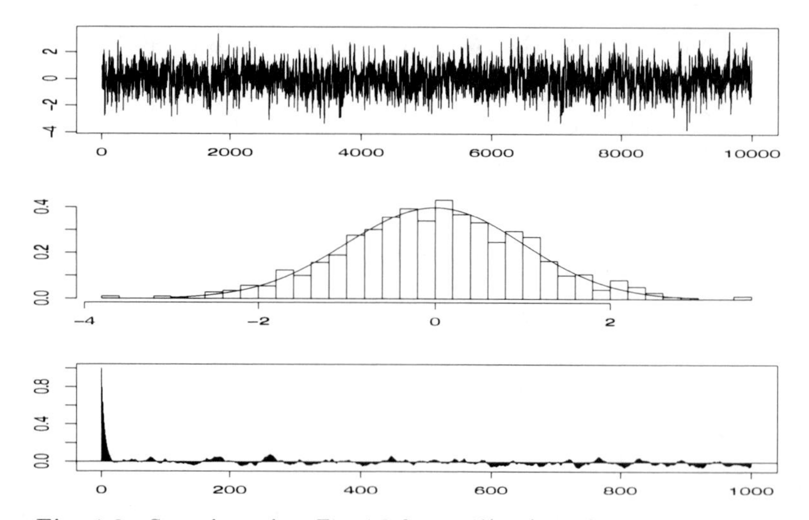

Fig. 4.3. Same legend as Fig. 4.2 for a $\mathscr{N}(x,1)$ random walk proposal

Several MCMC algorithms can be mixed together within a single algorithm using either a circular or a random design. While this construction is often suboptimal (in that the inefficient algorithms in the mixture are still used on a regular basis), it almost always brings an improvement compared with its individual components. A special case where a mixed scenario is used is the *Metropolis-within-Gibbs* algorithm: When building a Gibbs sampler, it may happen that it is difficult or impossible to simulate from one or several of the conditional distributions. In that case, a single Metropolis step associated with this conditional distribution (as its target) can be used instead.[5]

4.3 The Probit Model

We now engage in a full discussion of the Bayesian processing of the probit model introduced in Sect. 4.1, taking special care to distinguish between the various types of prior modeling.

4.3.1 Flat Prior

If no prior information is available, we can resort (as usual!) to a default flat prior on β, $\pi(\beta) \propto 1$, and then obtain the posterior distribution

$$\pi(\beta|\mathbf{y}) \propto \prod_{i=1}^{n} \Phi(\mathbf{x}^{i\mathsf{T}}\beta)^{y_i} \left[1 - \Phi(\mathbf{x}^{i\mathsf{T}}\beta)\right]^{1-y_i} ,$$

which is nonstandard and must be simulated using, e.g., MCMC techniques. First, the log-likelihood function is computable, as shown by the following R code[6]:

```
probitll=function(beta,y,X){
# probit likelihood
 if (is.matrix(beta)==F) beta=as.matrix(t(beta))
 n=dim(beta)[1]
 pll=rep(0,n)
 for (i in 1:n){
   lF1=pnorm(X%*%beta[i,],log=T)
   lF2=pnorm(-X%*%beta[i,],log=T)
```

[5]We stress that we do not resort to an MH algorithm for the purpose of simulating exactly from the corresponding conditional since this would require an infinite number of iterations but rather that we use a *single* iteration of the MH algorithm as a substitute for the simulation from the conditional since the resulting MCMC algorithm is still associated with the same stationary distribution.

[6]The use of the is.matrix test ensures that the function can be computed at one point as well as on multiple points and thus allows for calls from plot and other graphical functions.

```
    pll[i]=sum(y*lF1+(1-y)*lF2)
    }
  pll
  }
```

A variety of Metropolis–Hastings algorithms have been proposed for obtaining samples from this posterior distribution. Here we consider a sampler that appears to work well when the number of predictors is reasonably small. This Metropolis–Hastings sampler is a random walk scheme that uses the maximum likelihood estimate $\hat{\beta}$ as a starting value and the asymptotic (Fisher) covariance matrix $\hat{\Sigma}$ of the maximum likelihood estimate as the covariance matrix for the proposal[7] density, $\tilde{\beta} \sim \mathscr{N}_k(\beta^{(t-1)}, \tau^2\hat{\Sigma})$.

Algorithm 4.7 Probit Metropolis–Hastings Sampler

Initialization: Compute the MLE $\hat{\beta}$ and the covariance matrix $\hat{\Sigma}$ corresponding to the asymptotic covariance of $\hat{\beta}$, and set $\beta^{(0)} = \hat{\beta}$.
Iteration $t \geq 1$:
1. Generate $\tilde{\beta} \sim \mathscr{N}_k(\beta^{(t-1)}, \tau^2\hat{\Sigma})$.
2. Compute

$$\rho(\beta^{(t-1)}, \tilde{\beta}) = \min\left(1, \pi(\tilde{\beta}|\mathbf{y})/\pi(\beta^{(t-1)}|\mathbf{y})\right) .$$

3. With probability $\rho(\beta^{(t-1)}, \tilde{\beta})$, take $\beta^{(t)} = \tilde{\beta}$;
   otherwise take $\beta^{(t)} = \beta^{(t-1)}$.

The R function glm is obviously quite helpful in setting the initialization step of Algorithm 4.7. The step used in the R code to scale the algorithm is based on

```
> mod=summary(glm(y~X,family=binomial(link="probit")))
```

with mod$coeff[,1] corresponding to $\hat{\beta}$ and mod$cov.unscaled to $\hat{\Sigma}$. The following code is then reproducing the above algorithm in R::

```
hmflatprobit=function(niter,y,X,scale){
 p=dim(X)[2]
 mod=summary(glm(y~-1+X,family=binomial(link="probit")))
 beta=matrix(0,niter,p)
 beta[1,]=as.vector(mod$coeff[,1])
 Sigma2=as.matrix(mod$cov.unscaled)
```

[7] A choice of parameters that depend on the data for the Metropolis–Hastings proposal is completely valid, both from an MCMC point of view (meaning that this is not a self-tuning algorithm) and from a Bayesian point of view (since the parameters of the proposal are not those of the prior).

```
for (i in 2:niter){
  tildebeta=rmnorm(1,beta[i-1,],scale*Sigma2)
  llr=probitll(tildebeta,y,X)-probitll(beta[i-1,],y,X)
  if (runif(1)<=exp(llr)) beta[i,]=tildebeta
  else beta[i,]=beta[i-1,]
  }
 beta
 }
```

It takes advantage of the multivariate normal generator `rmnorm`, part of the package `mnormt` that caters to the multivariate normal distribution.

For **bank**, using a probit modeling with no intercept over the four measurements, we tested three different scales, namely $\tau = 1, 0.1, 10$, by running Algorithm 4.7 over 10,000 iterations. Looking both at the raw sequences and at the autocorrelation graphs, it appears that the best mixing behavior is associated with $\tau = 1$. Figure 4.4 illustrates the output of the simulation run in that case.[8] Using a burn-in range of 1,000 iterations, the averages of the parameters over the last 9,000 iterations are equal to $-1.2193, 0.9540, 0.9795$, and 1.1481, respectively. A plug-in estimate of the predictive probability of a counterfeit banknote is therefore

$$\hat{p}_i = \Phi\left(-1.2193x_{i1} + 0.9540x_{i2} + 0.9795x_{i3} + 1.1481x_{i4}\right).$$

For instance, according to this equation, a banknote of length 214.9 mm, left-edge width 130.1 mm, right-edge width 129.9 mm, and bottom margin width 9.5 mm is counterfeited with probability

$$\Phi\left(-1.1293 \times 214.9 + \ldots + 1.1481 \times 9.5\right) \approx 0.5917.$$

While the plug-in representation above gives an immediate evaluation of the predictive probability, a better approximation to this probability function is provided by the average over the iterations of the current predictive probabilities, $\Phi\left(\beta_1^{(t)}x_{i1} + \beta_2^{(t)}x_{i2} + \beta_3^{(t)}x_{i3} + \beta_4^{(t)}x_{i4}\right)$. It is easily derived from the output of the `hmflatprobit` function.

4.3.2 Noninformative G-Priors

Following the principles discussed in earlier chapters (see, e.g., Chap. 3), a flat prior on $\boldsymbol{\beta}$ is not appropriate for comparison purposes since we cannot validate the corresponding Bayes factors. In a variable selection setup, we thus need to replace the flat prior with, e.g., a hierarchical prior,

[8] We do not include the graphs for the other values of τ, but the curious reader can check that there is indeed a clear difference with the case $\tau = 1$.

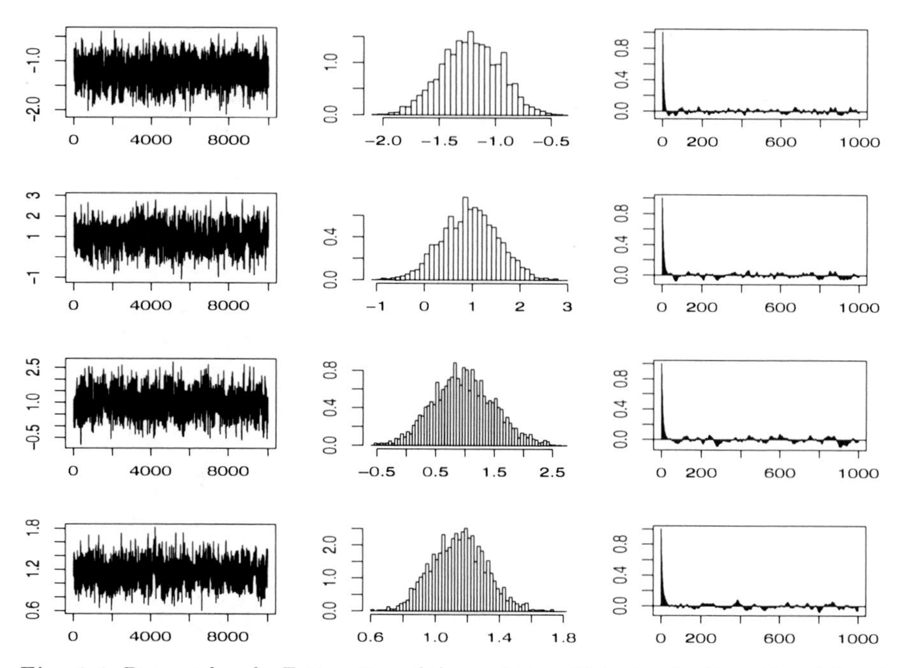

Fig. 4.4. Dataset **bank**: Estimation of the probit coefficients via Algorithm 4.7 and a flat prior. *Left*: β_i's ($i = 1, \ldots, 4$); *center*: histogram over the last 9,000 iterations; *right*: autocorrelation over the last 9,000 iterations

$$\boldsymbol{\beta}|\sigma^2 \sim \mathscr{N}_k\left(\mathbf{0}_k, \sigma^2(\mathbf{X}^\mathsf{T}\mathbf{X})^{-1}\right) \quad \text{and} \quad \pi(\sigma^2) \propto \sigma^{-2},$$

inspired by the normal linear regression model.[9] Integrating out σ^2 in this joint prior then leads to

$$\pi(\boldsymbol{\beta}) \propto |\mathbf{X}^\mathsf{T}\mathbf{X}|^{1/2}\Gamma(k/2)\left(\boldsymbol{\beta}^\mathsf{T}(\mathbf{X}^\mathsf{T}\mathbf{X})\boldsymbol{\beta}\right)^{-k/2}\pi^{-k/2},$$

which is clearly improper. Nonetheless, if we consider the *same* hierarchical prior for a submodel associated with a subset of the predictor variables in $\mathbf{X}$, associated with the *same* variance factor σ^2, the marginal distribution of $\mathbf{y}$ then depends on the *same* unknown multiplicative constant as the full model, and this constant cancels in the corresponding Bayes factor. This is exactly the same idea as for Zellner's noninformative G-prior, see Sect. 3.4.3.

The corresponding posterior distribution of $\boldsymbol{\beta}$ is

$$\pi(\boldsymbol{\beta}|\mathbf{y}) \propto |\mathbf{X}^\mathsf{T}\mathbf{X}|^{1/2}\Gamma(k/2)\left(\boldsymbol{\beta}^\mathsf{T}(\mathbf{X}^\mathsf{T}\mathbf{X})\boldsymbol{\beta}\right)^{-k/2}\pi^{-k/2}$$

[9]Note that the matrix $\mathbf{X}^\mathsf{T}\mathbf{X}$ is *not* the Fisher information matrix outside of the normal model. However, the (genuine) Fisher information matrix usually involves a function of $\boldsymbol{\beta}$ that prevents its use as a prior (inverse) covariance matrix on $\boldsymbol{\beta}$.

$$\times \prod_{i=1}^{n} \Phi(\mathbf{x}^{i\mathsf{T}}\boldsymbol{\beta})^{y_i} \left[1 - \Phi(\mathbf{x}^{i\mathsf{T}}\boldsymbol{\beta})\right]^{1-y_i} . \tag{4.4}$$

Note that we need to keep the "constant" terms $|\mathbf{X}^\mathsf{T}\mathbf{X}|^{1/2}$, $\Gamma(k/2)$, and $\pi^{-k/2}$, in this expression because they vary among submodels. To omit these terms would thus result in a bias in the computation of the Bayes factors.

Contrary to the linear regression setting and as for the flat prior in Sect. 4.3.1, neither the posterior distribution of $\boldsymbol{\beta}$ nor the marginal distribution of $\mathbf{y}$ can be derived analytically. We can however use exactly the same Metropolis–Hastings sampler as in Sect. 4.3.1, namely a random walk proposal based on the estimated Fisher information matrix for its scale and the MLE $\hat{\boldsymbol{\beta}}$ as its starting value.

For **bank**, the corresponding approximate Bayes estimate of $\boldsymbol{\beta}$ is given by

$$\mathbb{E}^\pi[\boldsymbol{\beta}|\mathbf{y}] \approx (-1.1552, 0.9200, 0.9121, 1.0820),$$

which slightly differs from the estimate found in Sect. 4.3.1 for the flat prior. This approximation was obtained by running the MH algorithm with scale $\tau^2 = 1$ over 10,000 iterations and averaging over the last 9,000 iterations. Figure 4.5 gives an assessment of the convergence of the MH scheme that does not vary very much compared with the previous figure.

We now address the specific problem of approximating the marginal distribution of $\mathbf{y}$ toward providing approximations to the Bayes factor and thus achieve the Bayesian equivalent of standard software to identify significant variables in the probit model. The marginal distribution of $\mathbf{y}$ is

$$f(\mathbf{y}) \propto |\mathbf{X}^\mathsf{T}\mathbf{X}|^{1/2}\pi^{-k/2}\Gamma(k/2) \int \left(\boldsymbol{\beta}^\mathsf{T}(\mathbf{X}^\mathsf{T}\mathbf{X})\boldsymbol{\beta}\right)^{-k/2}$$
$$\times \prod_{i=1}^{n} \Phi(\mathbf{x}^{i\mathsf{T}}\boldsymbol{\beta})^{y_i} \left[1 - \Phi(\mathbf{x}^{i\mathsf{T}}\boldsymbol{\beta})\right]^{1-y_i} \, \mathrm{d}\boldsymbol{\beta},$$

which cannot be computed in closed form. We thus propose to use as a generic proxy an importance sampling approximation to this integral based on a normal approximation $\mathscr{N}_k(\hat{\boldsymbol{\beta}}, 2\hat{V})$ to $\pi(\boldsymbol{\beta}|\mathbf{y})$, where $\hat{\boldsymbol{\beta}}$ is the MCMC approximation of $\mathbb{E}^\pi[\boldsymbol{\beta}|\mathbf{y}]$ and $\hat{V}$ is the MCMC approximation[10] of $\mathbb{V}(\boldsymbol{\beta}|\mathbf{y})$. The corresponding estimate of the marginal distribution of $\mathbf{y}$ is then, up to a constant,

[10] The factor 2 in the covariance matrix allows some amount of overdispersion, which is always welcomed in importance sampling settings, if only for variance finiteness purposes.

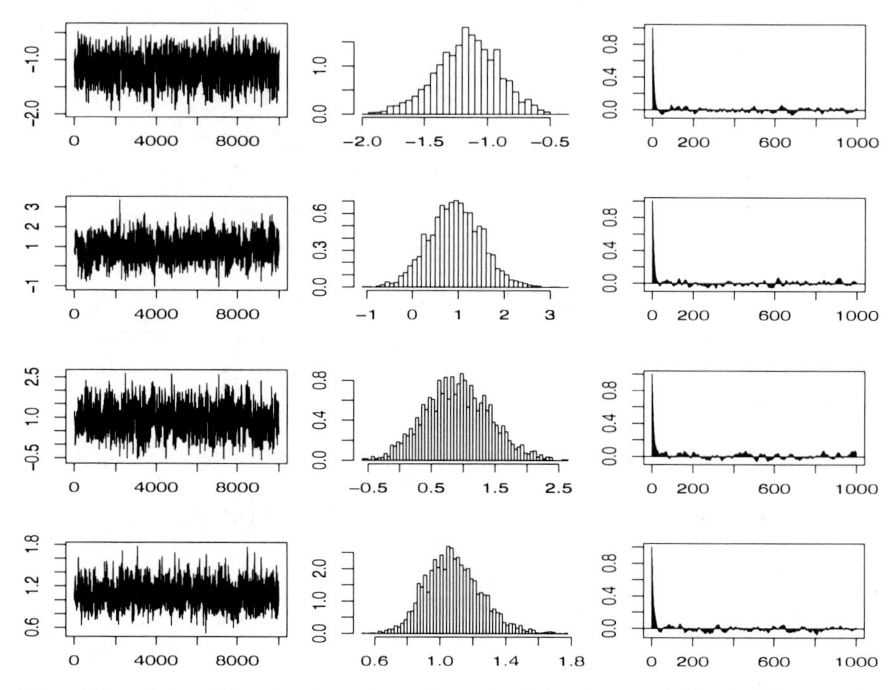

Fig. 4.5. Dataset **bank**: Same legend as Fig. 4.4 using an MH algorithm and a G-prior on $\boldsymbol{\beta}$

$$\frac{|\mathbf{X}^\top\mathbf{X}|^{1/2}}{\pi^{k/2}M}\sum_{m=1}^{M}\left(\boldsymbol{\beta}^{(m)\top}(\mathbf{X}^\top\mathbf{X})\boldsymbol{\beta}^{(m)}\right)^{-k/2}\prod_{i=1}^{n}\Phi(\mathbf{x}^{i\top}\boldsymbol{\beta}^{(m)})^{y_i}\left[1-\Phi(\mathbf{x}^{i\top}\boldsymbol{\beta}^{(m)})\right]^{1-y_i}$$

$$\times\,|\hat{V}|^{1/2}(4\pi)^{k/2}\,e^{(\boldsymbol{\beta}^{(m)}-\hat{\boldsymbol{\beta}})^\top\hat{V}^{-1}(\boldsymbol{\beta}^{(m)}-\hat{\boldsymbol{\beta}})/4}\,, \tag{4.5}$$

where the $\boldsymbol{\beta}^{(m)}$'s are simulated from the $\mathscr{N}_k(\hat{\boldsymbol{\beta}}, 2\,\hat{V})$ importance distribution.

If we consider a linear restriction on $\boldsymbol{\beta}$ such as $H_0: R\boldsymbol{\beta} = r$, with $r \in \mathbb{R}^q$ and R a $q \times k$ matrix of rank q, the submodel is associated with the likelihood

$$\ell(\boldsymbol{\beta}^0|\mathbf{y}) \propto \prod_{i=1}^{n}\Phi(\mathbf{x}_0^{i\top}\boldsymbol{\beta}^0)^{y_i}\left[1-\Phi(\mathbf{x}_0^{i\top}\boldsymbol{\beta}^0)\right]^{1-y_i}\,,$$

where $\boldsymbol{\beta}^0$ is $(k-q)$-dimensional and $\mathbf{X}_0$ and $\mathbf{x}_0$ are linear transforms of $\mathbf{X}$ and $\mathbf{x}$ of dimensions $(n, k-q)$ and $(k-q)$, respectively. Under the G-prior

$$\boldsymbol{\beta}^0|\sigma^2 \sim \mathscr{N}_{k-q}\left(\mathbf{0}_{k-q}, \sigma^2(\mathbf{X}_0^\top\mathbf{X}_0)^{-1}\right) \quad \text{and} \quad \pi(\sigma^2) \propto \sigma^{-2}\,,$$

the marginal distribution of $\mathbf{y}$ is of the same type as in the unconstrained case, namely,

$$f(\mathbf{y}) \propto |\mathbf{X}_0^\mathsf{T}\mathbf{X}_0|^{1/2}\pi^{-(k-q)/2}\Gamma\{(k-q)/2\} \int \left\{(\boldsymbol{\beta}^0)^\mathsf{T}(\mathbf{X}_0^\mathsf{T}X_0)\boldsymbol{\beta}^0\right\}^{-(k-q)/2}$$

$$\times \prod_{i=1}^{n} \Phi(\mathbf{x}_0^{i\mathsf{T}}\boldsymbol{\beta}^0)^{y_i}\left[1-\Phi(\mathbf{x}_0^{i\mathsf{T}}\boldsymbol{\beta}^0)\right]^{1-y_i}\,\mathrm{d}\boldsymbol{\beta}^0\,.$$

Once again, if we first run an MCMC sampler for the posterior of $\boldsymbol{\beta}^0$ for this submodel, it provides both parameters of a normal importance distribution and thus allows an approximation of the marginal distribution of $\mathbf{y}$ in the submodel in all ways similar to (4.5).

For **bank**, if we want to test the null hypothesis $H_0 : \beta_1 = \beta_2 = 0$, we obtain the Bayes factor $B_{10}^\pi = 8916.0$ via the importance sampling approximation of (4.5). We use the following R commands, which again borrow functions like `dmnorm` and `rmnorm` from the package `mnormt`,

```
# full model
mkprob=apply(noinfprobit,2,mean)
vkprob=var(noinfprobit)
simk=rmnorm(100000,mkprob,2*vkprob)
usk=probitnoinflpost(simk,y,X[,2:5])-
    dmnorm(simk,mkprob,2*vkprob,log=T)
# null model
noinfprobit0=hmnoinfprobit(10000,y,X[,4:5],1)
mk0=apply(noinfprobit0,2,mean)
vk0=var(noinfprobit0)
simk0=rmnorm(100000,mk0,2*vk0)
usk0=probitnoinflpost(simk0,y,X[,4:5])-
     dmnorm(simk0,mk0,2*vk0,log=T)
# Bayes factor
bf0probit=mean(exp(usk))/mean(exp(usk0))
```

Using Jeffreys' scale of evidence, since $\log_{10}(B_{10}^\pi) = 3.950$, the posterior distribution is strongly against H_0.

More generally, we can produce a Bayesian regression output, programmed in R, that mimics the standard software output for generalized linear models. Along with the estimates of the β_i's, given by their posterior expectation, we include the posterior variances of the β_i's, also derived from the MCMC sample, and the log Bayes factors $\log_{10}\left(B_{10}^i\right)$ corresponding to the null hypotheses $H_0 : \beta_i = 0$. As above, the Bayes factors are computed by importance sampling based on 100,000 simulations. The stars are related to Jeffreys' scale of evidence.

For **bank**, the corresponding outcome is

```
               Estimate  Post. var.  log10(BF)

X1             -1.1552    0.0631      4.5844 (****)
X2              0.9200    0.3299     -0.2875
X3              0.9121    0.2595     -0.0972
X4              1.0820    0.0287     15.6765 (****)

evidence against H0: (****) decisive, (***) strong,
(**) substantial, (*) poor
```

Although these Bayes factors cannot be used simultaneously, an informal conclusion is that the significant variables for the identification of counterfeited banknotes are X_1 and X_4.

4.3.3 About Informative Prior Analyses

In the setting of probit (and other generalized linear) models, it is unrealistic to expect practitioners to come up with precise prior information about the parameters $\boldsymbol{\beta}$. There exists nonetheless an amenable approach to prior information through what is called the *conditional mean family of prior distributions.* The intuition behind this approach is that prior beliefs about the probabilities p_i can be assessed to some extent by the practitioners for *particular values* of the explanatory variables $x_{1i}, \ldots, x_{ki}$. Once this information is taken into account, a corresponding prior can be derived for the parameter vector $\boldsymbol{\beta}$. This technique is certainly one of the easiest methods of incorporating subjective prior information into the processing of the binary regression problem, especially because it appeals to practitioners for whom the $\boldsymbol{\beta}$'s have, at best, a virtual meaning.

Starting with k explanatory variables, we derive the subjective prior information from k different values[11] of the covariate vector, denoted by $\tilde{\mathbf{x}}^1, \ldots, \tilde{\mathbf{x}}^k$. For each of these values, the practitioner is asked to specify two things:

1. a prior guess g_i at the probability of success p_i associated with $\mathbf{x}^i$; and
2. an assessment of her or his certainty about that guess translated as a number K_i of equivalent "prior observations."[12] This question can be expressed as "On how many imaginary observations did you build this guess?"

Both quantities can be turned into a formal prior density on $\boldsymbol{\beta}$ by imposing a beta prior distribution on p_i with parameters $K_i g_i$ and $K_i(1 - g_i)$ since the mean of a $\mathscr{B}e(a, b)$ distribution is $a/(a + b)$. If we make the additional

[11] The theoretical motivation for setting the number of covariate vectors equal to the dimension of $\boldsymbol{\beta}$ will be made clear below.

[12] This technique is called the device of *imaginary observations* and was proposed by the Italian statistician Bruno de Finetti for prior elicitation.

assumption that the k probabilities $p_1, \ldots, p_k$ are a priori independent (which clearly does not hold since they all depend on the same $\boldsymbol{\beta}$!), their joint density is

$$\pi(p_1, \ldots, p_k) \propto \prod_{i=1}^{k} p_i^{K_i g_i - 1}(1 - p_i)^{K_i(1-g_i)-1} . \tag{4.6}$$

Now, if we relate the probabilities p_i to the parameter $\boldsymbol{\beta}$, conditional on the covariate vectors $\tilde{\mathbf{x}}^1, \ldots, \tilde{\mathbf{x}}^k$, by $p_i = \Phi(\tilde{\mathbf{x}}^{i\mathsf{T}}\boldsymbol{\beta})$, we conclude that the corresponding distribution on $\boldsymbol{\beta}$ is

$$\pi(\boldsymbol{\beta}) \propto \prod_{i=1}^{k} \Phi(\tilde{\mathbf{x}}^{i\mathsf{T}}\boldsymbol{\beta})^{K_i g_i - 1} \left[1 - \Phi(\tilde{\mathbf{x}}^{i\mathsf{T}}\boldsymbol{\beta})\right]^{K_i(1-g_i)-1} \varphi(\tilde{\mathbf{x}}^{i\mathsf{T}}\boldsymbol{\beta}) .$$

This change of variable explains why we needed exactly k different covariate vectors in the prior assessment.

This intuitive approach to prior modeling is also interesting from a computational point of view since the corresponding posterior distribution

$$\begin{aligned}\pi(\boldsymbol{\beta}|\mathbf{y}) \propto & \prod_{i=1}^{n} \Phi(\mathbf{x}^{i\mathsf{T}}\boldsymbol{\beta})^{y_i} \left[1 - \Phi(\mathbf{x}^{i\mathsf{T}}\boldsymbol{\beta})\right]^{1-y_i} \\ & \times \prod_{j=1}^{k} \Phi(\tilde{\mathbf{x}}^{j\mathsf{T}}\boldsymbol{\beta})^{K_j g_j - 1} \left[1 - \varphi(\tilde{\mathbf{x}}^{j\mathsf{T}}\boldsymbol{\beta})\right]^{K_j(1-g_j)-1} \Phi(\tilde{\mathbf{x}}^{j\mathsf{T}}\boldsymbol{\beta})\end{aligned}$$

is of almost exactly the same type as the posterior distributions in both noninformative modelings above. The main difference stands in the product of the Jacobian terms $\varphi(\tilde{\mathbf{x}}^{j\mathsf{T}}\boldsymbol{\beta})$ $(1 \leq j \leq k)$, but

$$\prod_{j=1}^{k} \varphi(\tilde{\mathbf{x}}^{j\mathsf{T}}\boldsymbol{\beta}) \propto \exp\left\{ -\sum_{j=1}^{k} (\tilde{\mathbf{x}}^{j\mathsf{T}}\boldsymbol{\beta})^2/2 \right\} = \exp\left\{ -\boldsymbol{\beta}^{\mathsf{T}} \left[\sum_{j=1}^{k} \tilde{\mathbf{x}}^j \tilde{\mathbf{x}}^{j\mathsf{T}} \right] \boldsymbol{\beta}/2 \right\}$$

means that, if we forget about the -1's in the exponents, this posterior distribution corresponds to a regular posterior distribution for the probit model when adding to the observations $(y_1, \mathbf{x}^1), \ldots, (y_n, \mathbf{x}^n)$ the pseudo-observations[13] $(g_1, \tilde{\mathbf{x}}^1), \ldots, (g_1, \tilde{\mathbf{x}}^1), \ldots, (g_k, \tilde{\mathbf{x}}^k), \ldots, (g_k, \tilde{\mathbf{x}}^k)$, where each pair $(g_i, \tilde{\mathbf{x}}^i)$ is repeated K_i times and when using the G-prior

$$\boldsymbol{\beta} \sim \mathscr{N}_k \left(\mathbf{0}_k, \left[\sum_{j=1}^{k} \tilde{\mathbf{x}}^j \tilde{\mathbf{x}}^{j\mathsf{T}} \right]^{-1} \right) .$$

Therefore, Algorithm 4.7 need not be adapted to this case.

[13] Note that the fact that the g_j's do not take their values in $\{0, 1\}$ but rather in $(0, 1)$ does not create any difficulty in the implementation of Algorithm 4.7.

4.4 The Logit Model

We now reproduce some of the developments of the previous section in the case of the logit model, as defined in Sect. 4.1.2, not because there exist notable differences with either the processing or the conclusions of the probit model but rather because there is hardly any difference! For instance, Algorithm 4.7 can also be used for this model, while based on the same proposal, by simply modifying the definition of $\pi(\boldsymbol{\beta}|\mathbf{y})$, since the likelihood is now

$$\ell(\boldsymbol{\beta}|\mathbf{y}) = \exp\left\{\sum_{i=1}^{n} y_i\, \mathbf{x}^{i\mathrm{T}}\boldsymbol{\beta}\right\} \Big/ \prod_{i=1}^{n}\left[1+\exp(\mathbf{x}^{i\mathrm{T}}\boldsymbol{\beta})\right]. \tag{4.7}$$

The R function that computes the log-likelihood of the logit model is

```
logitll=function(beta,y,X){
 if (is.matrix(beta)==F) beta=as.matrix(t(beta))
 n=dim(beta)[1]
 pll=rep(0,n)
 for (i in 1:n){
   lF1=plogis(X%*%beta[i,],log=T)
   lF2=plogis(-X%*%beta[i,],log=T)
   pll[i]=sum(y*lF1+(1-y)*lF2)
   }
 pll
 }
```

That both models can be processed in a very similar manner means, for instance, that they can be easily compared when one is uncertain about which link function to adopt. The Bayes factor used in the comparison of the probit and logit models is directly derived from the importance sampling experiments described for the probit model. Note also that, while the values of the parameter $\boldsymbol{\beta}$ differ between the two models, a subjective prior modeling as in Sect. 4.3.3 can be conducted simultaneously for both models, the only difference occurring for the change of variables from $(p_1, \ldots, p_k)$ to $\boldsymbol{\beta}$.

If we use a flat prior on $\boldsymbol{\beta}$, the posterior distribution proportional to (4.7) can be inserted directly in Algorithm 4.7 to produce a sample approximately distributed from this posterior (assuming it exists, which means observing a sufficiently large and diverse sample). The corresponding R code is

```
hmflatlogit=function(niter,y,X,scale){
 p=dim(X)[2]
 mod=summary(glm(y~-1+X,family=binomial(link="logit")))
 beta=matrix(0,niter,p)
 beta[1,]=as.vector(mod$coeff[,1])
 Sigma2=as.matrix(mod$cov.unscaled)
 for (i in 2:niter){
   tildebeta=rmvn(1,beta[i-1,],scale*Sigma2)
```

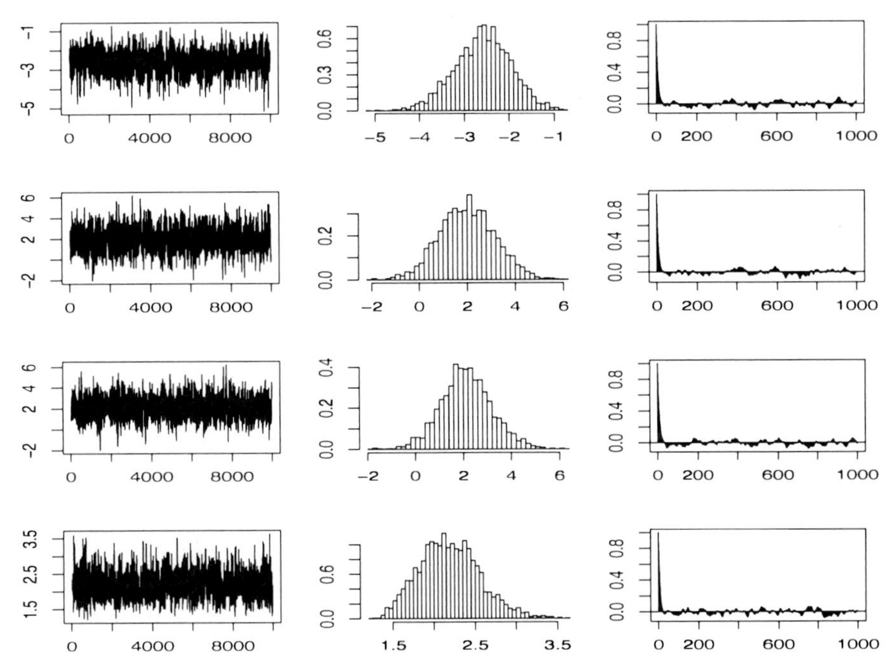

Fig. 4.6. Dataset **bank**: Estimation of the logit coefficients via Algorithm 4.7 under a flat prior. *Left*: β_i's ($i = 1, \dots, 4$); *center*: histogram over the last 9,000 iterations; *right*: autocorrelation over the last 9,000 iterations

```
    llr=logitll(tildebeta,y,X)-logitll(beta[i-1,],y,X)
       if (runif(1)<=exp(llr)) beta[i,]=tildebeta
    else beta[i,]=beta[i-1,]
    }
  beta
  }
```

For **bank**, Fig. 4.6 summarizes the results of running Algorithm 4.7 with the scale factor equal to $\tau = 1$: There is no clear difference between these graphs and those of earlier figures, except for a slight increase in the skewness of the histograms of the β_i's. (Obviously, this does not necessarily reflect a different convergence behavior but possibly a different posterior behavior since we are not dealing with the *same* posterior distribution.) The MH approximation—based on the last 9,000 iterations—of the Bayes estimate of β is equal to $(-2.5888, 1.9967, 2.1260, 2.1879)$. We can note the numerical difference between these values and those produced by the probit model. The sign and the relative magnitudes of the components are, however, very similar. For comparison purposes, consider the plug-in estimate of the predictive probability of a counterfeit banknote,

$$\hat{p}_i = \frac{\exp\left(-2.5888x_{i1} + 1.9967x_{i2} + 2.1260x_{i3} + 2.1879x_{i4}\right)}{1 + \exp\left(-2.5888x_{i1} + 1.9967x_{i2} + 2.1260x_{i3} + 2.1879x_{i4}\right)}.$$

Using this approximation, a banknote of length 214.9 mm, of left-edge width 130.1 mm, of right-edge width 129.9 mm, and of bottom margin width 9.5 mm is counterfeited with probability

$$\frac{\exp\left(-2.5888 \times 130.1 + \ldots + 2.1879 \times 9.5\right)}{1 + \exp\left(-2.5888 \times 130.1 + \ldots + 2.1879 \times 9.5\right)} \approx 0.5963\,.$$

This estimate of the probability is therefore very close to the estimate derived from the probit modeling, which was equal to 0.5917 (especially if we take into account the uncertainties associated both with the MCMC experiments and with the plug-in shortcut).

For model comparison purposes and the computation of Bayes factors, we can also use the same G-prior as for the probit model and thus multiply (4.7) by $|\mathbf{X}^{\mathsf{T}}\mathbf{X}|^{1/2}\Gamma(k/2)\left(\boldsymbol{\beta}^{\mathsf{T}}(\mathbf{X}^{\mathsf{T}}\mathbf{X})\boldsymbol{\beta}\right)^{-k/2}\pi^{-k/2}$. The MH implementation obviously remains the same.

For **bank**, Fig. 4.7 once more summarizes the output of the MH scheme over 10,000 iterations. Since we observe the same skewness in the histograms as in Fig. 4.6, this feature is most certainly due to the corresponding posterior distribution rather than to a deficiency in the convergence of the algorithm.)

We can repeat the test of the null hypothesis $H_0 : \beta_1 = \beta_2 = 0$ already done for the probit model and then obtain an approximate Bayes factor of $B^{\pi}_{10} = 16972.3$, with the same conclusion as earlier (although with twice as large an absolute value. We can also take advantage of the output software programmed for the probit model to produce the following summary:

```
              Estimate  Post. var.  log10(BF)

X1            -2.3970   0.3286       4.8084 (****)
X2             1.6978   1.2220      -0.2453
X3             2.1197   1.0094      -0.1529
X4             2.0230   0.1132      15.9530 (****)

evidence against H0: (****) decisive, (***) strong,
(**) substantial, (*) poor
```

Therefore, the most important covariates are again X_1 and X_4.

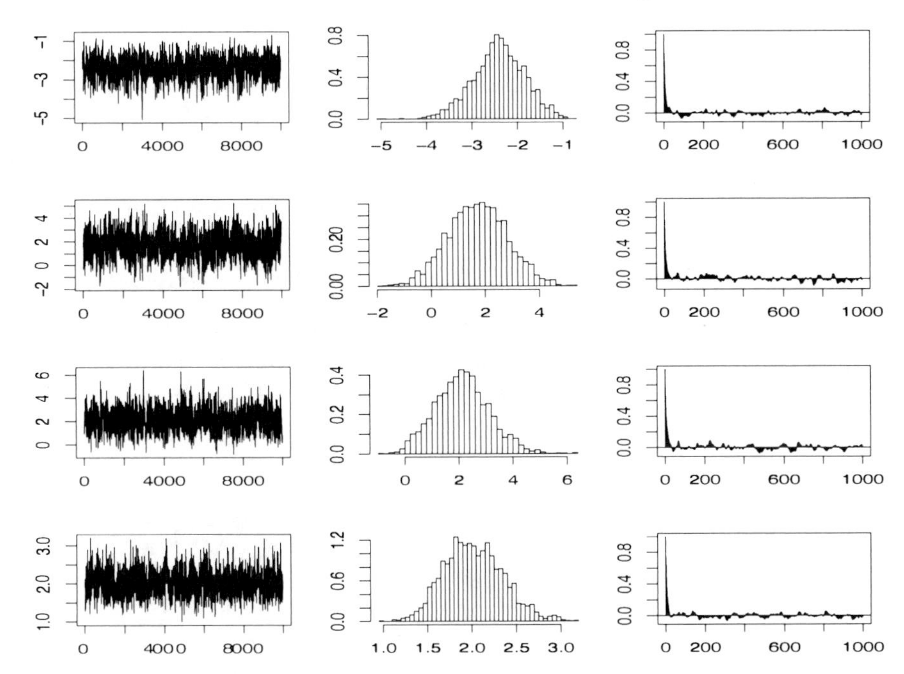

Fig. 4.7. Dataset **bank**: Same legend as Fig. 4.6 using an MH algorithm and a G-prior on β

4.5 Log-Linear Models

We conclude this chapter with an application of generalized linear modeling to the case of factors, already mentioned in Sect. 3.1. A standard approach to the analysis of associations (or dependencies) between *categorical* variables (that is, variables that take a finite number of values) is to use *log-linear models*. These models are special cases of generalized linear models connected to the Poisson distribution, and their name stems from the fact that they have traditionally been based on the logarithmic link function.

4.5.1 Contingency Tables

In such models, a sufficient statistic is the *contingency table*, which is a multiple-entry table made up of the cross-classified counts for the different categorical variables. There is much literature on contingency tables, including for instance Whittaker (1990) and Agresti (1996), because the corresponding models are quite handy both in the social sciences and in survey processing, where the observables are always reduced to a finite number of values.

The **airquality** dataset was obtained from the New York State Department of Conservation (ozone data) and from the American National Weather Service (meteorological data) and is part of the datasets contained in R (Chambers et al., 1983) and available as

```
> air=data(airquality)
```

This dataset involves two repeated measurements over 111 consecutive days of 1973, namely the mean ozone u (in parts per billion) from 1 pm to 3 pm at Roosevelt Island, the maximum daily temperature v (in degrees F) at La Guardia Airport, and, in addition, the month w (coded from 5 for May to 9 for September). If we discretize the measurements u and v into dichotomous variables (using the empirical median as the cutting point), we obtain the following three-way contingency table of counts per combination of the three (discretize) factors:

```
                 month  5  6  7  8  9
ozone      temp
[1,31]     [57,79] 17  4  2  5 18
           (79,97]  0  2  3  3  2
(31,168] [57,79]  6  1  0  3  1
           (79,97]  1  2 21 12  8
```

This contingency table thus has $5 \times 2 \times 2 = 20$ entries deduced from the number of categories of the three factors, among which some are zero because the corresponding combination of the three factors has not been observed in the study.

Each term in the table being an integer, it can then in principle be modeled as a Poisson variable. If we denote the counts by $\mathbf{y} = (y_1, \ldots, y_n)$, where $i = 1, \ldots, n$ is an arbitrary way of indexing the cells of the table, we can thus assume that $y_i \sim \mathscr{P}(\mu_i)$. Obviously, the likelihood

$$\ell(\mu|\mathbf{y}) = \prod_{i=1}^{n} \frac{1}{\mu_i!} \mu_i^{y_i} \exp(-\mu_i)\,,$$

where $\mu = (\mu_1, \ldots, \mu_n)$, shows that the model is *saturated*, namely that no structure can be exhibited because there are as many parameters as there are entries in the table. To exhibit any structure, we need to constrain the μ_i's and do so via a GLM whose covariate matrix $\mathbf{X}$ is directly derived from the contingency table itself. If some entries are structurally equal to zero (as for instance when crossing "number of pregnancies" with "male indicators"), these entries should be removed from the model.

An R function that corresponds to this log-linear model log-likelihood is

```
loglinll=function(beta,y,X){
  if (is.matrix(beta)==FALSE) beta=as.matrix(t(beta))
  n=dim(beta)[1]
  pll=rep(0,n)
  for (i in 1:n){
    lF=exp(X%*%beta[i,])
    pll[i]=sum(dpois(y,lF,log=T))
    }
  pll
  }
```

with again the use of `is.matrix` and `as.matrix` to allow for matricial calls to the `loglinll` function.

When we constrain the mean parameters μ_i of a log-linear model to satisfy

$$\log(\mu_i) = \mathbf{x}^{i\mathsf{T}}\boldsymbol{\beta}\,,$$

the covariate vector $\mathbf{x}^i$ is rather peculiar in that it is constituted *only* of indicators. The so-called *incidence matrix* $\mathbf{X}$ with rows equal to the $\mathbf{x}^i$'s is thus such that its elements are all zeros or ones. Given a contingency table, the choice of indicator variables to include in $\mathbf{x}^i$ can vary, depending on what is deemed (or found) to be an important relation between some categorical variables. For instance, suppose that there are three categorical variables, u, v, and w as in airquality, and that u takes I values, v takes J values, and w takes K values. If we only include the indicators for the values of the three categorical variables in $\mathbf{X}$, we have

$$\log(\mu_\tau) = \sum_{b=1}^{I} \beta_b^u \mathbb{I}_b(u_\tau) + \sum_{b=1}^{J} \beta_b^v \mathbb{I}_b(v_\tau) + \sum_{b=1}^{K} \beta_b^w \mathbb{I}_b(w_\tau)\,;$$

that is, $(1 \le i \le I,\ 1 \le j \le J,\ 1 \le k \le K)$,

$$\log(\mu_{l(i,j,k)}) = \beta_i^u + \beta_j^v + \beta_k^w$$

$(1 \le i \le I,\ 1 \le j \le J,\ 1 \le k \le K)$, where $l(i,j,k)$ corresponds to the index of the (i,j,k) entry in the table, namely the case when $u = i$, $v = j$, and $w = k$. Similarly, the saturated log-linear model corresponds to the use of one indicator per entry of the table; that is $1 \le i \le I,\ 1 \le j \le J,\ 1 \le k \le K)$,

$$\log(\mu_{l(i,j,k)}) = \beta_{ijk}^{uvw}\,.$$

For comparative reasons that will very soon become apparent, and by analogy with analysis of variance (ANOVA) conventions, we can also overparameterize this representation as

$$\log(\mu_{l(i,j,k)}) = \lambda + \lambda_i^u + \lambda_j^v + \lambda_k^w + \lambda_{ij}^{uv} + \lambda_{ik}^{uw} + \lambda_{jk}^{vw} + \lambda_{ijk}^{uvw}\,, \tag{4.8}$$

where λ appears as the overall or reference average effect, λ_i^u appears as the marginal discrepancy (against the reference effect λ) when $u = i$, λ_{ij}^{uv} as the interaction discrepancy (against the added effects $\lambda + \lambda_i^u + \lambda_j^v$) when $(u, v) = (i, j)$, etc.

Using the representation (4.8) is quite convenient because it allows a straightforward parameterization of the nonsaturated models, which then appear as submodels of (4.8) where some groups of parameters are null. For example,

1. if both categorical variables v and w are irrelevant, then

$$\log(\mu_{l(i,j,k)}) = \lambda + \lambda_i^u \, ;$$

2. if all three categorical variables are mutually independent, then

$$\log(\mu_{l(i,j,k)}) = \lambda + \lambda_i^u + \lambda_j^v + \lambda_k^w \, ;$$

3. if u and v are associated but are both independent of w, then

$$\log(\mu_{l(i,j,k)}) = \lambda + \lambda_i^u + \lambda_j^v + \lambda_k^w + \lambda_{ij}^{uv} \, ;$$

(iv) if u and v are conditionally independent given w, then

$$\log(\mu_{l(i,j,k)}) = \lambda + \lambda_i^u + \lambda_j^v + \lambda_k^w + \lambda_{ik}^{uw} + \lambda_{jk}^{vw} \, ; \quad \text{and}$$

(v) if there is no three-factor interaction, then

$$\log(\mu_{l(i,j,k)}) = \lambda + \lambda_i^u + \lambda_j^v + \lambda_k^w + \lambda_{ij}^{uv} + \lambda_{ik}^{uw} + \lambda_{jk}^{vw} \, ,$$

which appears as the most complete submodel (or as the global model if the saturated model is not considered at all).

This representation naturally embeds log-linear modeling within a model choice perspective in that it calls for a selection of the most parsimonious submodel that remains compatible with the observations. This is clearly equivalent to a variable-selection problem of a special kind in the sense that *all* indicators related with the same association must remain or vanish *at once*. This specific feature means that there are much fewer submodels to consider than in a regular variable-selection problem.

As stressed above, the representation (4.8) is not identifiable. Although the following is not strictly necessary from a Bayesian point of view (since the Bayesian approach can handle nonidentifiable settings and still estimate properly identifiable quantities), it is customary to impose identifiability constraints on the parameters as in the ANOVA model. A common convention is to set to zero the parameters corresponding to the first category of each variable, which is equivalent to removing the indicator (or *dummy variable*) of the first category for each variable (or group of variables). For instance, for a 2×2 contingency table with two variables u and v, both having two categories, say 1 and 2, the constraint could be

$$\lambda_1^u = \lambda_1^v = \lambda_{11}^{uv} = \lambda_{12}^{uv} = \lambda_{21}^{uv} = 0\,.$$

For notational convenience, we assume below that $\boldsymbol{\beta}$ is the vector of the parameters once the identifiability constraint has been applied and that $\mathbf{X}$ is the indicator matrix with the corresponding columns removed.

4.5.2 Inference Under a Flat Prior

Even when using a noninformative flat prior on $\boldsymbol{\beta}$, $\pi(\boldsymbol{\beta}) \propto 1$, the posterior distribution

$$\begin{aligned}
\pi(\boldsymbol{\beta}|\mathbf{y}) &\propto \prod_{i=1}^n \left\{\exp(\mathbf{x}^{i\mathsf{T}}\boldsymbol{\beta})\right\}^{y_i} \exp\{-\exp(\mathbf{x}^{i\mathsf{T}}\boldsymbol{\beta})\} \\
&= \exp\left\{\sum_{i=1}^n y_i\,\mathbf{x}^{i\mathsf{T}}\boldsymbol{\beta} - \sum_{i=1}^n \exp(\mathbf{x}^{i\mathsf{T}}\boldsymbol{\beta})\right\} \\
&= \exp\left\{\left(\sum_{i=1}^n y_i\,\mathbf{x}^i\right)^{\mathsf{T}} \boldsymbol{\beta} - \sum_{i=1}^n \exp(\mathbf{x}^{i\mathsf{T}}\boldsymbol{\beta})\right\}
\end{aligned}$$

is nonstandard and must be approximated by an MCMC algorithm. While the shape of this density differs from the posterior densities in the probit and logit cases, we can once more implement Algorithm 4.7 based on the normal Fisher approximation of the likelihood (whose parameters are again derived using the R `glm()` function as in

```
> mod=summary(glm(y~-1+X,family=poisson()))
```

which provides $\hat{\boldsymbol{\beta}}$ as `mod$coeff[,1]` and $\hat{\Sigma}$ as `mod$cov.unscaled`).

For **airquality**, we first consider the most general nonsaturated model, as described in Sect. 4.5.1. Taking into account the identifiability constraints, there are therefore

$$1+(2-1)+(2-1)+(5-1)+(2-1)\times(2-1)+(2-1)\times(5-1)+(2-1)\times(5-1)\,,$$

i.e., 16, free parameters in the model (to be compared with the 20 counts in the contingency table). Given the dimension of the simulated parameter, it is impossible to provide a complete picture of the convergence properties of the algorithm, and we represented in Fig. 4.8 the traces and histograms for the marginal posterior distributions of the parameters β_i based on 10,000 iterations using a scale factor equal to $\tau^2 = 0.5$. (This value was obtained by trial and error, producing a smooth trace for all parameters. Larger values of τ required a larger number of iterations since the acceptance rate was lower, as the reader can check using the `BCoRe` package.) Note that some of the traces represented in Fig. 4.8 show periodic patterns that indicate that more iterations could be necessary. However, the corresponding histograms remain

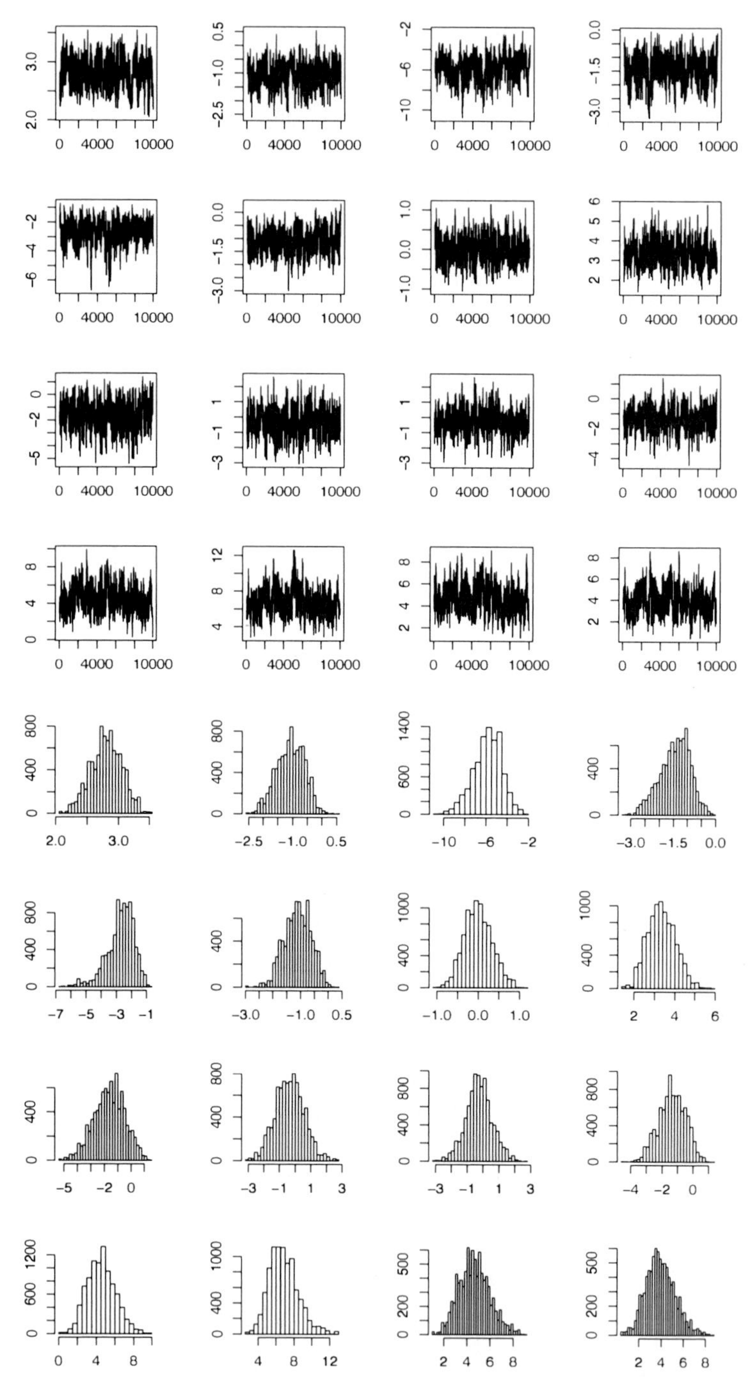

Fig. 4.8. Dataset airquality: Traces (*top*) and histograms (*bottom*) of the simulations from the posterior distributions of the components of $\boldsymbol{\beta}$ using a flat prior and a random walk Metropolis–Hastings algorithm with scale factor $\tau^2 = 0.5$ (same order row-wise as in Table 4.1)

Table 4.1. Dataset airquality: Bayes estimates of the parameter $\boldsymbol{\beta}$ using a random walk MH algorithm with scale factor $\tau^2 = 0.5$

Effect	Post. mean	Post. var.
λ	2.8041	0.0612
λ_2^u	–1.0684	0.2176
λ_2^v	–5.8652	1.7141
λ_2^w	–1.4401	0.2735
λ_3^w	–2.7178	0.7915
λ_4^w	–1.1031	0.2295
λ_5^w	–0.0036	0.1127
λ_{22}^{uv}	3.3559	0.4490
λ_{22}^{uw}	-1.6242	1.2869
λ_{23}^{uw}	–0.3456	0.8432
λ_{24}^{uw}	–0.2473	0.6658
λ_{25}^{uw}	–1.3335	0.7115
λ_{22}^{vw}	4.5493	2.1997
λ_{23}^{vw}	6.8479	2.5881
λ_{24}^{vw}	4.6557	1.7201
λ_{25}^{vw}	3.9558	1.7128

quite stable over iterations. Both the approximated posterior means and the posterior variances for the 16 parameters as deduced from the MCMC run are given in Table 4.1. A few histograms in Fig. 4.8 are centered at 0, signaling a potential lack of significance for the corresponding β_i's.

4.5.3 Model Choice and Significance of the Parameters

If we try to compare different levels of association (or interaction), or if we simply want to test the significance of some parameters β_i, the flat prior is once again inappropriate. The G-prior alternative proposed for the probit and logit models is still available, though, and we can thus replace the posterior distribution of the previous section with

$$\pi(\boldsymbol{\beta}|\mathbf{y}) \propto |\mathbf{X}^\mathsf{T}\mathbf{X}|^{1/2}\Gamma(k/2)\left(\boldsymbol{\beta}^\mathsf{T}(\mathbf{X}^\mathsf{T}\mathbf{X})\boldsymbol{\beta}\right)^{-k/2}\pi^{-k/2}$$
$$\exp\left\{\left(\sum_{i=1}^n y_i\,\mathbf{x}^i\right)^\mathsf{T}\boldsymbol{\beta} - \sum_{i=1}^n \exp(\mathbf{x}^{i\mathsf{T}}\boldsymbol{\beta})\right\} \qquad (4.9)$$

as an alternative posterior.

Table 4.2. Dataset airquality: Metropolis–Hastings approximations of the posterior means under the G-prior

Effect	Post. mean	Post. var.
λ	2.7202	0.0603
λ_2^u	-1.1237	0.1981
λ_2^v	-4.5393	0.9336
λ_2^w	-1.4245	0.3164
λ_3^w	-2.5970	0.5596
λ_4^w	-1.1373	0.2301
λ_5^w	0.0359	0.1166
λ_{22}^{uv}	2.8902	0.3221
λ_{22}^{uw}	-0.9385	0.8804
λ_{23}^{uw}	0.1942	0.6055
λ_{24}^{uw}	0.0589	0.5345
λ_{25}^{uw}	-1.0534	0.5220
λ_{22}^{vw}	3.2351	1.3664
λ_{23}^{vw}	5.3978	1.3506
λ_{24}^{vw}	3.5831	1.0452
λ_{25}^{vw}	2.8051	1.0061

For **airquality** and the same model as in the previous analysis, namely the maximum nonsaturated model with 16 parameters, Algorithm 4.7 can be used with (4.9) as target and $\tau^2 = 0.5$ as the scale in the random walk. The result of this simulation over 10,000 iterations is presented in Fig. 4.9. The traces of the components of $\boldsymbol{\beta}$ show the same slow mixing as in Fig. 4.8, with similar occurrences of large deviances from the mean value that may indicate the weak identifiability of some of these parameters. Note also that the histograms of the posterior marginal distributions are rather close to those associated with the flat prior, as shown in Fig. 4.8. The MCMC approximations to the posterior means and the posterior variances are given in Table 4.2 for all 16 parameters, based on the last 9,000 iterations. While the first parameters are quite close to those provided by Table 4.1, the estimates of the interaction coefficients vary much more and are associated with much larger variances. This indicates that much less information is available within the contingency table about interactions, as can be expected.

If we now consider the very reason why this alternative to the flat prior was introduced, we are facing the same difficulty as in the probit case for the computation of the marginal density of $\mathbf{y}$. And, once again, the same solution applies: using an importance sampling experiment to approximate the integral works when the importance function is a multivariate normal (or t) distribution with mean (approximately) $\mathbb{E}[\boldsymbol{\beta}|\mathbf{y}]$ and covariance matrix (approximately) $2 \times \mathbb{V}(\boldsymbol{\beta}|\mathbf{y})$ using the Metropolis–Hastings approximations reported in Table 4.2. We can therefore approximate Bayes factors for testing all possible structures of the log-linear model.

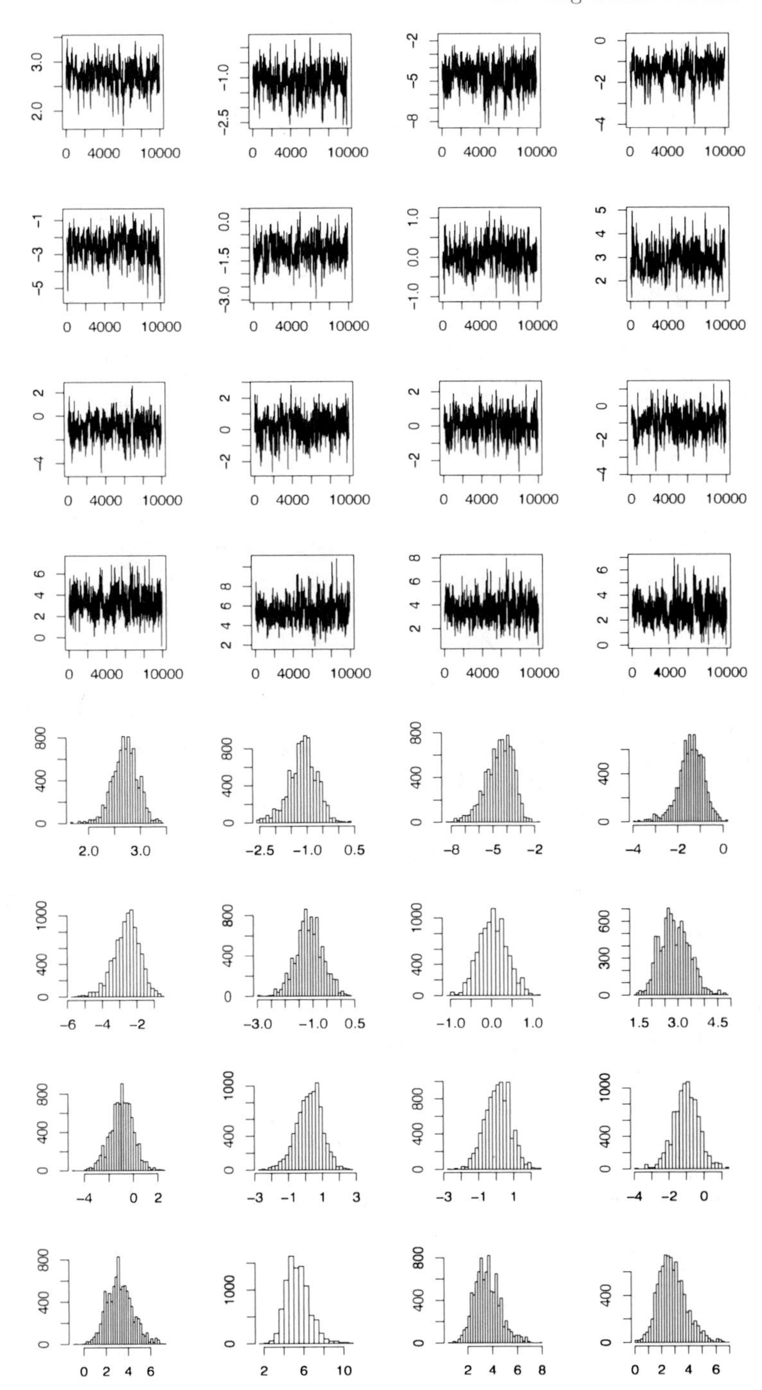

Fig. 4.9. Dataset airquality: Same legend as Fig. 4.8 for the posterior distribution (4.9) as target

For **airquality**, we illustrate this ability by testing the presence of two-by-two interactions between the three variables. We thus compare the largest non-saturated model with each submodel where one interaction is removed. An ANOVA-like output is

```
Effect log10(BF)

u:v      6.0983 (****)
u:w     -0.5732
v:w      6.0802 (****)

evidence against H0: (****) decisive, (***) strong,
(**) substantial, (*) poor
```

which means that the interaction between u and w (that is, ozone and month) is too small to be significant given all the other effects. (Note that it would be excessive to derive from this lack of significance a conclusion of independence between u and w because this interaction is conditional on all other interactions in the complete nonsaturated model.)

The above was obtained by the following R code: first we simulated an importance sample towards approximating the full model integrated likelihood

```
mklog=apply(noinfloglin,2,mean)
vklog=var(noinfloglin)
simk=rmnorm(100000,mklog,2*vklog)
usk=loglinnoinflpost(simk,counts,X)-
    dmnorm(simk,mklog,2*vklog,log=T)
```

then reproduced this computation for the three corresponding submodels, namely

```
noinfloglin1=hmnoinfloglin(10^4,counts,X[,-8],0.5)
mk1=apply(noinfloglin1,2,mean)
vk1=var(noinfloglin1)
simk1=rmnorm(100000,mk1,2*vk1)
usk1=loglinnoinflpost(simk1,counts,X[,-8])-
  dmnorm(simk1,mk1,2*vk1,log=T)
bf1loglin=mean(exp(usk))/mean(exp(usk1))
```

and the same pattern with

```
noinfloglin2=hmnoinfloglin(10^4,counts,cbind(X[,-(9:12)],0.5)
```

and

```
noinfloglin3=hmnoinfloglin(10^4,counts,X[,1:12],0.5)
```

4.6 Exercises

4.1 Show that, for the logistic regression model, the statistic $\sum_{i=1}^n y_i \mathbf{x}^i$ is sufficient when conditioning on the $\mathbf{x}^i$'s $(1 \le i \le n)$, and give the corresponding family of conjugate priors.

4.2 Show that the logarithmic link is the canonical link function in the case of the Poisson regression model.

4.3 Suppose $y_1, \ldots, y_k$ are independent Poisson $\mathscr{P}(\mu_i)$ random variables. Show that, conditional on $n = \sum_{i=1}^k y_i$,

$$\mathbf{y} = (y_1, \ldots, y_k) \sim \mathscr{M}_k(n; \alpha_1, \ldots, \alpha_k),$$

and determine the α_i's.

4.4 For π the density of an inverse normal distribution with parameters $\theta_1 = 3/2$ and $\theta_2 = 2$,

$$\pi(x) \propto x^{-3/2} \exp(-3/2x - 2/x) \mathbb{I}_{x>0},$$

write down and implement an independence MH sampler with a Gamma proposal with parameters $(\alpha, \beta) = (4/3, 1)$ and $(\alpha, \beta) = (0.5\sqrt{4/3}, 0.5)$.

4.5 Consider x_1, x_2, and x_3 iid $\mathscr{C}(\theta, 1)$, and $\pi(\theta) \propto \exp(-\theta^2/100)$. Show that the posterior distribution of θ, $\pi(\theta|x_1, x_2, x_3)$, is proportional to

$$\exp(-\theta^2/100)[(1 + (\theta - x_1)^2)(1 + (\theta - x_2)^2)(1 + (\theta - x_3)^2)]^{-1}$$

and that it is trimodal when $x_1 = 0$, $x_2 = 5$, and $x_3 = 9$. Using a random walk based on the Cauchy distribution $\mathscr{C}(0, \sigma^2)$, estimate the posterior mean of θ using different values of σ^2. In each case, monitor the convergence.

4.6 Estimate the mean of a $\mathscr{G}a(4.3, 6.2)$ random variable using

1. direct sampling from the distribution via the R command
```
> x=rgamma(n,4.3,scale=6.2)
```
2. Metropolis–Hastings with a $\mathscr{G}a(4, 7)$ proposal distribution;
3. Metropolis–Hastings with a $\mathscr{G}a(5, 6)$ proposal distribution.

In each case, monitor the convergence of the cumulated average.

4.7 For a standard normal distribution as target, implement a Hastings–Metropolis algorithm with a mixture of five random walks with variances $\sigma = 0.01, 0.1, 1, 10, 100$ and equal weights. Compare its output with the output of Fig. 4.3.

4.8 For the probit model under flat prior, find conditions on the observed pairs $(\mathbf{x}^i, y_i)$ for the posterior distribution above to be proper.

4.9 For the probit model under non-informative prior, find conditions on $\sum_i y_i$ and $\sum_i (1 - y_i)$ for the posterior distribution defined by (4.4) to be proper.

4.10 Include an intercept in the probit analysis of **bank** and run the corresponding version of Algorithm 4.7 to discuss whether or not the posterior variance of the intercept is high.

4.11 Using the latent variable representation of the probit model, introduce $z_i|\boldsymbol{\beta} \sim \mathscr{N}\left(\mathbf{x}^{i\mathsf{T}}\boldsymbol{\beta}, 1\right)$ $(1 \leq i \leq n)$ such that $y_i = \mathbb{I}_{z_i \leq 0}$. Deduce that

$$z_i|y_i, \boldsymbol{\beta} \sim \begin{cases} \mathscr{N}_+\left(\mathbf{x}^{i\mathsf{T}}\boldsymbol{\beta}, 1, 0\right) & \text{if} \quad y_i = 1\,, \\ \mathscr{N}_-\left(\mathbf{x}^{i\mathsf{T}}\boldsymbol{\beta}, 1, 0\right) & \text{if} \quad y_i = 0\,, \end{cases}$$

where $\mathscr{N}_+(\mu, 1, 0)$ and $\mathscr{N}_-(\mu, 1, 0)$ are the normal distributions with mean μ and variance 1 that are left-truncated and right-truncated at 0, respectively. Check that those distributions can be simulated using the R commands

```
> xp=qnorm(runif(1)*pnorm(mu)+pnorm(-mu))+mu
> xm=qnorm(runif(1)*pnorm(-mu))+mu
```

Under the flat prior $\pi(\boldsymbol{\beta}) \propto 1$, show that

$$\boldsymbol{\beta}|\mathbf{y}, \mathbf{z} \sim \mathscr{N}_k\left((\mathbf{X}^\mathsf{T}\mathbf{X})^{-1}\mathbf{X}^\mathsf{T}\mathbf{z}, (\mathbf{X}^\mathsf{T}\mathbf{X})^{-1}\right)\,,$$

where $\mathbf{z} = (z_1, \ldots, z_n)$, and derive the corresponding Gibbs sampler, sometimes called the *Albert–Chib* sampler. (*Hint*: A good starting point is the maximum likelihood estimate of $\boldsymbol{\beta}$.) Compare the application to **bank** with the output in Fig. 4.4. (*Note*: Account for differences in computing time.)

4.12 For the **bank** dataset and the probit model, compute the Bayes factor associated with the null hypothesis $H_0 : \beta_2 = \beta_3 = 0$.

4.13 In the case of the logit model—i.e., when $p_i = \exp \tilde{\mathbf{x}}^{i\mathsf{T}}\boldsymbol{\beta}/\{1 + \exp \tilde{\mathbf{x}}^{i\mathsf{T}}\boldsymbol{\beta}\}$ $(1 \leq i \leq k)$—derive the prior distribution on $\boldsymbol{\beta}$ associated with the prior (4.6) on $(p_1, \ldots, p_k)$.

4.14 Examine whether or not the sufficient conditions for propriety of the posterior distribution found in Exercise 4.9 for the probit model are the same for the logit model.

4.15 For the **bank** dataset and the logit model, compute the Bayes factor associated with the null hypothesis $H_0 : \beta_2 = \beta_3 = 0$ and compare its value with the value obtained for the probit model in Exercise 4.12.

4.16 Given a contingency table with four categorical variables, determine the number of submodels to consider.

4.17 In the case of a 2×2 contingency table with fixed total count $n = n_{11} + n_{12} + n_{21} + n_{22}$, we denote by $\theta_{11}, \theta_{12}, \theta_{21}, \theta_{22}$ the corresponding probabilities. If the prior on those probabilities is a Dirichlet $\mathscr{D}_4(1/2, \ldots, 1/2)$, give the corresponding marginal distributions of $\alpha = \theta_{11} + \theta_{12}$ and $\beta = \theta_{11} + \theta_{21}$. Deduce the associated Bayes factor if H_0 is the hypothesis of independence between the factors and if the priors on the margin probabilities α and β are those derived above.

5

Capture–Recapture Experiments

> He still couldn't be sure that
> he hadn't landed in a trap.
> —**Ian Rankin, *Resurrection Men*.**—

Roadmap

This chapter deals with a very special case of survey models. Surveys are used in many settings to evaluate some features of a given population, including its main characteristic, the *size* of the population. In the case of capture–recapture surveys, individuals are observed and identified either once or several times and the repeated observations can be used to draw inference on the population size and its dynamic characteristics. Along with the original model, we will also introduce extensions that can be seen as a first entry into hidden Markov chain models, detailed further in Chap. 6. In particular, we cover the generic *Arnason–Schwarz* model that is customarily used by biologists for open populations.

On the methodological side, we provide an introduction to the accept–reject method, which is the central simulation technique behind most standard random generators and relates to the Metropolis–Hastings methodology in many ways.

J.-M. Marin and C.P. Robert, *Bayesian Essentials with R*, Springer Texts in Statistics, DOI 10.1007/978-1-4614-8687-9_5,

5.1 Inference in a Finite Population

In this chapter, we consider the problem of estimating the unknown size, N, of a population, based on a *survey*; that is, on a partial observation of this population. To be able to evaluate a population size without going through the enumeration of all its members is obviously very appealing, both timewise and moneywise, especially when sampling those members has a perturbing effect on them.[1]

A primary type of survey (which we do not study in this chapter) is based on knowledge of the structure of the population. For instance, in a political survey about voting intentions, we build a sample of 1,000 individuals, say, such that the main sociological groups (farmers, civil servants, senior citizens, etc.) are represented in proportion in the sample. In that situation, there is no statistical inference, so to speak, except about the variability of the responses, which are in the simplest cases binomial variables.

Obviously, such surveys require primary knowledge of the population, which can be obtained either by a (costly) census, like those that states run every 5 or 10 years, or by a preliminary exploratory survey that aims at uncovering these hidden structures. This secondary type of survey is the purpose of this chapter, under the name of *capture–recapture* (or *capture–mark–recapture*) experiments, where a few individuals sampled at random from the population of interest bring some information about the characteristics of this population and in particular about its size.

The capture–recapture models were first used in biology and ecology to estimate the size of animal populations, such as herds of caribous (e.g., for culling) or of whales (e.g., for the International Whaling Commission to determine fishing quotas), cod populations, and the number of different species in a particular area. While our illustrative dataset will be related to a biological problem, we stress that these capture–recapture models apply in a much wider range of domains, such as, for instance,

- sociology and demography, where the estimation of the size of populations at risk is always delicate (e.g., homeless people, prostitutes, illegal migrants, drug addicts, etc.);
- official statistics for reducing the cost of a census[2] or improving its efficiency on delicate or rare subcategories (as in the U.S. census undercount procedure and the new French census);
- finance (e.g., in credit scoring, defaulting companies, etc.) and marketing (consumer habits, telemarketing, etc.);

[1] In the most extreme cases, sampling an individual may lead to its destruction, as for instance in forestry when estimating the volume of trees or in meat production when estimating the content of fat in meat.

[2] Even though a census is formally a deterministic process since it aims at the complete enumeration of a given population, it inevitably involves many random components at the selection, collection, and processing levels (Särndal et al., 2003).

- fraud detection (e.g., phone, credit card, etc.) and document authentication (historical documents, forgery, etc.); and
- software debugging, to determine an evaluation of the number of bugs in a computer program.

In these different examples, the size N of the whole population is unknown but samples (with fixed or random sizes) can easily be extracted from the population. For instance, in a computer program, the total number N of bugs is unknown but one can record the number n_1 of bugs detected in a given perusal. Similarly, the total number N of homeless people in a city like Philadelphia at a given time is not known but it is possible to count the number n_1 of homeless persons in a given shelter on a precise night, to record their ID, and to cross this sample with a sample of n_2 persons collected the night after in order to detect how many persons n_{12} were present in the shelter on both nights.

The dataset we consider throughout this chapter is called **eurodip** and is related to a population of birds called *European dippers (Cinclus cinclus)*. These birds are closely dependent on streams, feeding on underwater invertebrates, and their nests are always close to water. The capture–recapture data on the European dipper contained in **eurodip** covers 7 years (1981–1987 inclusive) of observations in a zone of 200 km^2 in eastern France. The data consist of markings and recaptures of breeding adults each year during the breeding period from early March to early June. Birds were at least 1 year old when initially banded. In **eurodip**, each row of seven digits corresponds to a capture–recapture story for a given dipper, 0 indicating an absence of capture that year and, in the case of a capture, 1, 2, or 3 representing the zone where the dipper is captured. For instance, the three lines from **eurodip**

```
1 0 0 0 0 0 0
1 3 0 0 0 0 0
0 2 2 2 1 2 2
```

indicate that the first dipper was only captured the first year in zone 1 and that the second dipper was captured in years 1981 and 1982 and moved from zone 1 to zone 3 between those years. The third dipper was captured every year but 1981 and moved between zones 1 and 2 during the remaining year.

In conclusion, we hope that the introduction above was motivating enough to convince the reader that population sampling models are deeply relevant in statistical practice. Besides, these models also provide an interesting application of Bayesian modeling and in particular they allow for the inclusion of often available prior information.

5.2 Sampling Models

5.2.1 The Binomial Capture Model

We start with the simplest model of all, namely the independent observation or *capture*[3] of n^+ individuals from a population of size N. For instance, a trap is positioned on a rabbit track for five hours and n^+ rabbits are found in the trap. While the population size $N \in \mathbb{N}$ is the parameter of interest, there exists a nuisance parameter, namely the probability $p \in [0, 1]$ with which each individual is captured. (This model assumes that catching the ith individual is independent of catching the jth individual.) For this model,

$$n^+ \sim \mathscr{B}(N, p)$$

and the corresponding likelihood is

$$\ell(N, p|n^+) = \binom{N}{n^+} p^{n^+} (1-p)^{N-n^+} \mathbb{I}_{N \geq n^+} .$$

Obviously, with a single observation n^+, we cannot say much on (N, p), but the posterior distribution is still well-defined. For instance, if we use the vague prior

$$\pi(N, p) \propto N^{-1} \mathbb{I}_{\mathbb{N}}(N) \mathbb{I}_{[0,1]}(p) ,$$

the posterior distribution of N is

$$\begin{aligned} \pi(N|n^+) &\propto \frac{N!}{(N-n^+)! n^+!} N^{-1} \mathbb{I}_{N \geq n^+} \mathbb{I}_{\mathbb{N}^*}(N) \int_0^1 p^{n^+} (1-p)^{N-n^+} \mathrm{d}p \\ &\propto \frac{(N-1)!}{(N-n^+)!} \frac{(N-n^+)!}{(N+1)!} \mathbb{I}_{N \geq n^+ \vee 1} \\ &= \frac{1}{N(N+1)} \mathbb{I}_{N \geq n^+ \vee 1} , \end{aligned} \tag{5.1}$$

where $n^+ \vee 1 = \max(n^+, 1)$. Note that this posterior distribution is defined even when $n^+ = 0$. If we use the (more informative) uniform prior

$$\pi(N, p) \propto \mathbb{I}_{\{1,\ldots,S\}}(N) \mathbb{I}_{[0,1]}(p) ,$$

the posterior distribution of N is

$$\pi(N|n^+) \propto \frac{1}{N+1} \mathbb{I}_{\{n^+ \vee 1,\ldots,S\}}(N) .$$

[3] We use the original terminology of *capture* and *individuals*, even though the sampling mechanism may be far from genuine capture, as in whale sightseeing or software bug detection.

For illustrative purposes, consider the case of year 1981 in **eurodip** (which is the first column in the file):

```
> data(eurodip)
> year81=eurodip[,1]
> nplus=sum(year81>0)
[1] 22
```

where $n^+ = 22$ dippers were thus captured. By using the binomial capture model and the vague prior $\pi(N, p) \propto N^{-1}$, the number of dippers N can be estimated by the posterior median. (Note that the mean of (5.1) does not exist, no matter what n^+ is.)

```
> N=max(nplus,1)
> rangd=N:(10^4*N)
> post=1/(rangd*(rangd+1))
> 1/sum(post)
[1] 22.0022
> post=post/sum(post)
> min(rangd[cumsum(post)>.5])
[1] 43
```

For this year 1981, the estimate of N is therefore 43 dippers. (See Exercise 5.1 for theoretical justifications as to why the sum of the probabilities is equal to n^+ and why the median is exactly $2n^+ - 1$.) If we use the ecological information that there cannot be more than 400 dippers in this region, we can take the prior $\pi(N, p) \propto \mathbb{I}_{\{1,\ldots,400\}}(N)\mathbb{I}_{[0,1]}(p)$ and estimate the number of dippers N by its posterior expectation:

```
> pbino=function(nplus){
+     prob=c(rep(0,max(nplus,1)-1),1/(max(nplus,1):400+1))
+     prob/sum(prob)
+     }
> sum((1:400)*pbino(nplus))
[1] 130.5237
```

5.2.2 The Two-Stage Capture–Recapture Model

A logical extension to the capture model above is the *capture–mark–recapture* model, which considers two capture periods plus a marking stage, as follows:

1. n_1 individuals from a population of size N are "captured", that is, sampled without replacement.
2. Those individuals are "marked", that is, identified by a numbered tag (for birds and fishes), a collar (for mammals), or another device (like the Social Security number for homeless people or a picture for whales), and they are then released into the population.

3. A second and similar sampling (once again without replacement) is conducted, with n_2 individuals captured.
4. m_2 individuals out of the n_2's bear the identification mark and are thus characterized as having been captured in both experiments.

If we assume a *closed population* (that is, a fixed population size N throughout the capture experiment), a constant capture probability p for all individuals, and complete independence between individuals and between captures, we end up with a product of binomial models,

$$n_1 \sim \mathscr{B}(N, p)\,, \quad m_2|n_1 \sim \mathscr{B}(n_1, p)\,,$$

and

$$n_2 - m_2|n_1, m_2 \sim \mathscr{B}(N - n_1, p)\,.$$

If

$$n^c = n_1 + n_2 \quad \text{and} \quad n^+ = n_1 + (n_2 - m_2)$$

denote the total number of captures over both periods and the total number of captured individuals, respectively, the corresponding likelihood $\ell(N, p|n_1, n_2, m_2)$ is

$$\begin{aligned}
&\binom{N-n_1}{n_2-m_2} p^{n_2-m_2}(1-p)^{N-n_1-n_2+m_2}\mathbb{I}_{\{0,\ldots,N-n_1\}}(n_2-m_2) \\
&\qquad \times \binom{n_1}{m_2} p^{m_2}(1-p)^{n_1-m_2}\binom{N}{n_1} p^{n_1}(1-p)^{N-n_1}\mathbb{I}_{\{0,\ldots,N\}}(n_1) \\
&\qquad \propto \frac{N!}{(N-n_1-n_2+m_2)!} p^{n_1+n_2}(1-p)^{2N-n_1-n_2}\mathbb{I}_{N\geq n^+} \\
&\qquad \propto \binom{N}{n^+} p^{n^c}(1-p)^{2N-n^c}\mathbb{I}_{N\geq n^+}\,,
\end{aligned}$$

which shows that (n^c, n^+) is a sufficient statistic. If we choose the prior $\pi(N, p) = \pi(N)\pi(p)$ such that $\pi(p)$ is a $\mathscr{U}([0,1])$ density, the conditional posterior distribution on p is such that

$$\pi(p|N, n_1, n_2, m_2) = \pi(p|N, n^c) \propto p^{n^c}(1-p)^{2N-n^c}\,;$$

that is,

$$p|N, n^c \sim \mathscr{B}e(n^c + 1, 2N - n^c + 1).$$

Unfortunately, the marginal posterior distribution of N is more complicated. For instance, if $\pi(N) = \mathbb{I}_{\mathbb{N}^*}(N)$, it satisfies

$$\pi(N|n_1, n_2, m_2) = \pi(N|n^c, n^+) \propto \binom{N}{n^+} B(n^c + 1, 2N - n^c + 1)\mathbb{I}_{N\geq n^+\vee 1}\,,$$

where $B(a, b)$ denotes the beta function This distribution is called a *beta-Pascal* distribution, but it is not very tractable. The same difficulty occurs if $\pi(N) = N^{-1}\mathbb{I}_{\mathbb{N}^*}(N)$.

The intractability in the posterior distribution $\pi(N|n_1, n_2, m_2)$ is due to the infinite summation resulting from the unbounded support of N. A feasible approximation is to replace the missing normalizing factor by a finite sum with a large enough bound on N, the bound being determined by a lack of perceivable impact on the sum. But the approximation errors due to the computations of terms such as $\binom{N}{n^+}$ or $B(n^c + 1, 2N - n^c + 1)$ can become a serious problem when n^+ is large. However,

```
> prob=lchoose((471570:10^7),471570)+lgamma(2*(471570:10^7)-
+ 582681+1)-lgamma(2*(471570:10^7)+2)
> range(prob)
[1] -7886469 -7659979
```

shows that relatively large populations are manageable.

If we have information about an upper bound S on N and use the corresponding uniform prior,

$$\pi(N) \propto \mathbb{I}_{\{1,\ldots,S\}}(N)\,,$$

the posterior distribution of N is thus proportional to

$$\pi(N|n^+) \propto \binom{N}{n^+} \frac{\Gamma(2N - n^c + 1)}{\Gamma(2N + 2)} \mathbb{I}_{\{n^+\vee 1,\ldots,S\}}(N)\,,$$

and, in this case, it is possible to calculate the posterior expectation of N with no approximation error.

For the first 2 years of the **eurodip** experiment, which correspond to the first two columns and the first 70 rows of the dataset, $n_1 = 22$, $n_2 = 60$, and $m_2 = 11$. Hence, $n^c = 82$ and $n^+ = 71$. Therefore, within the frame of the two-stage capture–recapture model[4] and the uniform prior $\mathscr{U}(\{1, \ldots, 400\}) \times \mathscr{U}([0, 1])$ on (N, p), the posterior expectation of N is derived as follows:

```
> n1=sum(eurodip[,1]>0)
> n2=sum(eurodip[,2]>0)
> m2=sum((eurodip[,1]>0) & (eurodip[,2]>0))
> nc=n1+n2
> nplus=nc-m2
> pcapture=function(T,nplus,nc){
+ #T is the number of capture episodes
+   lprob=lchoose(max(nplus,1):400,nplus)+
      lgamma(T*max(nplus,1):400-nc+1)-
+     lgamma(T*max(nplus,1):400+2)
+   prob=c(rep(0,max(nplus,1)-1),exp(lprob-max(lprob)))
```

[4]This analysis is based on the assumption that all birds captured in the second year were already present in the population during the first year.

```
+    prob/sum(prob)
+    }
> sum((1:400)*pcapture(2,nplus,nc))
[1] 165.2637
```

A simpler model used in capture–recapture settings is the hypergeometric model, also called the *Darroch model.* This model can be seen as a conditional version of the two-stage model when conditioning on both sample sizes n_1 and n_2 since (see Exercise 5.3)

$$m_2|n_1, n_2 \sim \mathscr{H}(N, n_2, n_1/N), \tag{5.2}$$

the hypergeometric distribution If we choose the uniform prior $\mathscr{U}(\{1, \ldots, 400\})$ on N, the posterior distribution of N is thus

$$\pi(N|m_2) \propto \binom{N - n_1}{n_2 - m_2} \Big/ \binom{N}{n_2} \mathbb{I}_{\{n^+ \vee 1, \ldots, 400\}}(N),$$

and posterior expectations can be computed numerically by simple summations.

For the first 2 years of the **eurodip** dataset and $S = 400$, the posterior distribution of N for the Darroch model is given by

$$\pi(N|m_2) \propto (n - n_1)!(N - n_2)!/\{(n - n_1 - n_2 + m_2)!N!\} \, \mathbb{I}_{\{71, \ldots, 400\}}(N),$$

the normalization factor being the inverse of

$$\sum_{k=71}^{400} (k - n_1)!(k - n_2)!/\{(k - n_1 - n_2 + m_2)!k!\}.$$

We thus have a closed-form posterior distribution and the posterior expectation of N is given by

```
pdarroch=function(n1,n2,m2){

  prob=c(rep(0,max(n1+n2-m2,1)-1),
       choose(n1,m2)*choose(max((n1+n2-m2),1):400-n1,n2-m2)/
       choose(max((n1+n2-m2),1):400,n2))
  prob/sum(prob)
  }
> sum((1:400)*pdarroch(n1,n2,m2))
[1] 137.5962
```

```
Table~5.1 shows the evolution of this
posterior expectation for different values of m_2,
obtained by

> for (i in 6:16) print(round(sum(pdarroch(n1,n2,i)*1:400)))
[1] 277
[1] 252
[1] 224
[1] 197
[1] 172
[1] 152
[1] 135
[1] 122
[1] 111
[1] 101
[1] 94
```

The number of recaptures is thus highly influential on the estimate of N. In parallel, Table 5.2 shows the evolution of the posterior expectation for different values of S (taken equal to 400 in the above). When S is large enough, say larger than $S = 250$, the estimate of N is quite stable, as expected.

Table 5.1. Dataset **eurodip**: Rounded posterior expectation of the dipper population size, N, under a uniform prior $\mathscr{U}(\{1,\ldots,400\})$

m_2	0	1	2	3	4	5	6	7	8	9	10	11	12	13	14	15
$\mathbb{E}^\pi[N\vert m_2]$	355	349	340	329	316	299	277	252	224	197	172	152	135	122	110	101

Table 5.2. Dataset **eurodip**: Rounded posterior expectation of the dipper population size, N, under a uniform prior $\mathscr{U}(\{1,\ldots,S\})$, for $m_2 = 11$

S	100	150	200	250	300	350	400	450	500
$\mathbb{E}^\pi[N\vert m_2]$	95	125	141	148	151	151	152	152	152

Leaving the Darroch model and getting back to the two-stage capture model with probability p of capture, the posterior distribution of (N, p) associated with the noninformative prior $\pi(N, p) = 1/N$ is proportional to

$$\frac{(N-1)!}{(N-n^+)!}\,p^{n^c}(1-p)^{2N-n^c}\,.$$

Thus, if $n^+ > 0$, both conditional posterior distributions are standard distributions since

$$p|n^c, N \sim \mathscr{B}e(n^c+1, 2N-n^c+1)$$
$$N-n^+|n^+, p \sim \mathscr{N}eg(n^+, 1-(1-p)^2),$$

the latter being a negative binomial distribution. Indeed, as a function of N,

$$\frac{(N-1)!}{(N-n^+)!}(1-p)^{2N-n^c} \propto \binom{N-1}{N-n^+}\left\{(1-p)^2\right\}^{N-n^+}\left\{1-(1-p)^2\right\}^{n^+}.$$

Therefore, while the marginal posterior in N is difficult to manage, the joint distribution of (N, p) can be approximated by a Gibbs sampler, as follows:

Algorithm 5.8 TWO-STAGE CAPTURE–RECAPTURE GIBBS SAMPLER

Initialization: Generate $p^{(0)} \sim \mathscr{U}([0,1])$.
Iteration i $(i \geq 1)$:
1. Generate $N^{(i)} - n^+ \sim \mathscr{N}eg(n^+, 1-(1-p^{(i-1)})^2)$.
2. Generate $p^{(i)} \sim \mathscr{B}e(n^c+1, 2N^{(i)}-n^c+1)$.

5.2.3 The T-Stage Capture–Recapture Model

A further extension to the two-stage capture–recapture model is to consider instead a series of T consecutive captures. In that case, if we denote by n_t the number of individuals captured at period t $(1 \leq t \leq T)$ and by m_t the number of recaptured individuals (with the convention that $m_1 = 0$), under the same assumptions as in the two-stage model, then $n_1 \sim \mathscr{B}(N, p)$ and, conditionally on the $j-1$ previous captures and recaptures $(2 \leq j \leq T)$,

$$m_j \sim \mathscr{B}\left(\sum_{t=1}^{j-1}(n_t - m_t), p\right) \quad \text{and} \quad n_j - m_j \sim \mathscr{B}\left(N - \sum_{t=1}^{j-1}(n_t - m_t), p\right).$$

The likelihood $\ell(N, p|n_1, n_2, m_2 \ldots, n_T, m_T)$ is thus

$$\binom{N}{n_1} p^{n_1}(1-p)^{N-n_1} \prod_{j=2}^{T}\left[\binom{N-\sum_{t=1}^{j-1}(n_t-m_t)}{n_j-m_j} p^{n_j-m_j+m_j}\right.$$
$$\left. \times (1-p)^{N-\sum_{t=1}^{j-1}(n_t-m_t)}\binom{\sum_{t=1}^{j-1}(n_t-m_t)}{m_j}(1-p)^{\sum_{t=1}^{j-1}(n_t-m_t)-m_j}\right]$$
$$\propto \frac{N!}{(N-n^+)!} p^{n^c}(1-p)^{TN-n^c}\mathbb{I}_{N \geq n^+}$$

if we denote the sufficient statistics as

$$n^+ = \sum_{t=1}^{T}(n_t - m_t) \quad \text{and} \quad n^c = \sum_{t=1}^{T} n_t,$$

the total numbers of captured individuals and captures over the T periods, respectively.

For a noninformative prior such as $\pi(N,p) = 1/N$, the joint posterior satisfies

$$\pi(N,p|n^+,n^c) \propto \frac{(N-1)!}{(N-n^+)!} p^{n^c}(1-p)^{TN-n^c} \mathbb{I}_{N\geq n^+\vee 1}\,.$$

Therefore, the conditional posterior distribution of p is

$$p|N,n^+,n^c \sim \mathscr{B}e(n^c+1, TN-n^c+1)$$

and the marginal posterior distribution of N

$$\pi(N|n^+,n^c) \propto \frac{(N-1)!}{(N-n^+)!}\,\frac{(TN-n^c)!}{(TN+1)!}\,\mathbb{I}_{N\geq n^+\vee 1}\,,$$

is computable. Note that the normalization coefficient can also be approximated by summation with an arbitrary precision unless N and n^+ are very large.

For the uniform prior $\mathscr{U}(\{1,\ldots,S\})$ on N and $\mathscr{U}([0,1])$ on p, the posterior distribution of N is then proportional to

$$\pi(N|n^+) \propto \binom{N}{n^+}\frac{(TN-n^c)!}{(TN+1)!}\,\mathbb{I}_{\{n^+\vee 1,\ldots,S\}}(N).$$

For the whole set of observations in **eurodip**, we have $T = 7$, $n^+ = 294$, and $n^c = 519$. Under the uniform prior with $S = 400$, the posterior expectation of N is given by

```
> sum((1:400)*pcapture(7,294,519))
[1] 372.7384
```

While this value seems dangerously close to the upper bound of 400 on N and thus leads us to suspect a strong influence of the upper bound S, the computation of the posterior expectation for $S = 2500$

```
> S=2500;T=7;nplus=294;nc=519
> lprob=lchoose(max(nplus,1):S,nplus)+
+    lgamma(T*max(nplus,1):S-nc+1)-lgamma(T*max(nplus,1):S+2)
> prob=c(rep(0,max(nplus,1)-1),exp(lprob-max(lprob)))
> sum((1:S)*prob)/sum(prob)
[1] 373.9939
```

leads to 373.99, which shows the limited impact of this hyperparameter S.

Using even a slightly more advanced sampling model may lead to genuine computational difficulties. For instance, consider a heterogeneous capture–recapture model where the individuals are captured at time $1 \leq t \leq T$ with probability p_t and where both the size N of the population and the probabilities p_t are unknown. The corresponding likelihood is

$$\ell(N, p_1, \ldots, p_T | n_1, n_2, m_2 \ldots, n_T, m_T) \propto \frac{N!}{(N-n^+)!} \prod_{t=1}^{T} p_t^{n_t} (1-p_t)^{N-n_t}.$$

If the associated prior on $(N, p_1, \ldots, p_T)$ is such that

$$N \sim \mathscr{P}(\lambda)$$

and $(1 \leq t \leq T)$,

$$\alpha_t = \log\left(\frac{p_t}{1-p_t}\right) \sim \mathscr{N}(\mu_t, \sigma^2),$$

where both σ^2 and the μ_t's are known,[5] the posterior distribution satisfies

$$\pi(\alpha_1, \ldots, \alpha_T, N | , n_1, \ldots, n_T) \propto \frac{N!}{(N-n^+)!} \frac{\lambda^N}{N!} \prod_{t=1}^{T} (1+e^{\alpha_t})^{-N} \tag{5.3}$$
$$\times \prod_{t=1}^{T} \exp\left\{\alpha_t n_t - \frac{1}{2\sigma^2}(\alpha_t - \mu_t)^2\right\}.$$

It is thus much less manageable from a computational point of view, especially when there are many capture episodes. A corresponding Gibbs sampler could simulate easily from the conditional posterior distribution on N since

$$N - n^+ | \alpha, n^+ \sim \mathscr{P}\left(\lambda \prod_{t=1}^{T} (1+e^{\alpha_t})\right),$$

but the conditionals on the α_t's $(1 \leq t \leq T)$ are less conventional,

$$\alpha_t | N, \mathbf{n} \sim \pi_t(\alpha_t | N, \mathbf{n}) \propto (1+e^{\alpha_t})^{-N} e^{\alpha_t n_t - (\alpha_t - \mu_t)^2 / 2\sigma^2},$$

and they require either an accept–reject algorithm (Sect. 5.4) or a Metropolis–Hastings algorithm in order to be simulated.

For the prior

$$\pi(N, p) \propto \frac{\lambda^N}{N!} \mathbb{I}_{\mathbb{N}}(N) \mathbb{I}_{[0,1]}(p),$$

the conditional posteriors are then

$$p | N, n^c \sim \mathscr{B}e(n^c + 1, TN - n^c + 1) \quad \text{and} \quad N - n^+ | p, n^+ \sim \mathscr{P}(\lambda(1-p)^T)$$

and a Gibbs sampler similar to the one developed in Algorithm 5.8 can easily be implemented, for instance via the code

[5]This assumption can be justified on the basis that each capture probability is only observed once on the tth round (and so cannot reasonably be associated with a noninformative prior).

```
> lambda=200
> nsimu=10^4
> p=rep(1,nsimu); N=p
> N[1]=2*nplus
> p[1]=rbeta(1,nc+1,T*lambda-nc+1)
> for (i in 2:nsimu){
+  N[i]=nplus+rpois(1,lambda*(1-p[i-1])^T)
+  p[i]=rbeta(1,nc+1,T*N[i]-nc+1)
+  }
```

For **eurodip**, we used this Gibbs sampler and obtained the results illustrated by Fig. 5.1. When the chain is initialized at the (unlikely) value $N^{(0)} = \lambda = 200$ (which is the prior expectation of N), the stabilization of the chain is quite clear: It only takes a few iterations to converge toward the proper region that supports the posterior distribution. We can thus visually confirm the convergence of the algorithm and approximate the Bayes estimators of N and p by the Monte Carlo averages

```
> mean(N)
[1] 326.9831
> mean(p)
[1] 0.2271828
```

The precision of these estimates can be assessed as in a regular Monte Carlo experiment, but the variance estimate is biased because of the correlation between the simulations. A simple way to assess this effect is to call R function `acf()` for each component θ_i of the parameter, as

$$\nu = 1 + 2 \sum_{t=1}^{\infty} \text{cor}(\theta_i^{(1)}, \theta_i^{(t+1)})$$

evaluates the loss of efficiency due to the correlation. The corresponding *effective sample size*, given by $T_{\text{ess}} = T/\nu$, provides the equivalent size of an iid sample. For instance,

```
> 1/(1+2*sum(acf(N)$acf[-1]))
[1] 0.599199
> 1/(1+2*sum(acf(p)$acf[-1]))
[1] 0.6063236
```

shows that the current Gibbs sampler offers an efficiency of 60% compared with an iid sample from the posterior distribution.

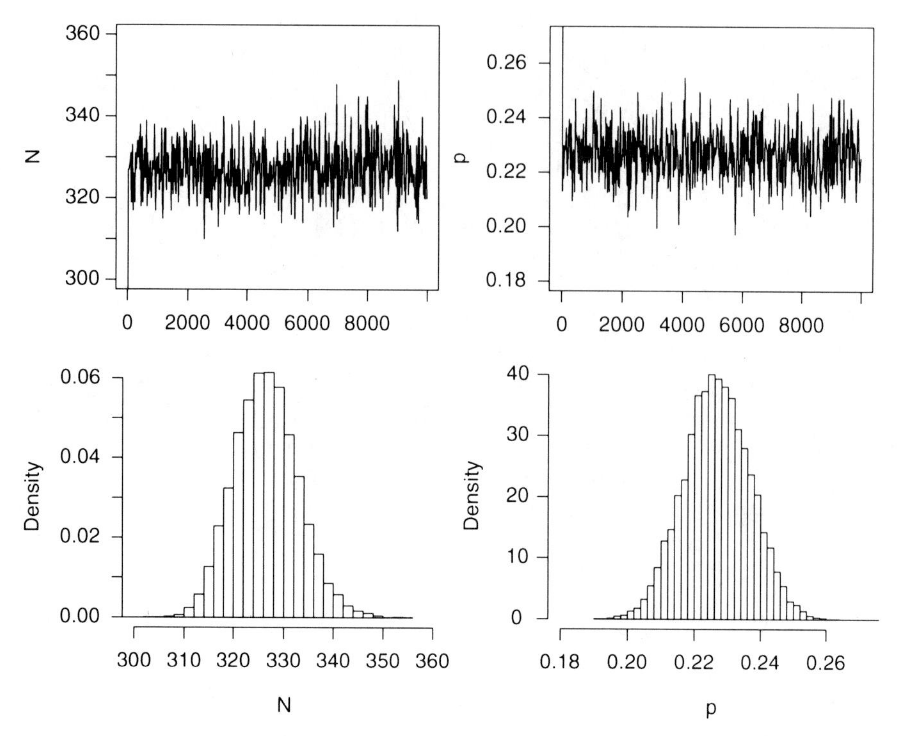

Fig. 5.1. Dataset **eurodip**: Representation of the Gibbs sampling output for the parameters p (*first column*) and N (*second column*)

5.3 Open Populations

Moving towards more realistic settings, we now consider the case of an *open population* model, where the population size does not remain fixed over the experiment but, on the contrary, there is a probability q for each individual to leave the population at each time (or, more accurately, between any two capture episodes). Given that the associated likelihood involves unobserved indicators (namely, indicators of survival; see Exercise 5.14), we study here a simpler model where only the individuals captured during the first capture experiment are marked and subsequent recaptures are registered. For three successive capture experiments, we thus have

$$n_1 \sim \mathscr{B}(N, p)\,, \quad r_1 | n_1 \sim \mathscr{B}(n_1, q)\,, \quad r_2 | n_1, r_1 \sim \mathscr{B}(n_1 - r_1, q)\,,$$

for the distributions of the first capture population size and of the numbers of individuals who vanished between the first and second, and the second and third experiments, respectively, and

$$c_2 | n_1, r_1 \sim \mathscr{B}(n_1 - r_1, p), \quad c_3 | n_1, r_1, r_2 \sim \mathscr{B}(n_1 - r_1 - r_2, p)\,,$$

for the number of recaptured individuals during the second and the third experiments, respectively. Here, only n_1, c_2, and c_3 are observed. The numbers of individuals removed at stages 1 and 2, r_1 and r_2, are not available and must therefore be simulated, as well as the parameters N, p, and q.[6] The likelihood $\ell(N,p,q,r_1,r_2|n_1,c_2,c_3)$ is given by

$$\binom{N}{n_1}p^{n_1}(1-p)^{N-n_1}\binom{n_1}{r_1}q^{r_1}(1-q)^{n_1-r_1}\binom{n_1-r_1}{c_2}p^{c_2}(1-p)^{n_1-r_1-c_2}$$
$$\times\binom{n_1-r_1}{r_2}q^{r_2}(1-q)^{n_1-r_1-r_2}\binom{n_1-r_1-r_2}{c_3}p^{c_3}(1-p)^{n_1-r_1-r_2-c_3}$$

and, if we use the prior $\pi(N,p,q)\propto N^{-1}\mathbb{I}_{[0,1]}(p)\mathbb{I}_{[0,1]}(q)$, the associated conditionals are

$$\begin{aligned}
\pi(p|N,q,\mathscr{D}^*) &\propto p^{n^+}(1-p)^{u_+}\,,\\
\pi(q|N,p,\mathscr{D}^*) &\propto q^{r_1+r_2}(1-q)^{2n_1-2r_1-r_2}\,,\\
\pi(N|p,q,\mathscr{D}^*) &\propto \frac{(N-1)!}{(N-n_1)!}(1-p)^N\mathbb{I}_{N\geq n_1}\,,\\
\pi(r_1|p,q,n_1,c_2,c_3,r_2) &\propto \frac{(n_1-r_1)!\,q^{r_1}(1-q)^{-2r_1}(1-p)^{-2r_1}}{r_1!(n_1-r_1-r_2-c_3)!(n_1-c_2-r_1)!}\,,\\
\pi(r_2|p,q,n_1,c_2,c_3,r_1) &\propto \frac{q^{r_2}[(1-p)(1-q)]^{-r_2}}{r_2!(n_1-r_1-r_2-c_3)!}\,,
\end{aligned}$$

where $\mathscr{D}^*=(n_1,c_2,c_3,r_1,r_2)$ and

$$\begin{aligned}
&u_1=N-n_1,\ u_2=n_1-r_1-c_2,\ u_3=n_1-r_1-r_2-c_3\,,\\
&n^+=n_1+c_2+c_3,\ u_+=u_1+u_2+u_3
\end{aligned}$$

(u stands for *unobserved*, even though these variables can be computed conditional on the remaining unknowns). Therefore, the full conditionals are

$$\begin{aligned}
p|N,q,\mathscr{D}^* &\sim \mathscr{B}e(n^++1,u_++1)\,,\\
q|N,p,\mathscr{D}^* &\sim \mathscr{B}e(r_1+r_2+1,2n_1-2r_1-r_2+1)\,,\\
N-n_1|p,q,\mathscr{D}^* &\sim \mathscr{N}eg(n_1,p)\,,\\
r_2|p,q,n_1,c_2,c_3,r_1 &\sim \mathscr{B}\left(n_1-r_1-c_3,\frac{q}{q+(1-q)(1-p)}\right)\,,
\end{aligned}$$

which are very easily simulated, while r_1 has a less conventional distribution. However, this difficulty is minor since, in our case, n_1 is not extremely

[6]From a theoretical point of view, r_1 and r_2 are *missing variables* rather than true parameters. This obviously does not change anything either for simulation purposes or for Bayesian inference.

large. It is thus possible to compute the probability that r_1 is equal to each of the values in $\{0, 1, \ldots, \min(n_1 - r_2 - c_3, n_1 - c_2)\}$. This means that the corresponding Gibbs sampler can be implemented as well.

```
gibbscap1=function(nsimu,n1,c2,c3){

N=p=q=r1=r2=rep(0,nsimu)
N[1]=round(n1/runif(1))
r1[1]=max(c2,c3)+round((n1-c2)*runif(1))
r2[1]=round((n1-r1[1]-c3)*runif(1))
nplus=n1+c2+c3
for (i in 2:nsimu){
  uplus=N[i-1]-r1[i-1]-c2+n1-r1[i-1]-r2[i-1]-c3
  p[i]=rbeta(1,nplus+1,uplus+1)
  q[i]=rbeta(1,r1[i-1]+r2[i-1]+1,2*n1-2*r1[i-1]-r2[i-1]+1)
  N[i]=n1+rnbinom(1,n1,p[i])
  rbar=min(n1-r2[i-1]-c3,n1-c2)
  pq=q[i]/((1-q[i])*(1-p[i]))^2
  pr=lchoose(n1-c2,0:rbar)+(0:rbar)*log(pq)+
     lchoose(n1-(0:rbar),r2[i-1]+c3)
  r1[i]=sample(0:rbar,1,prob=exp(pr-max(pr)))
  r2[i]=rbinom(1,n1-r1[i]-c3,q[i]/(q[i]+(1-q[i])*(1-p[i])))
  }
  list(N=N,p=p,q=q,r1=r1,r2=r2)
}
```

We stress that R is quite helpful in simulating from unusual distributions and in particular from those with finite support. For instance, the conditional distribution of r_1 above can be simulated using the following representation of $\mathbb{P}(r_1 = k|p, q, n_1, c_2, c_3, r_2)$ $(0 \le k \le \bar{r} = \min(n_1 - r_2 - c_3, n_1 - c_2))$,

$$\binom{n_1 - c_2}{k} \left\{ \frac{q}{(1-q)^2(1-p)^2} \right\}^k \binom{n_1 - k}{r_2 + c_3}, \tag{5.4}$$

up to a normalization constant, since the binomial coefficients and the power in k can be computed for all values of k at once, thanks to the matrix capabilities of R, through the command `lchoose`. The above quantity corresponding to

```
pr=lchoose(n=n1-c2,k=0:rbar) + (0:rbar)*log(q1)
          + lchoose(n=n1-(0:rbar),k=r2+c3)
```

is the whole vector of the log-probabilities, with $q_1 = q/(1-q)^2(1-p)^2$.

↯ In most computations, it is safer to use logarithmic transforms to reduce the risk of running into overflow or underflow error messages. For instance, in the example above, the probability vector can be recovered by

```
pr=exp(pr-max(pr))/sum(exp(pr-max(pr)))
```

while a direct computation of exp(pr) may well produce an Inf value that invalidates the remaining computations.[7]

Once the probabilities are transformed as in the previous R code, a call to the R command

```
> sample(0:mm,n,prob=exp(pr-max(pr)))
```

is sufficient to provide n simulations of r_1. The production of a large Gibbs sample is immediate:

```
> system.time(gibbscap1(10^5,22,11,6))
   user  system elapsed
 12.816   0.000  12.830
```

Even a large value such as $n_1 = 1612$ used below does not lead to computing difficulties since we can run 10,000 iterations of the corresponding Gibbs sampler in a few seconds on a laptop:

```
> system.time(gibbscap1(10^4,1612,811,236))
   user  system elapsed
 10.245   0.028  10.294
```

For **eurodip**, we have $n_1 = 22$, $c_2 = 11$, and $c_3 = 6$. We obtain the Gibbs output

```
> gg=gibbscap1(10^5,22,11,6)
```

summarized in Fig. 5.2. The sequences for all components are rather stable and their mixing behavior (i.e., the speed of exploration of the support of the target) is satisfactory, even though we can still detect a trend in the first three rows. Since r_1 and r_2 are integers with only a few possible values, the last two rows show apparently higher jumps than the three other parameters. The MCMC approximations to the posterior expectations of N and p are equal

```
> mean(gg$N)
[1] 57.52955
> mean(gg$p)
[1] 0.3962891
```

respectively.

Given the large difference between n_1 and c_2 and the proximity between c_2 and c_3, high values of q are rejected, and the difference can be attributed

[7]This recommendation also applies to the computation of likelihoods that tend to take absolute values that exceed the range of the computer representation of real numbers, while only the relative values are relevant for Bayesian computations. Using a transform such as exp(loglike-max(loglike)) thus helps in reducing the risk of overflows.

with high likelihood to a poor capture rate. One should take into account the fact that there are only three observations for a model that involves three true parameters plus two missing variables. Figure 5.3 gives another insight into the posterior distribution by representing the joint distribution of the sample of (r_1, r_2)'s

```
> plot(jitter(gg$r1,factor=1),jitter(g2$r2,factor=1),cex=0.5,
+ xlab=expression(r[1]),ylab=expression(r[2]))
```

using for representation purposes the R function jitter(), which moves each point by a tiny random amount. There is a clear positive correlation between r_1 and r_2, despite the fact that r_2 is simulated on an $(n_1 - c_3 - r_1)$ scale. The mode of the posterior is $(r_1, r_2) = (0, 0)$, which means that it is likely that no dipper died or left the observation area over the 3-year period.

5.4 Accept–Reject Algorithms

In Chap. 2, we mentioned standard random number generators used for the most common distributions and presented importance sampling (Algorithm 2.2) as a possible alternative when such generators are not available. While MCMC algorithms always offer a solution when facing nonstandard distributions, there often exists a possibility that is in fact used in most of the standard random generators and which we now present. It also relates to the independent Metropolis–Hastings algorithm of Sect. 4.2.2.

Given a density g that is defined on an arbitrary space (of any dimension), a fundamental identity is that simulating X distributed from $g(x)$ is completely equivalent to simulating (X, U) uniformly distributed on the set

$$\mathscr{S} = \{(x, u) : 0 < u < g(x)\}$$

(this is called the *Fundamental Theorem of Simulation* in Robert and Casella, 2004, Chap. 3). The reason for this equivalence is simply that

$$\int_0^\infty \mathbb{I}_{0<u<g(x)} \, \mathrm{d}u = g(x) \, .$$

Since $\mathscr{S}$ usually has complex features, direct simulation from the uniform distribution on $\mathscr{S}$ is most often impossible (Exercise 5.16). The idea behind the accept–reject method is to find a simpler set $\mathscr{G}$ that contains $\mathscr{S}$, $\mathscr{S} \subset \mathscr{G}$, and then to simulate uniformly on this set $\mathscr{G}$ until the value belongs to $\mathscr{S}$. In practice, this means that one needs to find an upper bound on g; that is, another density f and a constant M such that

$$g(x) \leq M f(x) \tag{5.5}$$

on the support of the density g. (Note that $M > 1$ necessarily.) Implementing the following algorithm then leads to a simulation from g.

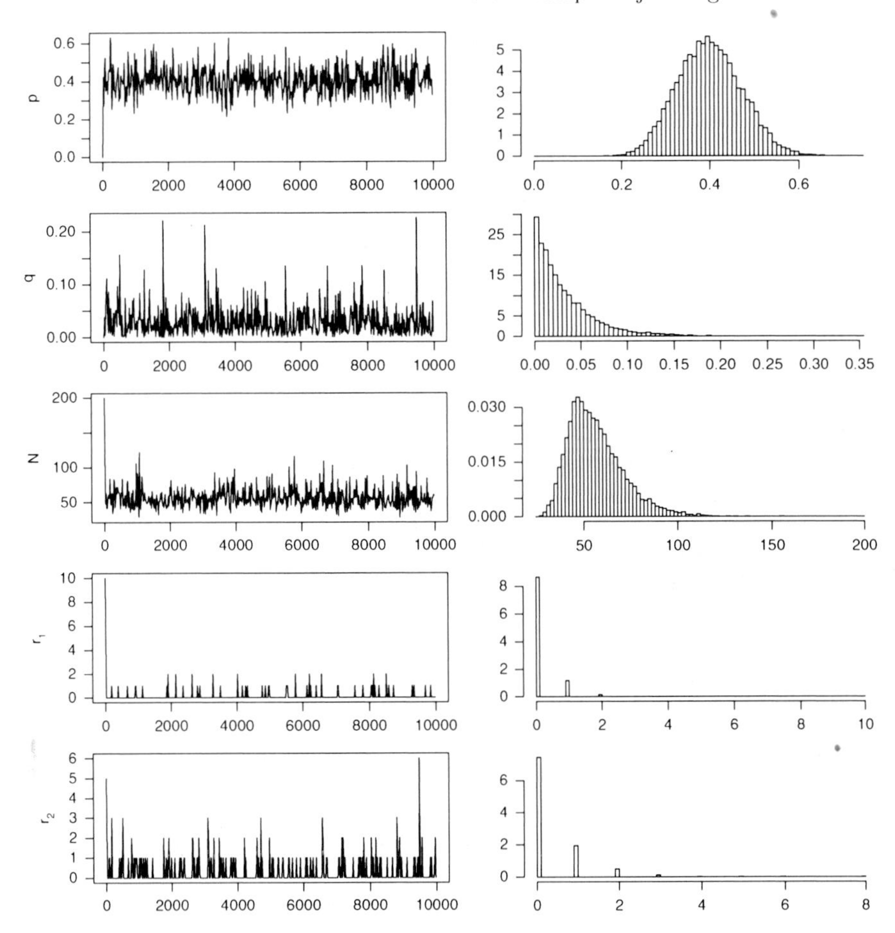

Fig. 5.2. Dataset **eurodip**: Representation of the Gibbs sampling output for the five parameters of the open population model, based on 10,000 iterations, with raw plots *(first column)* and histograms *(second column)*

Algorithm 5.9 Accept–Reject Sampler

1. Generate $X \sim f$, $U \sim \mathscr{U}_{[0,1]}$.
2. Accept $Y = x$ if $u \leq g(x)/(Mf(x))$.
3. Return to 1 otherwise.

This method provides a random generator for densities g that are known up to a multiplicative factor, which is a feature that occurs particularly often in Bayesian calculations since the posterior distribution is usually specified up to a normalizing constant.

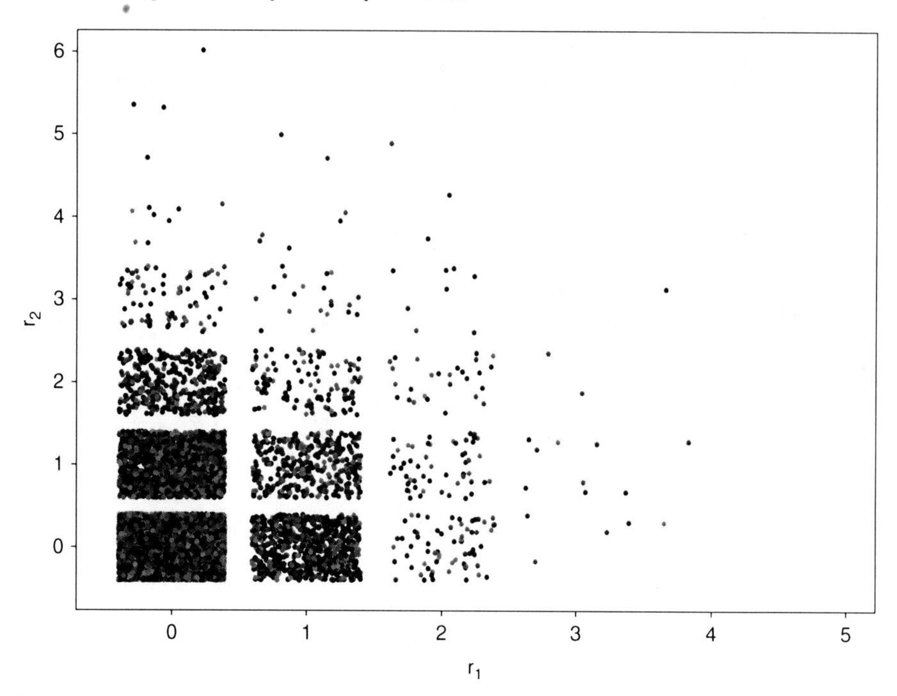

Fig. 5.3. Dataset **eurodip**: Representation of the Gibbs sampling output of the (r_1, r_2)'s by a jitterplot: to translate the density of the possible values of (r_1, r_2) on the $\mathbb{N}^2$ grid, each simulation has been randomly moved using the R jitter procedure and colored at random using *grey levels* to help distinguish the various simulations

For the open population model, we found the full conditional distribution of r_1 to be rather non-standard, as shown by (5.4). Rather than using an exhaustive enumeration of all probabilities $\mathbb{P}(m_1 = k) = g(k)$ and then sampling from this distribution, we can instead try to use a proposal based on a binomial upper bound. Take for instance f that corresponds to the binomial distribution $\mathscr{B}(\bar{r}, q_2)$ with

$$q_2 = q/\{q + (1-q)^2(1-p)^2\}\,.$$

The ratio $g(k)/f(k)$ is proportional to

$$\frac{\binom{n_1-c_2}{k}\binom{n_1-k}{r_2+c_3}}{\binom{\bar{r}}{k}} \propto \frac{(n_1-k)!}{(\max(n_1-c_2, n_1-r_2-c_3)-k)!}\,,$$

which is decreasing in k. The ratio is therefore bounded by

$$\frac{\binom{n_1-c_2}{0}\binom{n_1-0}{r_2+c_3}}{\binom{\bar{r}}{0}} = \frac{(n_1-c_2)!}{(r_2+c_3)!(n_1-r_2-c_3)!}$$

(up to the same normalizing constant). Note that this is *not* the constant M introduced in Algorithm 5.9 because we use unnormalized densities (the bound M may therefore also depend on q_2). Therefore we cannot derive the average acceptance rate from this ratio and we have to use a Monte Carlo experiment to check whether or not the method is really efficient (see Exercise 5.20).

If we use the values from **eurodip**—that is, $n_1 = 22$, $c_2 = 11$ and $c_3 = 6$, with $r_2 = 1$ and $q_1 = 0.1$—, we can use R functions like

```
thresh=function(k,n1,c2,c3,r2,barr){
  choose(n1-c2,k)*choose(n1-k,c3+r2)/choose(barr,k)
  }

ardipper=function(nsimu=1,n1,c2,c3,r2,q2){

  barr=min(n1-c2,n1-r2-c3)
  boundM=thresh(0,n1,c2,c3,r2,barr)
  echan=1:nsimu
  for (i in 1:nsimu){
    test=TRUE
    while (test){
      y=rbinom(1,size=barr,prob=q2)
      test=(runif(1)>thresh(y,n1,c2,c3,r2,barr))
      }
    echan[i]=y
    }
  echan
}
```

the average of the acceptance ratios $g(k)/Mf(k)$ is equal to 0.12. This is a relatively small value since it corresponds to a rejection rate of about 9/10. The simulation process could thus be a little slow, although

```
> system.time(ardipper(10^5,n1=22,c2=11,c3=6,r2=1,q1=.1))
   user  system elapsed
  8.148   0.024   8.1959
```

shows this is not the case. (Note that the code `ardipper` provided here does not produce the rejection rate. It has to be modified for this purpose.) An histogram of accepted values is shown in Fig. 5.4.

Obviously, this method is not hassle-free. For complex densities g, it may prove impossible to find a density f such that $g(x) \leq Mf(x)$ and M is small enough. However, there exists a large class of univariate distributions for which a generic choice of f is possible (see Robert and Casella, 2004, Chap. 2).

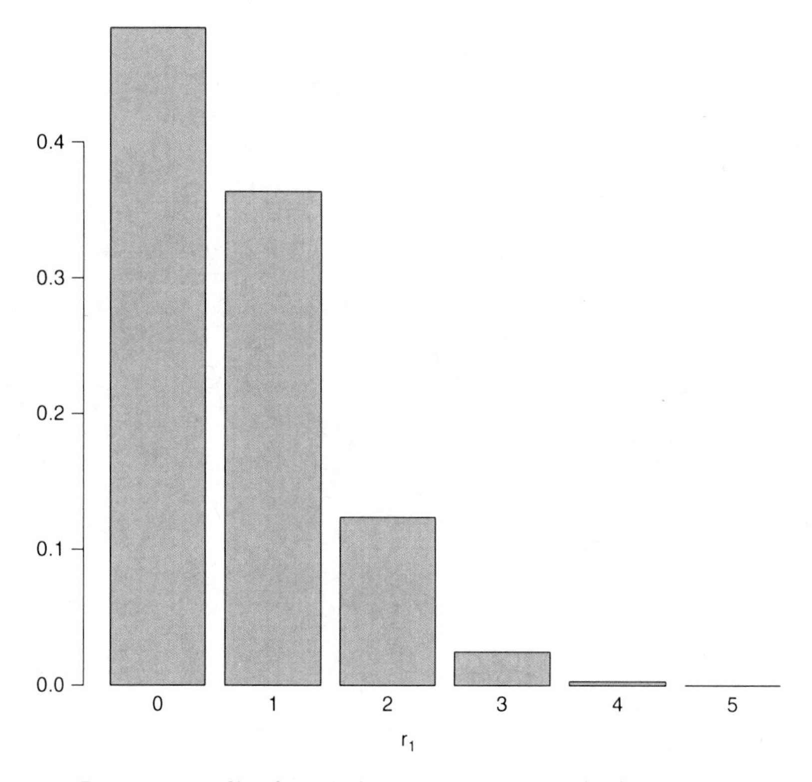

Fig. 5.4. Dataset **eurodip**: Sample from the distribution (5.4) obtained by accept–reject and based on the simulation of 10,000 values from a $\mathscr{B}(n_1, q_1)$ distribution for $n_1 = 22$, $c_2 = 11$, $c_3 = 6$, $r_2 = 1$, and $q_1 = 0.1$

5.5 The Arnason–Schwarz Capture–Recapture Model

We consider in this final section a more advanced capture–recapture model based on the realistic assumption that, in most capture–recapture experiments, we can tag individuals one by one; that is, we can distinguish each individual at the time of its first capture and thus follow its capture history. For instance, when tagging mammals and birds, differentiated tags can be used, so that there is only *one* individual with tag, say, 23131932.[8]

The *Arnason–Schwarz model* thus considers a capture–recapture experiment as a collection of individual histories. For each individual that has been

[8]In a capture–recapture experiment used in Dupuis (1995), a population of lizards was observed in the south of France (Lozère). When it was found that plastic tags caused necrosis on those lizards, the biologists in charge of the experiment decided to cut a phalange of one of the fingers of the captured lizards to identify them later. While the number of possibilities, 2^{20}, is limited, it is still much larger than the number of captured lizards in this study. Whether or not the lizards appreciated this ability to classify them is not known.

captured at least once during the experiment, individual characteristics of interest are registered at each capture. For instance, this may include location, weight, sexual status, pregnancy occurrence, social status, and so on. The probabilistic modeling includes this categorical decomposition by adding what we will call *movement* probabilities to the survival probabilities already used in the Darroch open population model of Sect. 5.2.2. From a theoretical point of view, this is a first example of a (partially) hidden Markov model, a structure studied in detail in Chap. 7. In addition, the model includes the possibility that individuals vanish from the population between two capture experiments. (This is thus another example of an open population model.)

As in **eurodip**, the interest that drives the capture–recapture experiment may be to study the movements of individuals within a zone $\mathfrak{K}$ divided into $k = 3$ strata denoted by $1, 2, 3$. (This structure is generic: Zones are not necessarily geographic and can correspond to anything from social status, to HIV stage, to university degree.) For instance, four consecutive rows of possible **eurodip** (individual) capture–recapture histories look as follows:

45	0 3 0 0 0 0 0
46	0 2 2 2 2 1 1
47	0 2 0 0 0 0 0
48	2 1 2 1 0 0 0

where 0 denotes a failure to capture. This means that, for dipper number 46, the first location was not observed but this dipper was captured for all the other experiments. For dippers number 45 and 47, there was no capture after the second time and thus one or both of them could be dead (or outside the range of the capture area) at the time of the last capture experiment. We also stress that the Arnason–Schwarz model often assumes that individuals that were not part of the population on the first capture experiments can be identified as such.[9] We thus have *cohorts* of individuals that entered the study in the first year, the second year, and so on.

5.5.1 Modeling

A description of the basic Arnason–Schwarz model involves two types of variables for each individual i ($i = 1, \ldots, n$) in the population: first, a variable that describes the location of this individual,

$$\mathbf{z}_i = (z_{(i,t)}, t = 1, .., \tau),$$

where τ is the number of capture periods; and, second, a binary variable that describes the capture history of this individual,

$$\mathbf{x}_i = (x_{(i,t)}, t = 1, .., \tau).$$

[9] This is the case, for instance, with newborns or new mothers in animal capture experiments.

We[10] assume that $z_{(i,t)} = r$ means the animal i is alive in stratum r at time t and that $z_{(i,t)} = \dagger$ denotes the case when the animal i is dead at time t. The variable $\mathbf{z}_i$ is sometimes called the *migration* process of individual i by analogy with the special case where one is considering animals moving between geographical zones, like some northern birds in spring and fall. Note that $\mathbf{x}_i$ is entirely observed, while $\mathbf{z}_i$ is not. For instance, we may have

$$\mathbf{x}_i = 1\ 1\ 0\ 1\ 1\ 1\ 0\ 0\ 0$$

and

$$\mathbf{z}_i = 1\ 2\ \cdot\ 3\ 1\ 1\ \cdot\ \cdot\ \cdot\,,$$

for which a possible completed $\mathbf{z}_i$ is

$$\mathbf{z}_i = 1\ 2\ 1\ 3\ 1\ 1\ 2\ \dagger\ \dagger\,,$$

meaning that the animal died between the seventh and the eighth capture events. In particular, the Arnason–Schwarz model assumes that dead animals are never observed (although this type of assumption can easily be modified when processing the model, in what are called *tag-recovery experiments*). Therefore $z_{(i,t)} = \dagger$ always corresponds to $x_{(i,t)} = 0$.

Moreover, we assume that the $(\mathbf{x}_i, \mathbf{z}_i)$'s $(i = 1, \ldots, n)$ are independent and that each random vector $\mathbf{z}_i$ is a Markov chain taking values in $\mathfrak{K} \cup \{\dagger\}$ with uniform initial probability on $\mathfrak{K}$ (unless there is prior information to the contrary). The parameters of the Arnason–Schwarz model are thus of two kinds: the capture probabilities

$$p_t(r) = \mathbb{P}\left(x_{(i,t)} = 1 | z_{(i,t)} = r\right)$$

on the one hand and the transition probabilities

$$q_t(r,s) = \mathbb{P}\left(z_{(i,t+1)} = s | z_{(i,t)} = r\right) \quad r \in \mathfrak{K}, s \in \mathfrak{K} \cup \{\dagger\}, \quad q_t(\dagger, \dagger) = 1$$

on the other hand. We derive two further sets of parameters, $\varphi_t(r) = 1 - q_t(r, \dagger)$ the *survival* probabilities and $\psi_t(r,s)$ the interstrata *movement* probabilities, defined as

$$q_t(r,s) = \varphi_t(r) \times \psi_t(r,s) \quad r \in \mathfrak{K}, s \in \mathfrak{K}\,.$$

The likelihood corresponding to the complete observation of the $(\mathbf{x}_i, \mathbf{z}_i)$'s, $\ell(p_1, \ldots, p_T, q_1, \ldots, q_T | (\mathbf{x}_1, \mathbf{z}_1), \ldots, (\mathbf{x}_n, \mathbf{z}_n))$, is then given by

$$\prod_{i=1}^{n} \left[\prod_{t=1}^{T} p_t(z_{(i,t)})^{x_{(i,t)}} \{1 - p_t(z_{(i,t)})\}^{1-x_{(i,t)}} \times \prod_{t=1}^{T-1} q_t(z_{(i,t)}, z_{(i,t+1)}) \right], \quad (5.6)$$

[10]Covariates registered once or at each time will not be used here, although they could be introduced via a generalized linear model as in Chap. 4, so we abstain from adding further notations in an already dense section.

up to a constant. The complexity of the likelihood corresponding to the data actually observed is due to the fact that the $\mathbf{z}_i$'s are not fully observed, hence that (5.6) would have to be summed over all possible values of the missing components of the $\mathbf{z}_i$'s. This complexity can be bypassed by a simulation alternative described below in Sect. 5.5.2.

The prior modeling corresponding to these parameters will depend on the information that is available about the population covered by the capture–recapture experiment. For illustration's sake, consider the use of conjugate priors

$$p_t(r) \sim \mathscr{B}e(a_t(r), b_t(r)), \qquad \varphi_t(r) \sim \mathscr{B}e(\alpha_t(r), \beta_t(r)),$$

where the hyperparameters, $a_t(r)$, $b_t(r)$ and so on, depend on both time t and location r, and

$$\psi_t(r) \sim \mathscr{D}ir(\gamma_t(r)),$$

a Dirichlet distribution, where $\psi_t(r) = (\psi_t(r,s); s \in \mathfrak{K})$ with

$$\sum_{s \in \mathfrak{K}} \psi_t(r,s) = 1,$$

and $\gamma_t(r) = (\gamma_t(r,s); s \in \mathfrak{K})$. The determination of these (numerous) hyperparameters is also case-dependent and varies from a noninformative modeling, where all hyperparameters are taken to be equal to 1 or 1/2, to a very informative setting where exact values of these hyperparameters can be chosen from the prior information. The following example is an illustration of the latter.

Table 5.3. Prior information about the capture and survival parameters of the Arnason–Schwarz model, represented by prior expectation and prior confidence interval, for a capture–recapture experiment on the migrations of lizards (*source*: Dupuis, 1995)

	Episode	2	3	4	5	6
p_t	Mean	0.3	0.4	0.5	0.2	0.2
	95% cred. int.	$[0.1, 0.5]$	$[0.2, 0.6]$	$[0.3, 0.7]$	$[0.05, 0.4]$	$[0.05, 0.4]$

	Site	A		B,C	
	Episode	$t = 1, 3, 5$	$t = 2, 4$	$t = 1, 3, 5$	$t = 2, 4$
$\varphi_t(r)$	Mean	0.7	0.65	0.7	0.7
	95% cred. int.	$[0.4, 0.95]$	$[0.35, 0.9]$	$[0.4, 0.95]$	$[0.4, 0.95]$

Example 5.1. For the capture–recapture experiment described in Footnote 8 on the migrations of lizards between three adjacent zones, there are six capture episodes. The prior information provided by the biologists on the capture and survival probabilities, p_t (which are assumed to be zone independent) and $\varphi_t(r)$, is given by Table 5.3. While this may seem very artificial, this

construction of the prior distribution actually happened that way because the biologists in charge were able to quantify their beliefs and intuitions in terms of prior expectation and prior confidence interval. (The differences in the prior values on p_t are due to differences in capture efforts, while the differences between the group of episodes 1, 3 and 5, and the group of episodes 2 and 4 are due to the fact that the odd indices correspond to spring and the even indices to fall and mortality is higher over the winter.) Moreover, this prior information can be perfectly translated in a collection of beta priors by the R divide-and-conquer function

```
probet=function(a,b,c,alpha){

  coc=(1-c)/c
  pbeta(b,alpha,alpha*coc)-pbeta(a,alpha,alpha*coc)
  }

solbeta=function(a,b,c,prec=10^(-3)){

  coc=(1-c)/c
  detail=alpha=1
  while (probet(a,b,c,alpha)<.95) alpha=alpha+detail
  while (abs(probet(a,b,c,alpha)-.95)>prec){

    alpha=max(alpha-detail,detail/10)
    detail=detail/10
    while (probet(a,b,c,alpha)<.95) alpha=alpha+detail
    }
  list(alpha=alpha,beta=alpha*coc)
  }
```

(see Exercise 5.23 for details). Repeated calls to `solbeta` as in

```
> solbeta(.1,.5,.3,10^(-4))
[1]  5.45300 12.72367
```

then leads to the hyperparameters given in Table 5.4. ◀

Table 5.4. Hyperparameters of the beta priors corresponding to the information contained in Table 5.3 (*source*: Dupuis, 1995)

Episode	2	3	4	5	6
Dist.	$\mathscr{B}e(5, 13)$	$\mathscr{B}e(8, 12)$	$\mathscr{B}e(12, 12)$	$\mathscr{B}e(3.5, 14)$	$\mathscr{B}e(3.5, 14)$

Site	A		B	
Episode	$t = 1, 3, 5$	$t = 2, 4$	$t = 1, 3, 5$	$t = 2, 4$
Dist.	$\mathscr{B}e(6.0, 2.5)$	$\mathscr{B}e(6.5, 3.5)$	$\mathscr{B}e(6.0, 2.5)$	$\mathscr{B}e(6.0, 2.5)$

5.5.2 Gibbs Sampler

Given the presence of missing data in the Arnason–Schwarz model, a Gibbs sampler is a natural solution to handle the complexity of the likelihood. It needs to include simulation of the missing components in the vectors $\mathbf{z}_i$ in order to simulate the parameters from the full conditional distribution

$$\pi(\theta|\mathbf{x},\mathbf{z}) \propto \ell(\theta|\mathbf{x},\mathbf{z}) \times \pi(\theta)\,,$$

Algorithm 5.10 Arnason–Schwarz Gibbs Sampler
Iteration l ($l \geq 1$):

1. **Parameter simulation**
   Simulate $\theta^{(l)} \sim \pi(\theta|\mathbf{z}^{(l-1)},\mathbf{x})$ as $(t=1,\ldots,\tau)$,

$$p_t^{(l)}(r)|\mathbf{x},\mathbf{z}^{(l-1)} \sim \mathscr{B}e\left(a_t(r)+u_t(r), b_t(r)+v_t^{(l)}(r)\right),$$

$$\varphi_t^{(l)}(r)|\mathbf{x},\mathbf{z}^{(l-1)} \sim \mathscr{B}e\left(\alpha_t(r)+\sum_{j\in\mathfrak{K}} w_t^{(l)}(r,j), \beta_t(r)+w_t^{(l)}(r,\dagger)\right),$$

$$\psi_t^{(l)}(r)|\mathbf{x},\mathbf{z}^{(l-1)} \sim \mathscr{D}ir\left(\gamma_t(r,s)+w_t^{(l)}(r,s); s\in\mathfrak{K}\right),$$

   where

$$w_t^{(l)}(r,s)=\sum_{i=1}^n \mathbb{I}_{\left(z_{(i,t)}^{(l-1)}=r,\, z_{(i,t+1)}^{(l-1)}=s\right)},$$

$$u_t^{(l)}(r)=\sum_{i=1}^n \mathbb{I}_{\left(x_{(i,t)}=1,\, z_{(i,t)}^{(l-1)}=r\right)},$$

$$v_t^{(l)}(r)=\sum_{i=1}^n \mathbb{I}_{\left(x_{(i,t)}=0,\, z_{(i,t)}^{(l-1)}=r\right)}.$$

2. **Missing location simulation**
   Generate the unobserved $z_{(i,t)}^{(l)}$'s from the full conditional distributions

$$\mathbb{P}(z_{(i,1)}^{(l)}=s|x_{(i,1)},z_{(i,2)}^{(l-1)},\theta^{(l)}) \propto q_1^{(l)}(s,z_{(i,2)}^{(l-1)})(1-p_1^{(l)}(s)),$$

$$\mathbb{P}(z_{(i,t)}^{(l)}=s|x_{(i,t)},z_{(i,t-1)}^{(l)},z_{(i,t+1)}^{(l-1)},\theta^{(l)}) \propto q_{t-1}^{(l)}(z_{(i,t-1)}^{(l)},s) \times q_t(s,z_{(i,t+1)}^{(l-1)})(1-p_t^{(l)}(s)),$$

$$\mathbb{P}(z_{(i,\tau)}^{(l)}=s|x_{(i,\tau)},z_{(i,\tau-1)}^{(l)},\theta^{(l)}) \propto q_{\tau-1}^{(l)}(z_{(i,\tau-1)}^{(l)},s)(1-p_\tau(s)^{(l)}).$$

where $\mathbf{x}$ and $\mathbf{z}$ denote the collections of the vectors of capture indicators and locations, respectively. This is thus a particular case of *data augmentation*, where the missing data $\mathbf{z}$ are simulated at each step t in order to reconstitute a complete sample $(\mathbf{x}, \mathbf{z}^{(t)})$ for which conjugacy applies. In the setting of the Arnason–Schwarz model, we can simulate the full conditional distributions both of the parameters and of the missing components. The Gibbs sampler is as follows:

Note that simulating the missing locations in the $\mathbf{z}_i$'s conditionally on the other locations and on the parameters is not a very complex task because of the good conditioning properties of these vectors (which stem from their Markovian nature). As shown in Step 2 of Algorithm 5.10, the full conditional distribution of $z_{(i,t)}$ only depends on the previous and next locations $z_{(i,t-1)}$ and $z_{(i,t+1)}$ (and obviously on the fact that it is not observed; that is, that $x_{(i,t)} = 0$). The corresponding part of the R code is based on a `latent` matrix containing the current values of both the observed and missing locations:

```
for (i in 1:n){

  if (z[i,1]==0) latent[i,1]=sample(1:(m+1),1,
     prob=q[,latent[i,2]]*(1-c(p[s,],0)))
  for (t in ((2:(T-1))[z[i,-c(1.T)]==0]))
    latent[i,t]=sample(1:(m+1),1,
      prob=q[latent[i,t-1],]*q[,latent[i,t+1]]*(1-c(p[s,],0)))
  if (z[i,T]==0) latent[i,T]=sample(1:(m+1),1,
     prob=q[latent[i,T-1],]*(1-c(p[s,],0)))
  }
```

(The convoluted range for the inner loop replaces an `if (z[i,t]==0)`.) When the number of states $s \in \mathfrak{K}$ is moderate, it is straightforward to simulate from such a distribution.

Take $\mathfrak{K} = \{1, 2\}$, $n = 4$, $m = 8$ and assume that, for $\mathbf{x}$, we have the following histories:

$$\begin{array}{c|cccccccc}
1 & 1 & 1 & \cdot & \cdot & 1 & \cdot & \cdot & \cdot \\
2 & 1 & \cdot & 1 & \cdot & 1 & \cdot & 2 & 1 \\
3 & 2 & 1 & \cdot & 1 & 2 & \cdot & \cdot & 1 \\
4 & 1 & \cdot & \cdot & 1 & 2 & 1 & 1 & 2
\end{array}$$

Assume also that all (prior) hyperparameters are taken equal to 1. Then one possible instance of a simulated $\mathbf{z}$ is

$$\begin{array}{cccccccc}
1 & 1 & 1 & 2 & 1 & 1 & 2 & \dagger \\
1 & 1 & 1 & 2 & 1 & 1 & 1 & 2 \\
2 & 1 & 2 & 1 & 2 & 1 & 1 & 1 \\
1 & 2 & 1 & 1 & 2 & 1 & 1 & 2
\end{array}$$

and it leads to the following simulation of the parameters:

$$\begin{aligned}
p_4^{(l)}(1)|\mathbf{x}, \mathbf{z}^{(l-1)} &\sim \mathscr{B}e(1+2, 1+0)\,, \\
\varphi_7^{(l)}(2)|\mathbf{x}, \mathbf{z}^{(l-1)} &\sim \mathscr{B}e(1+0, 1+1)\,, \\
\psi_2^{(l)}(1,2)|\mathbf{x}, \mathbf{z}^{(l-1)} &\sim \mathscr{B}e(1+1, 1+2)\,,
\end{aligned}$$

in the Gibbs sampler, where the hyperparameters are therefore derived from the (partly) simulated history above. Note that because there are only two possible states, the Dirichlet distribution simplifies into a beta distribution.

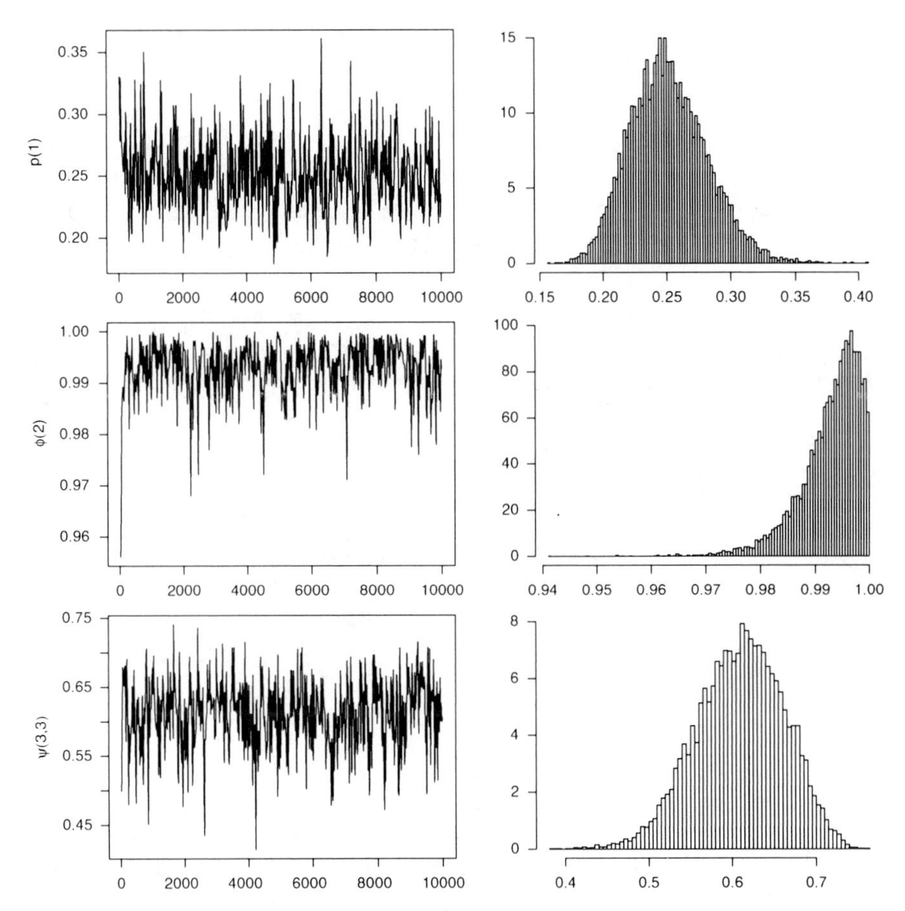

Fig. 5.5. Dataset **eurodip**: Representation of the Gibbs sampling output for some parameters of the Arnason–Schwarz model, based on 10,000 iterations, with raw plots (*first column*) and histograms (*second column*)

For **eurodip**, Lebreton et al. (1992) argue that the capture and survival rates should be constant over time. If we assume that the movement probabilities are also time independent, we are left with $3+3+3\times 2=12$ parameters. Figure 5.5 gives the Gibbs output for the parameters $p(1)$, $\varphi(2)$, and $\psi(3,3)$ using noninformative priors with $a(r)=b(r)=\alpha(r)=\beta(r)=\gamma(r,s)=1$. The simulation of the parameters is obtained by the following piece of R code, where s is the current index of the Gibbs iteration in the R code below:

```
for (r1 in 1:m){

  for (r2 in 1:(m+1))
    omega[r2]=sum(latent[,1:(T-1)]==r1 & latent[,2:T]==r2)
  u=sum(z!=0 & latent==r1)


  v=sum(z==0 & latent==r1)
  p[s,r1]=rbeta(1,1+u,1+v)
  phi[s,r1]=rbeta(1,1+sum(omega[1:m]),1+omega[m+1])
  psi[r1,,s]=rdirichlet(1,rep(1,m)+omega[1:m])
  }
```

The transition probabilities $q_t(r,s)$ are then reconstructed from the survival and movement probabilities, with the special case of the m+1 column corresponding to the absorbing † state:

```
tt=matrix(rep(phi[s,],m),m,byrow=T)
q=rbind(tt*psi[,,s],rep(0,m))
q=cbind(q,1-apply(q,1,sum))
```

The convergence of the Gibbs sampler to the region of interest occurs very quickly, even though we can spot an approximate periodicity in the raw plots on the left-hand side. The MCMC approximations of the estimates of $p(1)$, $\varphi(2)$, and $\psi(3,3)$, the empirical mean over the last 8,000 simulations, are equal to 0.25, 0.99, and 0.61, respectively.

5.6 Exercises

5.1 Show that the posterior distribution $\pi(N|n^+)$ given by (5.1), while associated with an improper prior, is defined for all values of n^+. Show that the normalization factor of (5.1) is $n^+ \vee 1$, and deduce that the posterior median is equal to $2(n^+ \vee 1)-1$. Discuss the relevance of this estimator and show that it corresponds to a Bayes estimate of p equal to $1/2$.

5.2 Under the same prior as in Sect. 5.2.1, derive the marginal posterior density of N in the case where $n_1^+ \sim \mathscr{B}(N,p)$ and

$$n_2^+, \ldots, n_k^+ \overset{\text{iid}}{\sim} \mathscr{B}(n_1^+, p)$$

are observed (the later are in fact recaptures). Apply to the sample

$$(n_1^+, n_2^+, \ldots, n_{11}^+) = (32, 20, 8, 5, 1, 2, 0, 2, 1, 1, 0)\,,$$

which describes a series of tag recoveries over 11 years.

5.3 Show that the conditional distribution of m_2 conditional on both sample sizes n_1 and n_2 is given by (5.2) and does not depend on p. Deduce the expectation $\mathbb{E}^\pi[m_2|n_1, n_2, N]$.

5.4 In order to determine the number N of buses in a town, a capture–recapture strategy goes as follows. We observe $n_1 = 20$ buses during the first day and keep track of their identifying numbers. Then we repeat the experiment the following day by recording the number of buses that have already been spotted on the previous day, say $m_2 = 5$, out of the $n_2 = 30$ buses observed the second day. For the Darroch model, give the posterior expectation of N under the prior $\pi(N) = 1/N$.

5.5 Show that the maximum likelihood estimator of N for the Darroch model is $\hat{N} = n_1/\left(m_2/n_2\right)$, and deduce that it is not defined when $m_2 = 0$.

5.6 Give the likelihood of the extension of Darroch's model when the capture–recapture experiments are repeated K times with capture sizes and recapture observations n_k $(1 \le k \le K)$ and m_k $(2 \le k \le K)$, respectively. (*Hint*: Exhibit first the two-dimensional sufficient statistic associated with this model.)

5.7 Give both conditional posterior distributions involved in Algorithm 5.8 in the case $n^+ = 0$.

5.8 Show that, for the two-stage capture model with probability p of capture, when the prior on N is a $\mathscr{P}(\lambda)$ distribution, the conditional posterior on $N - n^+$ is $\mathscr{P}(\lambda(1-p)^2)$.

5.9 Reproduce the analysis of **eurodip** summarized by Fig. 5.1 when switching the prior from $\pi(N,p) \propto \lambda^N/N!$ to $\pi(N,p) \propto N^{-1}$.

5.10 An extension of the T-stage capture–recapture model of Sect. 5.2.3 is to consider that the capture of an individual modifies its probability of being captured from p to q for future recaptures. Give the likelihood $\ell(N, p, q|n_1, n_2, m_2 \ldots, n_T, m_T)$.

5.11 Another extension of the two-stage capture–recapture model is to allow for mark loss.[11] If we introduce q as the probability of losing the mark, r as the probability of recovering a lost mark and k as the number of recovered lost marks, give the associated likelihood $\ell(N, p, q, r|n_1, n_2, m_2, k)$.

---

[11] Tags can be lost by marked animals, but the animals themselves could also be lost to recapture either by changing habitat or dying. Our current model assumes that the population is *closed*; that is, that there is no immigration, emigration, birth, or death within the population during the length of the study. These other kinds of extension are dealt with in Sects. 5.3 and 5.5.

5.12 Show that the conditional distribution of r_1 in the open population model of Sect. 5.3 is proportional to the product (5.4).

5.13 Show that the distribution of r_2 in the open population model of Sect. 5.3 can be integrated out from the joint distribution and that this leads to the following distribution on r_1:

$$\pi(r_1|p,q,n_1,c_2,c_3) \propto \frac{(n_1-r_1)!(n_1-r_1-c_3)!}{r_1!(n_1-r_1-c_2)!} \times \left(\frac{q}{(1-p)(1-q)[q+(1-p)(1-q)]}\right)^{r_1} .$$

Compare the computational cost of a Gibbs sampler based on this approach with a Gibbs sampler using the full conditionals.

5.14 Show that the likelihood associated with an open population as in Sect. 5.3 can be written as

$$\ell(N,p|\mathscr{D}^*) = \sum_{(\epsilon_{it},\delta_{it})_{it}} \prod_{t=1}^{T}\prod_{i=1}^{N} q_{\epsilon_{i(t-1)}}^{\epsilon_{it}} (1-q_{\epsilon_{i(t-1)}})^{1-\epsilon_{it}} \times p^{(1-\epsilon_{it})\delta_{it}}(1-p)^{(1-\epsilon_{it})(1-\delta_{it})},$$

where $q_0 = q$, $q_1 = 1$, and δ_{it} and ϵ_{it} are the capture and exit indicators, respectively. Derive the order of complexity of this likelihood; that is, the number of elementary operations necessary to compute it.[12]

5.15 In connection with the presentation of the accept–reject algorithm in Sect. 5.4, show that, for $M > 0$, if g is replaced with Mg in $\mathscr{S}$ and if (X, U) is uniformly distributed on $\mathscr{S}$, the marginal distribution of X is still g. Deduce that the density g only needs to be known up to a normalizing constant.

5.16 For the function $g(x) = (1+\sin^2(x))(2+\cos^4(4x))\exp[-x^4\{1+\sin^6(x)\}]$ on $[0, 2\pi]$, examine the feasibility of running a uniform sampler on the set $\mathscr{S}$ associated with the accept–reject algorithm in Sect. 5.4.

5.17 Show that the probability of acceptance in Step 2 of Algorithm 5.9 is $1/M$ and that the number of trials until a variable is accepted has a geometric distribution with parameter $1/M$. Conclude that the expected number of trials per simulation is M.

5.18 For the conditional distribution of α_t derived from (5.3), construct an accept–reject algorithm based on a normal bounding density f and study its performances for $N = 532$, $n_t = 118$, $\mu_t = -0.5$, and $\sigma^2 = 3$.

5.19 When uniform simulation on the accept–reject set $\mathscr{S}$ of Sect. 5.4 is impossible, construct a Gibbs sampler based on the conditional distributions of u and x. (*Hint*: Show that both conditionals are uniform distributions.) This special case of the Gibbs sampler is called the *slice sampler* (see Robert and Casella, 2004, Chap. 8). Apply to the distribution of Exercise 5.16.

[12] We will see in Chap. 7 a derivation of this likelihood that enjoys an $O(T)$ complexity.

5.20 Show that the normalizing constant M of a target density f can be deduced from the acceptance rate in the accept–reject algorithm (Algorithm 5.9) under the assumption that g is properly normalized.

5.21 Reproduce the analysis of Exercise 5.20 for the marginal distribution of r_1 computed in Exercise 5.13.

5.22 Modify the function `ardipper` used in Sect. 5.4 to return the acceptance rate as well as a sample from the target distribution.

5.23 Show that, given a mean and a 95% confidence interval in $[0,1]$, there exists at most one beta distribution $\mathscr{B}e(a,b)$ with such a mean and confidence interval.

5.24 Show that, for the Arnason–Schwarz model, groups of consecutive unknown locations are independent of one another, conditional on the observations. Devise a way to simulate these groups by blocks rather than one at a time; that is, using the joint posterior distributions of the groups rather than the full conditional distributions of the states.

6

Mixture Models

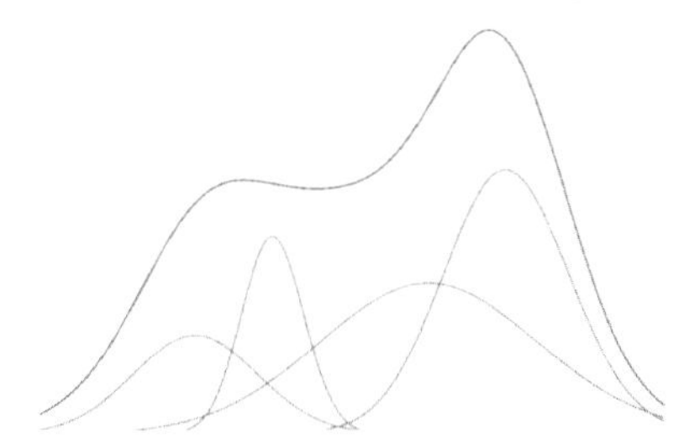

I must have missed something.
—**Ian Rankin, *The Hanging Garden*.**—

Roadmap

This chapter covers a class of models where a rather simple distribution is made more complex and less informative by a mechanism that mixes together several known or unknown distributions. This representation is naturally called a mixture of distributions, as illustrated above. Inference about the parameters of the elements of the mixtures and the weights is called mixture estimation, while recovery of the original distribution of each observation is called classification (or, more exactly, unsupervised classification to distinguish it from the supervised classification to be discussed in Chap. 8).

Both aspects almost always require advanced computational tools since even the representation of the posterior distribution may be complicated. Typically, Bayesian inference for these models was not correctly treated until the introduction of MCMC algorithms in the early 1990s. This chapter also covers the case of a mixture with an unknown number of components, for which a specific approximation of the Bayes factor was designed by Chib (1995).

J.-M. Marin and C.P. Robert, *Bayesian Essentials with R*, Springer Texts in Statistics, DOI 10.1007/978-1-4614-8687-9_6,

6.1 Missing Variable Models

In some cases, the complexity of a model originates from the fact that some piece of information about an original and more standard (simpler) model is *missing*. For instance, we have encountered a missing variable model in Chap. 5 with the Arnason–Schwarz model (Sect. 5.5), where the fact of ignoring the characteristics of the individuals outside their capture periods makes inference much harder. Similarly, we have seen in Chap. 4 that the probit model can be reinterpreted as a missing-variable model in that we only observe the sign of a normal variable.

Formally, all models that are defined via a marginalization mechanism, that is, such that the density of the observables $\mathbf{x}$, $f(\mathbf{x}|\theta)$, is given by an integral

$$f(\mathbf{x}|\theta) = \int_{\mathscr{Z}} g(\mathbf{x}, \mathbf{z}|\theta)\,\mathrm{d}\mathbf{z}\,, \tag{6.1}$$

can be considered as belonging to a *missing variable* (or *missing data*) model.[1]

This chapter focus on the case of the mixture model, which is *the* archetypical missing-variable model in that its simple representation (and interpretation) is mirrored by a need for complex processing. Later, in Chap. 7, we will also discuss *hidden Markov models* that add to the missing structure a temporal dependence dimension.

Although image analysis is the topic of Chap. 8, the dataset used in this chapter is derived from an image of a license plate, called **License** and not available in bayess, as

```
> image(license,col=grey(0:255/255),axes=FALSE,xlab="",
        ylab="")
```

represented in Fig. 6.1 (top). The actual histogram of the grey levels is concentrated on 256 values because of the poor resolution of the image, but we transformed the original data as

```
> license=scan("license.txt")
> license=jitter(license,10)
> datha=log((license-min(license)+.01)/
+ (max(license)+.01-license))
```

where `jitter` is used to randomize the dataset and avoid repetitions (as already described on page 156). The second line of code is a logit transform.

[1]This is not a definition in the mathematical sense since all densities can formally be represented that way. We thus stress that the model itself must be introduced that way. This point is not to be mistaken for a requirement that the variable $\mathbf{z}$ be meaningful for the data at hand. In many cases, for instance the probit model, the missing variable representation remains formal.

The transformed data used in this chapter has been stored in the file `datha.txt`.

```
> data(datha)
> datha=as.matrix(datha)
> hist(datha,nclas=200,xlab="",xlim=c(min(datha),max(datha)),
       ylab="",prob=TRUE,main="")
```

As seen from Fig. 6.1 (bottom), the resulting structure of the data is compatible with a sample from a mixture of several normal distributions (with at least two components). We point out at this early stage that mixture modeling is often used in image smoothing. Unsurprisingly, the current plate image would instead require feature recognition, for which this modeling does not help, because it requires spatial coherence and thus more complicated models that will be presented in Chap. 8.

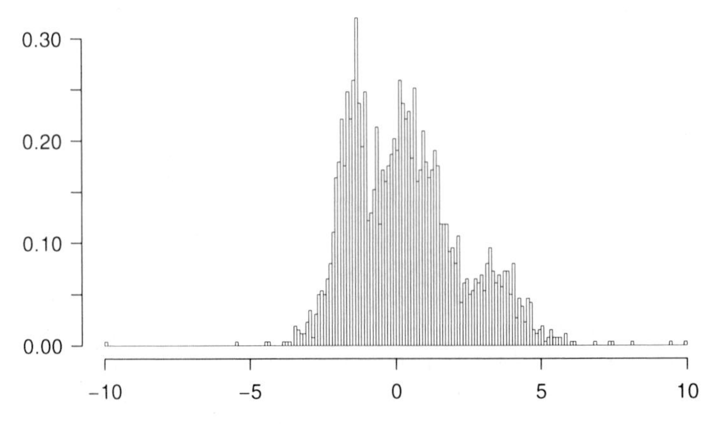

Fig. 6.1. Dataset **License**: (*top*) Image of a car license plate and (*bottom*) histogram of the transformed grey levels of the dataset

6.2 Finite Mixture Models

We now introduce the specific case of mixtures as it exemplifies the complexity of missing-variable models, both by its nature (in the sense that it is inherently linked with a missing variable) and by its processing, which also requires the incorporation of the missing structure.[2]

A *mixture distribution* is a convex combination

$$\sum_{j=1}^{k} p_j f_j(x), \qquad p_j \geq 0, \qquad \sum_{j=1}^{k} p_j = 1,$$

of k distributions f_j ($k > 1$). In the simplest situations, the f_j's are known and inference focuses either on the unknown proportions p_j or on the allocations of the points of the sample $(x_1, \ldots, x_n)$ to the components f_j, i.e. on the probability that x_i is generated from f_j by opposition to being generated from f_ℓ, say. In most cases, however, the f_j's are from a parametric family like the normal or Beta distributions, with unknown parameters θ_j, leading to the mixture model

$$\sum_{j=1}^{k} p_j f(x|\theta_j), \tag{6.2}$$

with parameters including both the weights p_j and the component parameters θ_j ($j = 1, \ldots, k$). It is actually relevant to distinguish the weights p_j from the other parameters in that they are solely associated with the missing-data structure of the model, while the others are related to the observations. This distinction is obviously irrelevant in the computation of the likelihood function or in the construction of the prior distribution, but it matters in the interpretation of the posterior output, for instance.

There are several motivations for considering mixtures of distributions as a useful extension to "standard" distributions. The most natural approach is to envisage a dataset as made of several latent (that is, missing, unobserved) strata or subpopulations. For instance, one of the earliest occurrences of mixture modeling can be found in Bertillon (1887),[3] where the bimodal structure of the heights of (military) conscripts in central France (Doubs) can be explained a posteriori by the aggregation of two populations of young men, one from the plains and one from the mountains. The mixture structure appears because the origin of each observation (that is, the allocation to a specific subpopulation or stratum) is lost. In the example of the military conscripts, this means that the geographical origin of each young man was not recorded.

---

[2]We will see later that the missing structure of a mixture actually *need* not be simulated but, for more complex missing-variable structures like hidden Markov models (introduced in Chap. 7), this completion cannot be avoided.

[3]The Frenchman Alphonse Bertillon is also the father of scientific police investigation. For instance, he originated the use of fingerprints in criminal investigations.

Depending on the setting, the inferential goal associated with a sample from a mixture of distributions may be either to reconstitute the groups by estimating the missing component z, an operation usually called classification (or *clustering*), to provide estimators for the parameters of the different groups, or even to estimate the number k of groups.

A completely different (if more involved) approach to the interpretation and estimation of mixtures is the *semiparametric* perspective. This approach considers that since very few phenomena obey probability laws corresponding to the most standard distributions, mixtures such as (6.2) can be seen as a good trade-off between fair representation of the phenomenon and efficient estimation of the underlying distribution. If k is large enough, there is theoretical support for the argument that (6.2) provides a good approximation (in some functional sense) to most distributions. Hence, a mixture distribution can be perceived as a type of (functional) basis approximation of unknown distributions, in a spirit similar to wavelets and splines, but with a more intuitive flavor (for a statistician at least). However, this chapter mostly focuses on the "parametric" case, that is, on situations when the partition of the sample into subsamples with different distributions f_j does make sense from the dataset or modelling point of view (even though the computational processing is the same in both cases). In other words, we consider settings where clustering the sample into strata or subpopulations is of interest.

6.3 Mixture Likelihoods and Posteriors

Let us consider an iid sample $\mathbf{x} = (x_1, \ldots, x_n)$ from model (6.2). The likelihood is such that

$$\ell(\boldsymbol{\theta}, \mathbf{p}|\mathbf{x}) = \prod_{i=1}^{n} \sum_{j=1}^{k} p_j \, f(x_i|\theta_j) \,.$$

This likelihood contains k^n terms when the inner sums are expanded. While this expansion is not necessary for computing the likelihood at a given value $(\boldsymbol{\theta}, \mathbf{p})$, a computation that is feasible in $\mathrm{O}(nk)$ operations as demonstrated by the representation in Fig. 6.2, it remains a necessary step in the understanding of the mixture structure. Alas, the computational difficulty in using the expanded version precludes analytic solutions for either maximum likelihood or Bayes estimators.

Example 6.1. Consider the simple case of a two-component normal mixture

$$p\,\mathscr{N}(\mu_1, 1) + (1-p)\,\mathscr{N}(\mu_2, 1)\,, \tag{6.3}$$

where the weight $p \neq 0.5$ is known. The likelihood surface can be computed by an R code as in the following `plotmix` function, which relies on the `image` function and a discretization of the (μ_1, μ_2) space into pixels. Given a sample sampl that is generated in the first lines of the function, the log-likelihood surface is computed by

```
pbar=1-p
mu1=mu2=seq(min(sampl),max(sampl),.1)
mo1=mu1%*%t(rep(1,length(mu2)))
mo2=rep(1,length(mu2))%*%t(mu2)
ca1=-0.5*mo1*mo1
ca2=-0.5*mo2*mo2
like=0*mo1
for (i in 1:n)
   like=like+log(p*exp(ca1+sampl[i]*mo1)+
   pbar*exp(ca2+sampl[i]*mo2))
like=like+.1*(ca1+ca2)
```

and plotted by

```
image(mu1,mu2,like,xlab=expression(mu[1]),
  ylab=expression(mu[2]),col=heat.colors(250))
contour(mu1,mu2,like,levels=seq(min(like),max(like),lengthl),
  add=TRUE,drawlabels=FALSE)
```

We note that the outcome of the plotmix function is the list list(sample=sampl,like=like), used in subsequent analyses of the data. For instance, this outcome, including the level sets obtained by contour, is provided in Fig. 6.2. In this case, the parameters are identifiable: μ_1 cannot be confused with μ_2 when p is different from 0.5. Nonetheless, the log-likelihood surface in this figure exhibits two modes, one being close to the true value of the parameters used to simulate the dataset and one corresponding to an inverse separation of the dataset into two groups.[4] ◀

For any prior $\pi(\boldsymbol{\theta}, \mathbf{p})$, the posterior distribution of $(\boldsymbol{\theta}, \mathbf{p})$ is available up to a multiplicative constant:

$$\pi(\boldsymbol{\theta}, \mathbf{p}|\mathbf{x}) \propto \left[\prod_{i=1}^{n}\sum_{j=1}^{k} p_j\, f(x_i|\theta_j)\right] \pi(\boldsymbol{\theta}, \mathbf{p}) \,. \tag{6.4}$$

While $\pi(\boldsymbol{\theta}, \mathbf{p}|\mathbf{x})$ can thus be computed for a given value of $(\boldsymbol{\theta}, \mathbf{p})$ at a cost of order $O(kn)$, we now explain why the derivation of the posterior characteristics, and in particular of posterior expectations of quantities of interest, is only possible in an exponential time of order $O(k^n)$.

To explain this difficulty in more detail, we consider the rather intuitive missing-variable representation of mixture models: With each x_i is associated

[4] To get a better understanding of this second mode, consider the limiting setting when $p = 0.5$. In that case, there are two equivalent modes of the likelihood, (μ_1, μ_2) and (μ_2, μ_1). As p moves away from 0.5, this second mode gets lower and lower compared with the other mode, but it still remains.

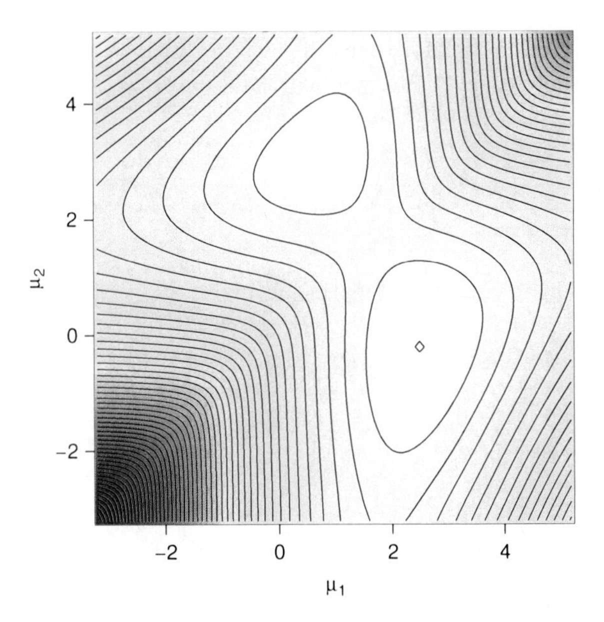

Fig. 6.2. R image representation of the log-likelihood of the mixture (6.3) for a simulated dataset of 500 observations and a true value $(\mu_1, \mu_2, p) = (2.5, 0, 0.7)$. Besides a mode (represented by a *diamond*) in the neighborhood of the true value, the R contour function exhibits an additional mode on the likelihood surface

a missing variable z_i that indicates "its" component, i.e. the index z_i of the distribution from which it was generated. Formally, this means that we have a hierarchical structure associated with the model:

$$z_i|\mathbf{p} \sim \mathscr{M}_k(p_1, \ldots, p_k)$$

and

$$x_i|z_i, \boldsymbol{\theta} \sim f(\cdot|\theta_{z_i}) \,.$$

The completed likelihood corresponding to the missing structure is such that

$$\ell(\boldsymbol{\theta}, \mathbf{p}|\mathbf{x}, \mathbf{z}) = \prod_{i=1}^{n} p_{z_i} \, f(x_i|\theta_{z_i})$$

and

$$\pi(\boldsymbol{\theta}, \mathbf{p}|\mathbf{x}, \mathbf{z}) \propto \left[\prod_{i=1}^{n} p_{z_i} \, f(x_i|\theta_{z_i})\right] \pi\left(\boldsymbol{\theta}, \mathbf{p}\right),$$

where $\mathbf{z} = (z_1, \ldots, z_n)$. If we denote by $\mathcal{Z} = \{1, \ldots, k\}^n$ the set of the k^n possible values of the vector $\mathbf{z}$, we can decompose $\mathcal{Z}$ into a partition of subsets

$$\mathcal{Z} = \cup_{j=1}^{\mathfrak{r}} \mathcal{Z}_j$$

as follows (see Exercise 6.2 for the value of $\mathfrak{r}$: For a given allocation size vector $(n_1, \ldots, n_k)$, where $n_1 + \ldots + n_k$, i.e. a given number of observations allocated to each component, we define the *partition sets*

$$\mathcal{Z}_j = \left\{ \mathbf{z} : \sum_{i=1}^{n} \mathbb{I}_{z_i=1}, \ldots, \sum_{i=1}^{n} \mathbb{I}_{z_i=k} \right\},$$

which consist of all allocations with the given allocation size $(n_1, \ldots, n_k)$. We label those partition sets with $j = j(n_1, \ldots, n_k)$ by using, for instance, the lexicographical ordering on the $(n_1, \ldots, n_k)$'s. (This means that $j = 1$ corresponds to $(n_1, \ldots, n_k) = (n, 0, \ldots, 0)$, $j = 2$ to $(n_1, \ldots, n_k) = (n - 1, 1, \ldots, 0)$, $j = 3$ to $(n_1, \ldots, n_k) = (n - 1, 0, 1, \ldots, 0)$, and so on). Using this partition, the posterior distribution of $(\boldsymbol{\theta}, \mathbf{p})$ can be written in closed form as

$$\pi(\boldsymbol{\theta}, \mathbf{p}|\mathbf{x}) = \sum_{\mathbf{z} \in \mathcal{Z}} \pi(\boldsymbol{\theta}, \mathbf{p}|\mathbf{x}, \mathbf{z}) = \sum_{i=1}^{\mathfrak{r}} \sum_{\mathbf{z} \in \mathcal{Z}_i} \omega(\mathbf{z}) \, \pi(\boldsymbol{\theta}, \mathbf{p}|\mathbf{x}, \mathbf{z}), \tag{6.5}$$

where $\omega(\mathbf{z})$ represents the marginal posterior probability of the allocation $\mathbf{z}$ conditional on the observations $\mathbf{x}$ (derived by integrating out the parameters $\boldsymbol{\theta}$ and $\mathbf{p}$). With this representation, a Bayes estimator of $(\boldsymbol{\theta}, \mathbf{p})$ can also be written in closed form as

$$\mathbb{E}^{\pi}[\boldsymbol{\theta}, p|\mathbf{x}] = \sum_{i=1}^{\mathfrak{r}} \sum_{\mathbf{z} \in \mathcal{Z}_i} \omega(\mathbf{z}) \, \mathbb{E}^{\pi}[\boldsymbol{\theta}, \mathbf{p}|\mathbf{x}, \mathbf{z}] .$$

Continuation of Example 6.1. In the special case of model (6.3), if we take two different independent normal priors on both means,

$$\mu_1 \sim \mathcal{N}(0, 4), \quad \mu_2 \sim \mathcal{N}(2, 4),$$

the posterior weight of a given allocation vector $\mathbf{z}$ is

$$\begin{aligned}\omega(\mathbf{z}) \propto & \sqrt{(n_1 + 1/4)(n - n_1 + 1/4)}\, p^{n_1}(n_1 - p)^{n-l} \\ & \times \exp\left\{-[(n_1 + 1/4)\hat{s}_1(\mathbf{z}) + n_1\{\bar{x}_1(\mathbf{z})\}^2/4]/2\right\} \\ & \times \exp\left\{-[(n - n_1 + 1/4)\hat{s}_2(\mathbf{z}) + (n - n_1)\{\bar{x}_2(\mathbf{z}) - 2\}^2/4]/2\right\},\end{aligned}$$

$$\bar{x}_1(\mathbf{z}) = \frac{1}{n_1} \sum_{i=1}^{n} \mathbb{I}_{z_i=1} x_i, \quad \bar{x}_2(\mathbf{z}) = \frac{1}{n - n_1} \sum_{i=1}^{n} \mathbb{I}_{z_i=2} x_i,$$

$$\hat{s}_1(\mathbf{z}) = \sum_{i=1}^{n} \mathbb{I}_{z_i=1} (x_i - \bar{x}_1(\mathbf{z}))^2, \quad \hat{s}_2(\mathbf{z}) = \sum_{i=1}^{n} \mathbb{I}_{z_i=2} (x_i - \bar{x}_2(\mathbf{z}))^2$$

(if we set $\bar{x}_1(\mathbf{z}) = 0$ when $n_1 = 0$ and $\bar{x}_2(\mathbf{z}) = 0$ when $n - n_1 = 0$). Implementing this derivation in R is quite straightforward:

```
omega=function(z,x,p){
  n=length(x)
  n1=sum(z==1);n2=n-n1
  if (n1==0) xbar1=0 else xbar1=sum((z==1)*x)/n1
  if (n2==0) xbar2=0 else xbar2=sum((z==2)*x)/n2
  ss1=sum((z==1)*(x-xbar1)^2)
  ss2=sum((z==2)*(x-xbar2)^2)
  return(sqrt((n1+.25)*(n2+.25))*p^n1*(1-p)^n2*
     exp(-((n1+.25)*ss1+(n2+.25)*ss2)/2)*
     exp(-(n1*xbar1^2+n2*xbar2)/8))
  }
```

leading for instance to

```
> omega(z=sample(1:2,4,rep=TRUE),
+ x=plotmix(n=4,plot=FALSE)$samp,p=.8)
[1] 0.0001781843
> omega(z=sample(1:2,4,rep=TRUE),
+ x=plotmix(n=4,plot=FALSE)$sample,p=.8)
[1] 5.152284e-09
```

Note that the `omega` function is not and cannot be normalized, so the values must be interpreted on a relative scale. ◀

The decomposition (6.5) makes a lot of sense from an inferential point of view. The posterior distribution simply considers each possible partition $\mathbf{z}$ of the dataset, then allocates a posterior probability $\omega(\mathbf{z})$ to this partition, and at last constructs a posterior distribution for the parameters conditional on this allocation. Unfortunately, the computational burden resulting from this decomposition is simply too intensive because there are k^n terms in the sum.

However, there exists a solution that overcomes this computational problem. It uses an MCMC approach that takes advantage of the missing-variable structure and removes the requirement to explore the k^n possible values of $\mathbf{z}$ by only looking at the most likely ones.

Although this is beyond the scope of the book, let us point out here that there also exists in the statistical literature a technique that predates MCMC simulation algorithms but still relates to the same missing-data structure and completion mechanism. It is called the *EM Algorithm*[5] and consists of an iterative but deterministic sequence of "E" (for *expectation*) and "M" (for *maximization*) steps that converge to a local maximum of the likelihood. At iteration t, the "E" step corresponds to the computation of the function

$$Q\{(\boldsymbol{\theta}^{(t)}, \mathbf{p}^{(t)}), (\boldsymbol{\theta}, \mathbf{p})\} = \mathbb{E}_{(\boldsymbol{\theta}^{(t)}, \mathbf{p}^{(t)})} \left[\log \ell(\boldsymbol{\theta}, \mathbf{p}|\mathbf{x}, \mathbf{z})|\mathbf{x}\right],$$

[5]In non-Bayesian statistics, the EM algorithm is certainly the most ubiquitous numerical method, even though it only applies to (real or artificial) missing variable models.

where the likelihood $\ell(\boldsymbol{\theta}, \mathbf{p}|\mathbf{x}, \mathbf{z})$ is the joint distribution of $\mathbf{x}$ and $\mathbf{z}$, while the expectation is computed under the conditional distribution of $\mathbf{z}$ given $\mathbf{x}$ and the value $(\boldsymbol{\theta}^{(t)}, \mathbf{p}^{(t)})$ for the parameter. The "M" step corresponds to the maximization of $Q((\boldsymbol{\theta}^{(t)}, \mathbf{p}^{(t)}), (\boldsymbol{\theta}, \mathbf{p}))$ in $(\boldsymbol{\theta}, \mathbf{p})$, with solution $(\boldsymbol{\theta}^{(t+1)}, \mathbf{p}^{(t+1)})$. As we will see in Sect. 6.4, the Gibbs sampler takes advantage of exactly the same conditional distribution. Further details on EM and its Monte Carlo versions (namely, when the "E" step is not analytically feasible) are given in Robert and Casella (2004, Chap. 5; 2009, Chap. 5).

6.4 MCMC Solutions

For the joint distribution (6.4), the full conditional distribution of $\mathbf{z}$ given $\mathbf{x}$ and the parameters is always available as

$$\pi(\mathbf{z}|\mathbf{x}, \boldsymbol{\theta}, \mathbf{p}) \propto \prod_{i=1}^{n} p_{z_i} f(x_i|\theta_{z_i})$$

and can thus be computed at a cost of $\mathrm{O}(n)$. Since, for standard distributions $f(\cdot|\theta)$, the full posterior conditionals are also easily simulated when using conjugate priors, this implies that the Gibbs sampler can be derived in this setting.[6]

If $\mathbf{p}$ and $\boldsymbol{\theta}$ are independent a priori, then, given $\mathbf{z}$, the vectors $\mathbf{p}$ and $\mathbf{x}$ are independent; that is,

$$\pi(\mathbf{p}|\mathbf{z}, \mathbf{x}) \propto \pi(\mathbf{p}) f(\mathbf{z}|\mathbf{p}) f(\mathbf{x}|\mathbf{z}) \propto \pi(\mathbf{p}) f(\mathbf{z}|\mathbf{p}) \propto \pi(\mathbf{p}|\mathbf{z})\,.$$

Moreover, in that case, $\boldsymbol{\theta}$ is also independent a posteriori from $\mathbf{p}$ given $\mathbf{z}$ and $\mathbf{x}$, with density $\pi(\boldsymbol{\theta}|\mathbf{z}, \mathbf{x})$. If we apply the Gibbs sampler in this problem, it involves the successive simulation of $\mathbf{z}$ and $(\mathbf{p}, \boldsymbol{\theta})$ conditional on one another and on the data:

[6] Historically, missing-variable models constituted one of the first instances where the Gibbs sampler was used by completing the missing variables by simulation under the name of *data augmentation* (see Tanner, 1996, and Robert and Casella, 2004, Chaps. 9 and 10).

Algorithm 6.11 Mixture Gibbs Sampler

Initialization: Choose $\mathbf{p}^{(0)}$ and $\boldsymbol{\theta}^{(0)}$ arbitrarily.
Iteration t $(t \geq 1)$:

1. For $i = 1, \ldots, n$, generate $z_i^{(t)}$ such that

$$\mathbb{P}\left(z_i = j | \boldsymbol{\theta}, \mathbf{p}\right) \propto p_j^{(t-1)} f\left(x_i | \theta_j^{(t-1)}\right) .$$

2. Generate $\mathbf{p}^{(t)}$ according to $\pi(\mathbf{p}|\mathbf{z}^{(t)})$.
3. Generate $\boldsymbol{\theta}^{(t)}$ according to $\pi(\boldsymbol{\theta}|\mathbf{z}^{(t)}, \mathbf{x})$.

The simulation of the p_j's is also generally obvious since there exists a conjugate prior (as detailed below). In contrast, the complexity in the simulation of the θ_j's will depend on the type of sampling density $f(\cdot|\theta)$ as well as the prior π.

The marginal (sampling) distribution of the z_i's is a multinomial distribution $\mathscr{M}_k(p_1, \ldots, p_k)$, which allows for a conjugate prior on $\mathbf{p}$, namely the Dirichlet distribution $\mathbf{p} \sim \mathscr{D}(\gamma_1, \ldots, \gamma_k)$, with density

$$\frac{\Gamma(\gamma_1 + \ldots + \gamma_k)}{\Gamma(\gamma_1) \cdots \Gamma(\gamma_k)} p_1^{\gamma_1} \cdots p_k^{\gamma_k}$$

on the simplex of $\mathbb{R}^k$,

$$\mathscr{S}_k = \left\{ (p_1, \ldots, p_k) \in [0,1]^k ; \sum_{j=1}^k p_j = 1 \right\} .$$

In this case, denoting $n_j = \sum_{l=1}^n \mathbb{I}_{z_l = j}$ $(1 \leq j \leq k)$ the allocation sizes, the posterior distribution of $\mathbf{p}$ given $\mathbf{z}$ is

$$\mathbf{p}|\mathbf{z} \sim \mathscr{D}(n_1 + \gamma_1, \ldots, n_k + \gamma_k) .$$

It is rather peculiar that, despite its importance for Bayesian statistics, the Dirichlet distribution is not available in R (at least in the standard stat package). It is however fairly straightforward to code, using a representation based on gamma variates, as shown below.

```
rdirichlet=function(n=1,par=rep(1,2)){

  k=length(par)
  mat=matrix(0,n,k)
  for (i in 1:n){
```

```
    sim=rgamma(k,shape=par,scale=1)
    mat[i,]=sim/sum(sim)
    }
  mat
  }
```

When the density $f(\cdot|\theta)$ also allows for conjugate priors, the simulation of $\boldsymbol{\theta}$ can be specified further since an independent conjugate prior on each θ_j leads to independent and conjugate posterior distributions on the θ_j's, given $\mathbf{z}$ and $\mathbf{x}$.

Continuation of Example 6.1. For the mixture (6.3), under independent normal priors $\mathcal{N}(\delta, 1/\lambda)$ (both $\delta \in \mathbb{R}$ and $\lambda > 0$ are fixed hyperparameters) on both μ_1 and μ_2, the parameters μ_1 and μ_2 are independent given $(\mathbf{z}, \mathbf{x})$, with conditional distributions

$$\mathcal{N}\left(\frac{\lambda\delta + n_1\bar{x}_1(\mathbf{z})}{\lambda + n_1}, \frac{1}{\lambda + n_1}\right) \quad \text{and} \quad \mathcal{N}\left(\frac{\lambda\delta + (n - n_1)\bar{x}_2(\mathbf{z})}{\lambda + n - n_1}, \frac{1}{\lambda + n - n_1}\right),$$

respectively. Similarly, the conditional posterior distribution of the z_i's given (μ_1, μ_2) is $(i = 1, \ldots, n)$

$$\mathbb{P}(z_i = 1|\mu_1, x_i) \propto p \exp\left(-0.5\,(x_i - \mu_1)^2\right).$$

We can thus construct an R function like the following one to generate a sample from the posterior distribution: assuming $\delta = 0$ and $\lambda = 1$,

```
gibbsmean=function(p,datha,niter=10^4){

  n=length(datha)
  z=rep(0,n); ssiz=rep(0,2)
  nxj=rep(0,2)
  mug=matrix(mean(datha),nrow=niter+1,ncol=2)

  for (i in 2:(niter+1)){
    for (t in 1:n){
      prob=c(p,1-p)*dnorm(datha[t],mean=mug[i-1,])
      z[t]=sample(c(1,2),size=1,prob=prob)
      }
    for (j in 1:2){
      ssiz[j]=1+sum(z==j)
      nxj[j]=sum(as.numeric(z==j)*datha)
      }
    mug[i,]=rnorm(2,mean=nxj/ssiz,sd=sqrt(1/ssiz))
    }
  mug
  }
```

which can be used as

```
> dat=plotmix()$sample
> simu=gibbsmean(0.7,dat)
> points(simu,pch=".")
```

to produce Figs. 6.3 and 6.4. This R code illustrates two possible behaviors of this algorithm if we use a simulated dataset of 500 points from the mixture $0.7\mathcal{N}(0,1) + 0.3\mathcal{N}(2.5,1)$, which corresponds to the level sets on both pictures. The starting point in both cases is located at the saddle point between the two modes, i.e. at an instable equilibrium. Depending on the very first (random) iterations of the algorithm, the final sample may end up located on the upper or on the lower mode. For instance, in Fig. 6.3, the Gibbs sample based on 10,000 iterations is in agreement with the likelihood surface, since the second mode discussed in Example 6.1 is much lower than the mode where the simulation output concentrates. In Fig. 6.4, the Gibbs sample ends up being trapped by this lower mode. ◀

Example 6.2. If we consider the more general case of a mixture of two normal distributions with all parameters unknown,

$$p\mathcal{N}(\mu_1, \sigma_1^2) + (1-p)\mathcal{N}(\mu_2, \sigma_2^2),$$

and for the conjugate prior distribution ($j = 1, 2$)

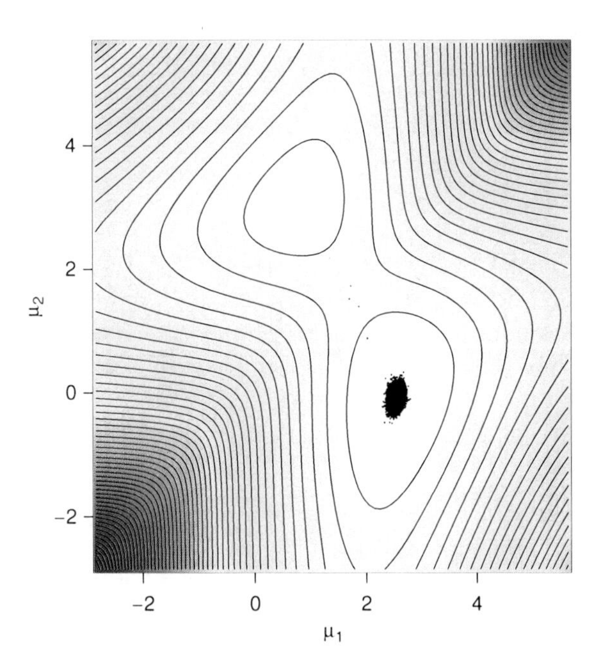

Fig. 6.3. Log-likelihood surface and the corresponding Gibbs sample for the model (6.3), based on 10,000 iterations

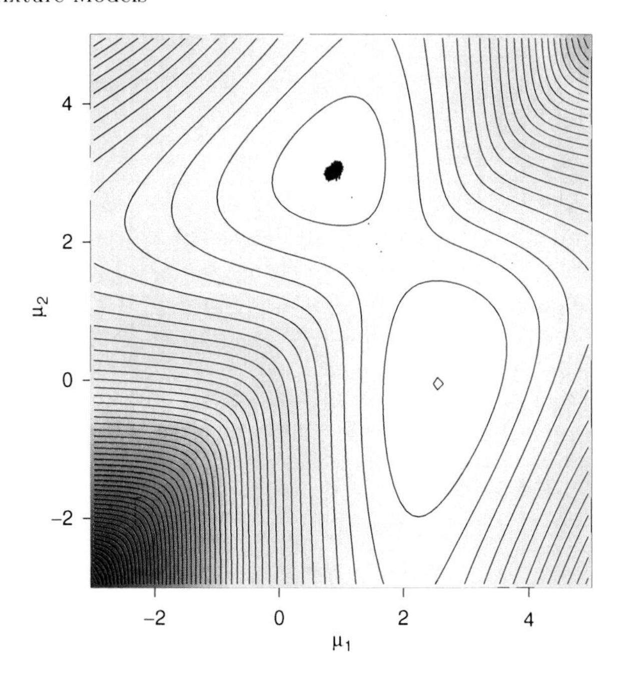

Fig. 6.4. Same legend as Fig. 6.3, with the same starting point located at the saddle point. In this instance, the Gibbs sample ends up around a lower mode

$$\mu_j|\sigma_j \sim \mathscr{N}(\xi_j, \sigma_j^2/l_j), \quad \sigma_j^2 \sim \mathscr{IG}(\nu_j/2, s_j^2/2), \quad p \sim \mathscr{Be}(\alpha, \beta),$$

the same decomposition conditional on $\mathbf{z}$ and straightforward (if dreary) algebra imply that

$$p|\mathbf{x}, \mathbf{z} \sim \mathscr{Be}(\alpha + n_1, \beta + n_2),$$

$$\mu_j|\sigma_j, \mathbf{x}, \mathbf{z} \sim \mathscr{N}\left(\xi_1(\mathbf{z}), \frac{\sigma_j^2}{n_j + l_j}\right), \quad \sigma_j^2|\mathbf{x}, \mathbf{z} \sim \mathscr{IG}((\nu_j + n_j)/2, s_j(\mathbf{z})/2),$$

where n_j is the number of z_i equal to j, $\bar{x}_j(\mathbf{z})$ and $\hat{s}_j^2(\mathbf{z})$ are the empirical mean and variance (biased) for the subsample with z_i equal to j, and

$$\xi_j(\mathbf{z}) = \frac{l_j \xi_j + n_j \bar{x}_j(\mathbf{z})}{l_j + n_j}, \quad s_j(\mathbf{z}) = s_j^2 + n_j \hat{s}_j^2(\mathbf{z}) + \frac{l_j n_j}{l_j + n_j}(\xi_j - \bar{x}_j(\mathbf{z}))^2.$$

The modification of the above R code is also straightforward and we do not reproduce it here to save space. The extension to more than two components is equally straightforward, as described below for **License**. ◂

If we model **License** by a $k = 3$ component normal mixture model, we start by deriving the prior distribution from the scale of the problem. Namely, we choose a $\mathscr{D}_3(1/2, 1/2, 1/2)$ prior for the weights (although picking parameters less than 1 in the Dirichlet prior has the potential drawback that it may allow very small weights for some components), a $\mathscr{N}(\overline{x}, \sigma_i^2)$ distribution on the means μ_i, and a $\mathscr{G}a(10, \hat{\sigma}^2)$ distribution on the precisions σ_i^{-2}, where $\overline{x}$ and $\hat{\sigma}^2$ are the empirical mean and variance of **License**, respectively. (This empirical choice of a prior is debatable on principle, as it depends on the dataset, but this is relatively harmless since it is equivalent to standardizing the dataset so that the empirical mean and variance are equal to 0 and 1, respectively.) If we define the parameter vector `mix` as a list,

```
> mix=list(k=k,p=p,mu=mu,sig=sig)
```

our R function

```
gibbsnorm=function(niter,mix)
```

is made of an initialization step:

```
n=length(datha);k=mix$k
z=rep(0,n) #missing data
nxj=rep(0,k)
ssiz=ssum=rep(0,k)
mug=sigg=prog=matrix(0,nrow=niter,ncol=k)
lopost=rep(0,niter) #log-posterior
lik=matrix(0,n,k)
prog[1,]=rep(1,k)/k;mug[1,]=rep(mix$mu,k)
sigg[1,]=rep(mix$sig,k)
#current log-likelihood
for (j in 1:k)
  lik[,j]=prog[1,j]*dnorm(x=datha,mean=mug[1,j],
       sd=sqrt(sigg[1,j]))
lopost[1]=sum(log(apply(lik,1,sum)))+
  sum(dnorm(mug[1,],mean(datha),sqrt(sigg[1,]),log=TRUE))-
  (10+1)*sum(log(sigg[1,]))-sum(var(datha)/sigg[1,])+
  .5*sum(log(prog[1,]))
```

and of the main loop for data completion and conditional parameter simulation:

```
for (i in 1:(niter-1)){
 for (t in 1:n){   #missing data completion
   prob=prog[i,]*dnorm(datha[t],mug[i,],sqrt(sigg[i,]))
   if (sum(prob)==0) prob=rep(1,k)/k
   z[t]=sample(1:k,1,prob=prob)
   }
```

```
  #conditional parameter simulation
  for (j in 1:k){
    ssiz[j]=sum(z==j)
    nxj[j]=sum(as.numeric(z==j)*datha)
    }
  mug[i+1,]=rnorm(k,(mean(datha)+nxj)/(ssiz+1),
            sqrt(sigg[i,]/(ssiz+1)))
  for (j in 1:k)
    ssum[j]=sum(as.numeric(z==j)*(datha-nxj[j]/ssiz[j])^2)
     sigg[i+1,]=1/rgamma(k,shape=.5*(20+ssiz),rate=var(datha)+
      .5*ssum+.5*ssiz/(ssiz+1)*(mean(datha)-nxj/ssiz)^2)
  prog[i+1,]=rdirichlet(1,par=ssiz+0.5)
  #current log-likelihood
  for (j in 1:k)
    lik[,j]=prog[i+1,j]*dnorm(x=datha,mean=mug[i+1,j],
     sd=sqrt(sigg[i+1,j]))
   lopost[i+1]=sum(log(apply(lik,1,sum)))+
  sum(dnorm(mug[i+1,],mean(datha),sqrt(sigg[i+1,]),log=TRUE))-
     (10+1)*sum(log(sigg[i+1,]))-sum(var(datha)/sigg[i+1,])+
     .5*sum(log(prog[i+1,]))
  }
```

returning all simulated values as a list

```
list(k=k,mu=mug,sig=sigg,p=prog,lopost=lopost)
```

The output of this R function, represented in Fig. 6.5 as an overlay of the **License** histogram is then produced by the R code

```
mix=list(k=3,mu=mean(datha),sig=var(datha))
simu=gibbsnorm(1000,mix)
hist(datha,prob=TRUE,main="",xlab="",ylab="",nclass=100)
x=y=seq(min(datha),max(datha),length=150)
yy=matrix(0,ncol=150,nrow=1000)
for (i in 1:150){
  yy[,i]=apply(simu$p*dnorm(x[i],mean=simu$mu,
  sd=sqrt(simu$sig)),1,sum)
  y[i]=mean(yy[,i])
  }
for (t in 501:1000)
  lines(x,yy[t,],col="gold")
lines(x,y,lwd=2.3,col="sienna2")
```

This output demonstrates that this crude prior modeling is sufficient to capture the modal features of the histogram as well as the tail behavior in a surprisingly small number of Gibbs iterations, despite the large sample size of

2,625 points. The range of the simulated densities represented in Fig. 6.5 reflects the variability of the posterior distribution, while the estimate of the density is obtained by averaging the simulated densities over the 500 iterations.[7]

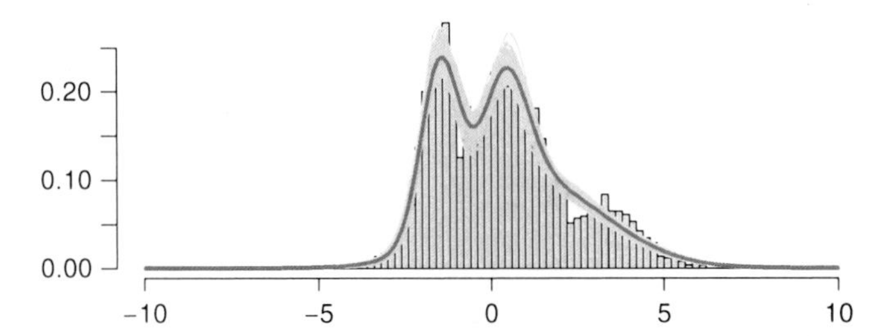

Fig. 6.5. Dataset **License**: Representation of 500 Gibbs iterations for the mixture estimation. (The accumulated *lines* correspond to the estimated mixtures at each iteration and the overlaid *curve* to the density estimate obtained by summation.)

The experiment produced in Example 6.1, page 184, gives a false sense of security about the performance of the Gibbs sampler because it hides the structural dependence of the sampler on its initial conditions. The fundamental feature of Gibbs sampling—its derivation from conditional distributions—implies that it is often restricted in the width of its moves and that, in some situations, this restriction may even jeopardize convergence. In the case of mixtures of distributions, conditioning on $\mathbf{z}$ implies that the proposals for $(\boldsymbol{\theta}, \mathbf{p})$ are quite concentrated and do not allow drastic changes in the allocations at the next step. To obtain a significant modification of $\mathbf{z}$ requires a considerable number of iterations once a stable position has been reached.[8] Figure 6.4 illustrates this phenomenon for the very same sample as in Fig. 6.3: A Gibbs sampler initialized at the saddlepoint may get close to the second mode in the very first iterations and is then unable to escape its (fatal) attraction, even after a large number of iterations, for the reason given above. It is quite interesting to see that this Gibbs sampler suffers from the same pathology as the EM algorithm. However, this is not immensely surprising given that it is based on a similar principle.

In general, there is very little one can do about improving the Gibbs sampler since its components are given by the joint distribution. The solutions are (a) to change the parameterization and thus the conditioning (see

[7]That this is a natural estimate of the model, compared with the "plug-in" density using the estimates of the parameters, will be explained more clearly in Sect. 6.5.

[8]In practice, the Gibbs sampler never leaves the vicinity of a given mode if the attraction of this mode is strong enough, for instance in the case of many observations.

Exercise 6.6), (b) to use tempering to facilitate exploration (see Sect. 6.7), or (c) to mix the Gibbs sampler with another MCMC algorithm.

To look for alternative MCMC algorithms is not a difficulty in this setting, given that the likelihood of mixture models is available in closed form, being computable in $O(kn)$ time, and the posterior distribution is thus known up to a multiplicative constant. We can therefore use any Metropolis–Hastings algorithm, as long as the proposal distribution q provides a correct exploration of the posterior surface, since the acceptance ratio

$$\frac{\pi(\boldsymbol{\theta}', \mathbf{p}'|\mathbf{x})}{\pi(\boldsymbol{\theta}, \mathbf{p}|\mathbf{x})} \frac{q(\boldsymbol{\theta}, \mathbf{p}|\boldsymbol{\theta}', \mathbf{p}')}{q(\boldsymbol{\theta}', \mathbf{p}'|\boldsymbol{\theta}, \mathbf{p})} \wedge 1$$

can be computed in $O(kn)$ time. For instance, we can use a random walk Metropolis–Hastings algorithm where each parameter is the mean of the proposal distribution for the new value, that is,

$$\widetilde{\xi}_j = \xi_j^{(t-1)} + u_j,$$

where $u_j \sim \mathscr{N}(0, \zeta^2)$ and ζ is chosen to achieve a reasonable acceptance rate.

Continuation of Example 6.1. For the posterior associated with (6.3), the Gaussian random walk proposal is

$$\widetilde{\mu_1} \sim \mathscr{N}\left(\mu_1^{(t-1)}, \zeta^2\right) \quad \text{and} \quad \widetilde{\mu_2} \sim \mathscr{N}\left(\mu_2^{(t-1)}, \zeta^2\right)$$

which leads to an acceptance probability of

$$r = \min\left\{1, \pi\left(\widetilde{\mu_1}, \widetilde{\mu_2} | x\right) / \pi\left(\mu_1^{(t-1)}, \mu_2^{(t-1)} \middle| x\right)\right\}.$$

The corresponding R function is then of the form

```
hmmean=function(dat,niter,var=1){
  mu=matrix(0,niter,2)
  mu[1,]=rnorm(2)
  for (i in 2:niter){
    muprop=rnorm(2,mu[i-1,],sqrt(var))
    bound=lpost(dat,muprop)-lpost(dat,mu[i-1,])
    if (runif(1)<=exp(bound)) mu[i,]=muprop else
      mu[i,]=mu[i-1,]
      }
    mu
    }
```

used as in

```
> dat=plotmix()$sample
> simu=hmmeantemp(dat,niter=10^4)
> points(simu,pch=".")
```

when `lpost` is the log-posterior density R function:

```
lpost=function(x,mu,p=0.7,delta=0,lambda=1){
  sum(log(p*dnorm(x,mu[1])+(1-p)*dnorm(x,mu[2])))+
  sum(log(dnorm(mu,delta,1/sqrt(lambda))))
  }
```

For the same simulated dataset as in Fig. 6.4, Fig. 6.6 shows how quickly this algorithm escapes the attraction of the spurious mode. After a few iterations of the algorithm, the chain drifts away from the poor mode and converges almost deterministically to the proper region of the posterior surface. The Gaussian random walk is scaled as $\zeta = 1$, although slightly smaller scales do work as well but would require more iterations to reach the proper modal regions. Too small a scale sees the same trapping phenomenon appear, as the chain does not have sufficient energy to escape the attraction of the current mode (see Example 6.1, page 199, and Fig. 6.8 below). Nonetheless, for a large enough scale, the Metropolis–Hastings algorithm overcomes the drawbacks of the Gibbs sampler. ◀

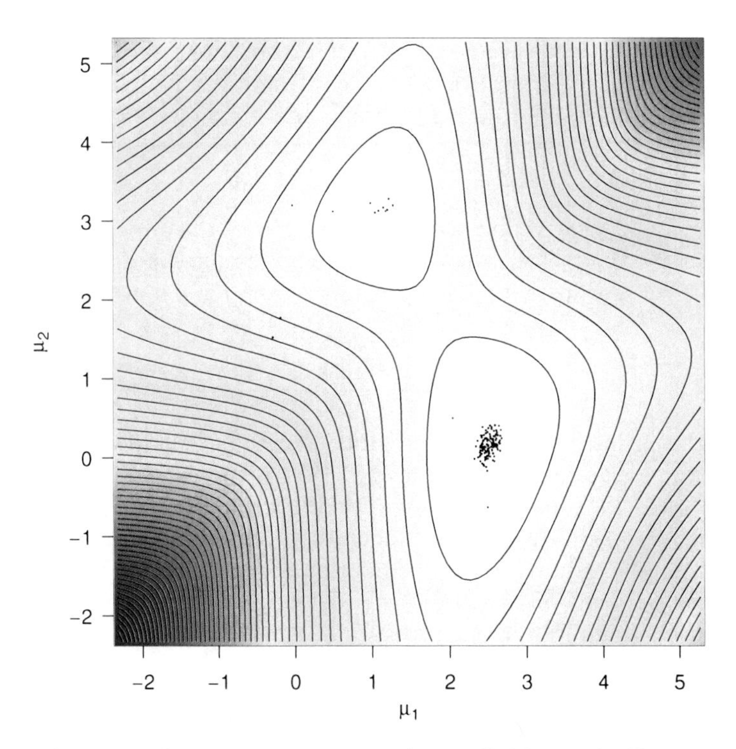

Fig. 6.6. Outcome of a 10,000 iteration random walk Metropolis–Hastings sample on the log-likelihood surface; the starting point is equal to $(1, 3)$. The scale ζ of the random walk is equal to 1

We must point out that, for constrained parameters, the unconstrained random walk Metropolis–Hastings proposal remains valid but is not efficient because when the chain $(\xi_j^{(t)})$ gets close to the boundary of the parameter space, it moves very slowly, given that the proposed values are often incompatible with the constraint and thus rejected at the Metropolis–Hastings acceptance step.

For instance, this lack of efficiency has an impact on the simulation of the weight vector $\mathbf{p}$ since $\sum_{i=1}^{k} p_k = 1$ in addition to positivity constraints. A practical resolution of this difficulty is to overparameterize the weights of (6.2) into

$$p_j = w_j \Big/ \sum_{l=1}^{k} w_l \,, \quad w_j > 0 \qquad (1 \le j \le k)\,.$$

Obviously, the w_j's are not identifiable, but this is not a difficulty from a simulation point of view and the p_j's remain identifiable (up to a permutation of indices). Perhaps paradoxically, using overparameterized representations often helps with the mixing of the corresponding MCMC algorithms since those algorithms are less constrained by the dataset or by the likelihood. The reader may have noticed that the w_j's are also constrained by a positivity requirement (just like the variances in a normal mixture or the scale parameters for a Gamma mixture), but this weaker constraint can be bypassed using the reparameterization $\eta_j = \log w_j$. The proposed random walk move on the w_j's is thus

$$\log(\widetilde{w_j}) = \log\left\{w_j^{(t-1)}\right\} + u_j\,,$$

where $u_j \sim \mathcal{N}(0, \zeta^2)$. An important difference from the original random walk Metropolis–Hastings algorithm is that the acceptance ratio also involves a Jacobian term. For instance, the acceptance ratio for a move from $w^{(t-1)}$ to $\tilde{w}$ is then

$$1 \wedge \frac{\pi(\tilde{w})}{\pi(w^{(t-1)})} \prod_{j=1}^{k} \frac{\widetilde{w_j}}{w_j^{(t-1)}}\,. \tag{6.6}$$

Note that, while being a fairly natural algorithm, the random walk Metropolis–Hastings algorithm usually falls victim to the curse of dimensionality since, obviously the same scale cannot perform well for every component of the parameter vector. In large or even moderate dimensions, a reparameterization of the parameter and preliminary estimation of the information matrix of the distribution are thus often necessary and must sometimes be completed by Gibbs steps operating in lower dimensions.

6.5 Label Switching Difficulty

A basic but extremely important feature of a mixture model is that it is invariant under permutations of the indices of the components. For instance, the normal mixtures $0.3\mathcal{N}(0,1)+0.7\mathcal{N}(2.5,1)$ and $0.7\mathcal{N}(2.5,1)+0.3\mathcal{N}(0,1)$

are exactly the same. Therefore, the $\mathscr{N}(2.5,1)$ distribution cannot be called the "first" component of the mixture! In other words, the component parameters θ_i are not identifiable *marginally* in the sense that θ_1 may be 2.5 as well as 0 in the example above. In this specific case, the pairs (θ_1, p) and $(\theta_2, 1-p)$ are exchangeable.

First, in a k-component mixture, the number of modes of the likelihood is of order $\mathrm{O}(k!)$ since if $((\theta_1,\ldots,\theta_k),(p_1,\ldots,p_k))$ is a local maximum of the likelihood function, so is $\tau(\boldsymbol{\theta},\mathbf{p}) = (\theta_{\tau(1)},\ldots,\theta_{\tau(k)},p_{\tau(1)},\ldots,p_{\tau(k)})$ for every permutation $\tau \in \mathfrak{S}_k$, the set of all permutations of $\{1,\ldots,k\}$. This makes maximization and even exploration of the posterior surface obviously harder because modes are separated by valleys that most samplers find difficult to cross.

Second, if an exchangeable prior is used on $(\boldsymbol{\theta},\mathbf{p})$ (that is, a prior invariant under permutation of the indices), all the posterior marginals on the θ_i's are identical, a fact which means for instance that the posterior expectation of θ_1 is identical to the posterior expectation of θ_2. Therefore, alternatives to posterior expectations must be considered to provide pertinent estimators.

Continuation of Example 6.1. In the special case of model (6.3), if we take *the same* normal prior on both μ_1 and μ_2, $\mu_1,\mu_2 \sim \mathscr{N}(0,10)$, say, the posterior weight conditional on $\mathbf{p}$ associated with an allocation $\mathbf{z}$ for which n_1 values are attached to the first component will simply be

$$\begin{aligned}\omega(\mathbf{z}) &\propto p^{n_1}(1-p)^{n-n_1} \int e^{-n_1(\mu_1-\bar{x}_1(\mathbf{z}))^2/2-(n-1)(\mu_2-\bar{x}_2(\mathbf{z}))^2/2}\,\mathrm{d}\pi(\mu_1)\,\mathrm{d}\pi(\mu_2)\\ &\quad\times\exp\left(-\left\{s_1^2(\mathbf{z})+s_2^2(\mathbf{z})\right\}/2\right)\\ &\propto \sqrt{(n_1+1/10)(n-n_1+1/10)}\,p^{n_1}(1-p)^{n-n_1}\exp\left(-\left\{s_1^2(\mathbf{z})+s_2^2(\mathbf{z})\right.\right.\\ &\quad\left.\left.+n_1\{\bar{x}_1(\mathbf{z})\}^2/(10n_1+1)+(n-n_1)\{\bar{x}_2(\mathbf{z})\}^2/(10(n-n_1)+1)\right\}/2\right),\end{aligned}$$

where $s_1^2(\mathbf{z})$ and $s_2^2(\mathbf{z})$ denote the sums of squares for both groups. ◀

For the Gibbs output of **License** discussed above, the exchangeability predicted by the theory is not observed at all, as shown in Fig. 6.7. This figure is derived from an R code repeating dual plots like

```
> simu=gibbsnorm(1000,mix)
> plot(simu$mu[,1],ylim=range(simu$mu),
+ ylab=expression(mu[i]),xlab="n",type="l",col="sienna3")
> lines(simu$mu[,2],col="gold4")
> lines(simu$mu[,3],col="steelblue")
> plot(simu$mu[,2],simu$p[,2],col="sienna3",
+ xlim=range(simu$mu),ylim=range(simu$p),
+ xlab=expression(mu[i]),ylab=expression(p[i]))
> points(simu$mu[,3],simu$p[,3],col="steelblue")
```

In Fig. 6.7, we see that each component is thus identified by its mean, and the posterior distributions of the means are very clearly distinct. Although this result has the appeal of providing distinct estimates for the three components, it suffers from the severe drawback that the Gibbs sampler has not explored the whole parameter space after 1,000 iterations. Running the algorithm for a much longer period does not solve this problem since the Gibbs sampler cannot simultaneously switch enough component allocations in this highly peaked setup. In other words, the algorithm is unable to explore more than one of the $3! = 6$ equivalent modes of the posterior distribution. Therefore, it is difficult to trust the estimates derived from such an output.

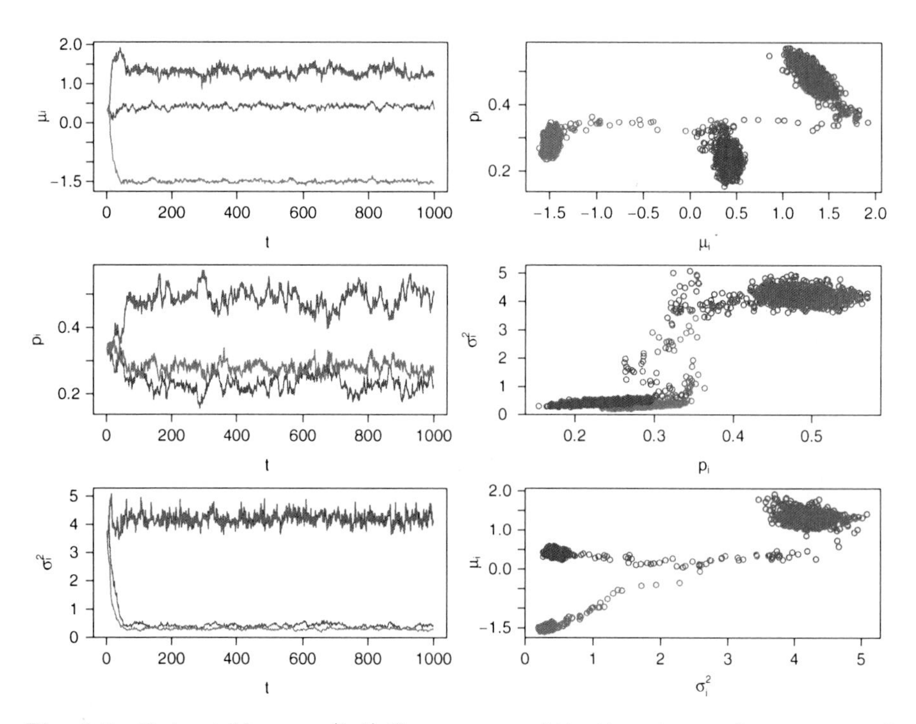

Fig. 6.7. Dataset **License**: (*left*) Convergence of the three types of parameters of the normal mixture, each component being identified by a different *grey level/color*; (*right*) 2×2 plot of the Gibbs sample for the three types of parameters of a normal mixture

This identifiability problem related to the exchangeability of the posterior distribution, often called "label switching," thus requires either an alternative prior modeling or a more tailored inferential approach. A naïve answer to the problem is to impose an *identifiability constraint* on the parameters, for instance defining the components by ordering the means (or the variances or the weights) in a normal mixture (see Exercise 6.3). From a Bayesian point of view, this amounts to truncating the original prior distribution, going from $\pi(\boldsymbol{\theta}, \mathbf{p})$ to

$$\pi(\boldsymbol{\theta}, \mathbf{p})\, \mathbb{I}_{\mu_1 \leq \ldots \leq \mu_k}$$

for instance. While this seems innocuous (given that the sampling distribution is the same with or without this indicator function), the introduction of an identifiability constraint has severe consequences on the resulting inference, both from a prior and from a computational point of view. When reducing the parameter space to its constrained part, the imposed truncation has no reason to respect the topology of either the prior or the likelihood. Instead of singling out one mode of the posterior, the constrained parameter space may then well include parts of several modes and the resulting posterior mean could, for instance, lie in a very low probability region between the modes, while the high posterior probability zones are located at the boundaries of this space.

In addition, the constraint may radically modify the prior modeling and come close to contradicting the prior information. For large values of k, the introduction of a constraint also has a consequence on posterior inference: With many components, the ordering of components in terms of one of the parameters of the mixture is unrealistic. Some components will be close in mean while others will be close in variance or in weight. This may even lead to very poor estimates of the parameters if the inappropriate ordering is chosen.

Note that while imposing a constraint that is not directly related to the modal regions of the target distribution may considerably reduce the efficiency of an MCMC algorithm, it must be stressed that the constraint does not need to be imposed *during* the simulation but can instead be imposed *after* simulation by reordering the MCMC output according to the constraint. For instance, if the constraint imposes an ordering of the means, once the simulation is over, the components can be relabeled for each MCMC iteration according to this constraint; that is, defining the first component as the one associated with the smallest simulated mean and so on. From this perspective, identifiability constraints have nothing to do with (or against) simulation.

An empirical resolution of the label switching problem that avoids imposing the constraints altogether consists of arbitrarily selecting one of the $k!$ modal regions of the posterior distribution once the simulation step is over and only then operate the relabeling in terms of proximity to this region.

Given an MCMC sample of size M, we can find a Monte Carlo approximation of the *maximum a posteriori* (MAP) estimator by taking $\boldsymbol{\theta}^{(i^*)}, \mathbf{p}^{(i^*)}$ such that

$$i^* = \arg \max_{i=1,\ldots,M} \pi\left\{(\boldsymbol{\theta}, \mathbf{p})^{(i)} | \mathbf{x}\right\} ;$$

that is, the simulated value that gives the maximal posterior density. (Note that π does not need its normalizing constant for this computation.) This value is quite likely to be in the vicinity of one of the $k!$ modes, especially if we run many simulations. The approximate MAP estimate will thus act as a *pivot* in the sense that it gives a good approximation to a mode and we can reorder the other iterations with respect to this mode.

Rather than selecting the reordering based on a Euclidean distance in the parameter space, we use a distance in the space of allocation probabilities. Indeed, the components of the parameter vary in different spaces, from the real line for the means to the simplex for the weights. Let $\mathfrak{S}_k$ be the k-permutation set and $\tau \in \mathfrak{S}_k$. We suggest to minimize in τ an entropy distance summing the relative entropies between the $\mathbb{P}(z_t = j|\boldsymbol{\theta}^{(i^*)}, \mathbf{p}^{(i^*)})$'s and the $\mathbb{P}(z_t = j|\tau\left\{(\boldsymbol{\theta}^{(i)}, \mathbf{p}^{(i)})\right\})$'s, namely

$$
\begin{aligned}
h(i,\tau) = \sum_{t=1}^{n}\sum_{j=1}^{k} & \mathbb{P}(z_t = j|\boldsymbol{\theta}^{(i^*)}, \mathbf{p}^{(i^*)}) \\
& \times \log\left\{\mathbb{P}(z_t = j|\boldsymbol{\theta}^{(i^*)}, \mathbf{p}^{(i^*)})/\mathbb{P}(z_t = j|\tau\left[(\boldsymbol{\theta}^{(i)}, \mathbf{p}^{(i)})\right])\right\}.
\end{aligned}
$$

The selection of the permutations reordering the MCMC output thus reads as follows:

Algorithm 6.12 PIVOTAL REORDERING
At iteration $i \in \{1, \ldots, M\}$:

1. Compute
$$\tau_i = \arg\min_{\tau\in\mathfrak{S}_k} h(i,\tau)\,,$$
2. Set $(\boldsymbol{\theta}^{(i)}, \mathbf{p}^{(i)}) = \tau_i\{(\boldsymbol{\theta}^{(i)}, \mathbf{p}^{(i)})\}$.

Thanks to this reordering, most iteration labels get switched to the same mode (when n gets large, this is almost a certainty), and the identifiability problem is thus solved. Therefore, after this reordering step, the Monte Carlo estimate of the posterior expectation $\mathbb{E}^\pi[\theta_i|\mathbf{x}]$,

$$\sum_{j=1}^{M} (\theta_i)^{(j)}/M\,,$$

can be used as in a standard setting because the reordering automatically gives different meanings to different components. Obviously, $\mathbb{E}^\pi[\theta_i|\mathbf{x}]$ (or its approximation) should also be compared with $\theta^{(i^*)}$ to check convergence.[9]

Using the Gibbs output `simu` of **License** (which is the `datha` of the following code) as in the previous illustration, the corresponding R code involves the determination of the MAP approximation

```
indimap=order(simu$lopost,decreasing=TRUE)[1]
map=list(mu=simu$mu[indimap,],sig=simu$sig[indimap,],
    p=simu$p[indimap,])
```

that is easily derived by storing the values of the log-likelihood densities in the Gibbs sampling function `gibbsnorm`. The corresponding (MAP) allocation probabilities for the data are then

```
lili=alloc=matrix(0,length(datha),3)
for (t in 1:length(datha)){
  lili[t,]=map$p*dnorm(datha[t],mean=map$mu,
  sd=sqrt(map$sig))
  lili[t,]=lili[t,]/sum(lili[t,])
  }
```

They are used as reference for the reordering:

```
ormu=orsig=orp=matrix(0,ncol=3,nrow=1000)
library(combinat)
perma=permn(3)
for (t in 1:1000){
  entropies=rep(0,factorial(3))
  for (j in 1:n){
    alloc[j,]=simu$p[t,]*dnorm(datha[j],mean=simu$mu[t,],
    sd=sqrt(simu$sig[t,]))
    alloc[j,]=alloc[j,]/sum(alloc[j,])
    for (i in 1:factorial(3))
     entropies[i]=entropies[i]+
     sum(lili[j,]*log(alloc[j,perma[[i]]]))
    }
  best=order(entropies,decreasing=TRUE)[1]
  ormu[t,]=simu$mu[t,perma[[best]]]
  orsig[t,]=simu$sig[t,perma[[best]]]
  orp[t,]=simu$p[t,perma[[best]]]
  }
```

[9] While this resolution seems intuitive enough, there is still a lot of debate in academic circles on whether or not label switching should be observed on an MCMC output and, in case it should, on which substitute to the posterior mean should be used.

An output comparing the original MCMC sample and the one corresponding to this reordering for the **License** dataset is then constructed. However, since the Gibbs sampler does not switch between the $k!$ modes in this case, the above reordering does not modify the labelling and we thus abstain from producing the corresponding graph as it is identical to Fig. 6.7.

6.6 Prior Selection

After[10] insisting in Chap. 2 that conjugate priors are not the only possibility for prior modeling, we seem to be using them quite extensively in this chapter! The fundamental reason for this is that, as explained below, it is not possible to use the standard alternative of noninformative priors on the components. Nonconjugate priors can be used as well (with Metropolis–Hastings steps) but are difficult to fathom when the components have no specific "real" meaning (as, for instance, when the mixture is used as a nonparametric proxy).

The representation (6.2) of a mixture model precludes the use of independent improper priors,

$$\pi(\boldsymbol{\theta}) = \prod_{j=1}^{k} \pi_j(\theta_j),$$

since if, for any $1 \leq j \leq k$,

$$\int \pi_j(\theta_i) \mathrm{d}\theta_j = \infty,$$

then, for every n,

$$\int \pi(\boldsymbol{\theta}, \mathbf{p}|\mathbf{x}) \mathrm{d}\boldsymbol{\theta} \mathrm{d}\mathbf{p} = \infty.$$

The reason for this inconvenient behavior is that among the k^n terms in the expansion (6.5) of $\pi(\boldsymbol{\theta}, \mathbf{p}|\mathbf{x})$, there are $(k-1)^n$ terms without *any* observation allocated to the ith component and thus there are $(k-1)^n$ terms with a conditional posterior $\pi(\theta_i|\mathbf{x}, \mathbf{z})$ that is equal to the prior $\pi_i(\theta_i)$.

The inability to use improper priors may be seen by some as a *marginalia*, a fact of little importance, since they argue that proper priors with large variances can be used instead. However, since mixtures are ill-posed problems,[11] this difficulty with improper priors is more of an issue, given that the

[10] This section may be skipped by most readers, as it only addresses the very specific issue of handling improper priors in mixture estimation.

[11] By nature, *ill-posed* problems are not precisely defined. They cover classes of models such as *inverse problems*, where the complexity of getting back from the data to the parameters is huge. They are not to be confused with nonidentifiable problems, though.

influence of a particular proper prior, no matter how large its variance, cannot be truly assessed. In other words, the prior gives a specific meaning to what distinguishes one component from another.

↯ Prior distributions must always be chosen with the utmost care when dealing with mixtures and their bearing on the resulting inference assessed by a sensitivity study. The fact that some noninformative priors are associated with undefined posteriors, no matter what the sample size, is a clear indicator of the complex nature of Bayesian inference for those models.

6.7 Tempering

The notion of *tempering* can be found in different areas under many different denominations, but it always comes down to the same intuition that governs simulated annealing (Chap. 8), namely that when you flatten a posterior surface, it is easier to move around, while if you sharpen it, it gets harder to do so except around peaks.

More formally, given a density $\pi(x)$, we can define an associated density $\pi_\alpha(x) \propto \pi(x)^\alpha$ for $\alpha > 0$ large enough (if α is too small, $\pi(x)^\alpha$ does not integrate). An important property of this family of distributions is that they all share the same modes. When $\alpha > 1$, the surface of π_α is more contrasted than the surface of π: Peaks are higher and valleys are lower. Increasing α to infinity results in a Dirac mass at the modes of π, and this is the principle behind simulated annealing. Conversely, lowering α to values less than 1 makes the surface smoother by lowering peaks and raising valleys. In a compact space, lowering α to 0 ends up with the uniform distribution.

This rather straightforward intuition can be exploited in several directions for simulation. For instance, a tempered version of π, π_α, can be simulated in a preliminary step to determine where the modal regions of π are. (Different values of α can be used in parallel to compare the results.) This preliminary exploration can then be used to build a more appropriate proposal. Alternatively, these simulations may be pursued and associated with appropriate importance weights. Note also that a regular Metropolis–Hastings algorithm may be used with π_α just as well as with π since the acceptance ratio is transformed into

$$\left(\frac{\pi(\boldsymbol{\theta}', \mathbf{p}'|\mathbf{x})}{\pi(\boldsymbol{\theta}, \mathbf{p}|\mathbf{x})}\right)^\alpha \frac{q(\boldsymbol{\theta}, \mathbf{p}|\boldsymbol{\theta}', \mathbf{p}')}{q(\boldsymbol{\theta}', \mathbf{p}'|\boldsymbol{\theta}, \mathbf{p})} \wedge 1 \tag{6.7}$$

in the case of the mixture parameters, with the same irrelevance of the normalizing constants.

Continuation of Example 6.1. If we consider once more the posterior associated with (6.3), we can check in Fig. 6.8 the cumulative effect of a small

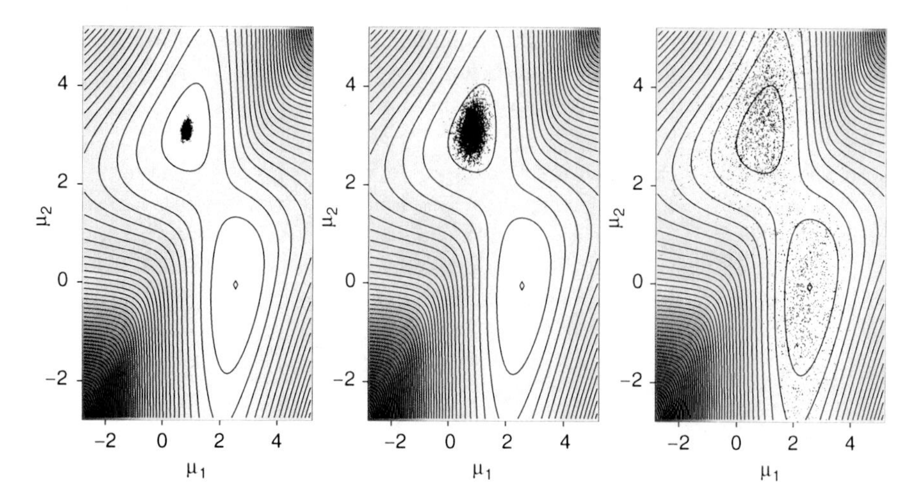

Fig. 6.8. Comparison of Metropolis–Hastings samples of 10^4 points started in the vicinity of the spurious mode for the target distributions π_α when $\alpha = 1, 0.1, 0.01$ (*from left to right*), π is the same as in Fig. 6.6, and the proposal is a random walk with variance 0.1 (the shape of log-likelihood does not changed)

variance for the random walk proposal (chosen here as 0.1) and a decrease in the power α. The R function used to produce this figure is

```
hmmeantemp=function(dat,niter=100,var=.1,alpha=1){
  mu=matrix(0,niter,2)
  mu[1,]=c(1,3)
  for (i in 2:niter){
    muprop=rnorm(2,mu[i-1,],sqrt(var))
    bound=lpost(dat,muprop)-lpost(dat,mu[i-1,])
    if (runif(1)<=exp(alpha*bound)) mu[i,]=muprop else
      mu[i,]=mu[i-1,]
    }
  mu
  }
```

It thus constitutes a very straightforward modification of the original Metropolis–Hastings algorithm. For the genuine target distribution π *(left)*, 10,000 iterations of the Metropolis–Hastings algorithm are not nearly sufficient to remove the attraction of the lower mode. When $\alpha = 0.1$, we can reasonably hope that a few thousand more iterations could bring the Markov chain toward the other mode. For $\alpha = 0.01$, only a few iterations suffice to switch modes, given that the saddle between both modes is not much lower than the modes themselves. (The best way to check this fact and to select α in practice is to run the R code!) ◀

6.8 Mixtures with an Unknown Number of Components

While the standard interpretation of mixtures gives each component a meaning, the semiparametric approach to mixtures only perceives components as base elements in a representation of an unknown density. In that perspective, the number k of components represents the degree of approximation, and it has no particular reason to be fixed in advance. Even from the traditional perspective, it may also happen that the number of homogeneous groups within the population of interest is unknown and that inference first seeks to determine this number. For instance, in a marketing study of Web-browsing behaviors, it may well be that the number of different behaviors is unknown. Also, for instance, in the analysis of financial stocks, the number of different patterns in the evolution of these stocks may be unknown to the analyst. For these different situations, it is thus necessary to extend the previous setting to include inference on the number k of components itself.

Inference on such a structure is somehow more complicated than on single models, especially when there are an infinite number of submodels, i.e. when k is not bounded, and it can be tackled from two different (or even opposite) perspectives. The first approach is to consider the variable dimension model as a whole and to estimate quantities that are meaningful for the whole model (such as moments or predictives) as well as quantities that only make sense for submodels (such as posterior probabilities of submodels and posterior moments of θ_k). From a Bayesian perspective, once a prior is defined on $\boldsymbol{\theta}$, the only difficulty is in finding an efficient way to explore the complex parameter space in order to produce these estimators. The second perspective on variable dimension models is to resort to *testing*, rather than estimation, by adopting a *model choice* stance. This requires choosing among all possible submodels the "best one" in terms of an appropriate criterion, usually through the Bayes factor (Sect. 2.3.2). The computational resolution of the comparison when the number of models is infinite requires MCMC exploration, while the variability of the resulting inference may be underestimated if the selection of the model is not accounted for in the assessment of the variability. Nonetheless, this is an approach often used in linear and generalized linear models (Chaps. 3 and 4) where subgroups of covariates are compared against a given dataset.

Mixtures with an unknown number of components are one particular instance of *variable dimension models*. Other cases include the selection of covariates among k possible covariates in a generalized linear model (Chap. 4) which can be seen as a collection of 2^k submodels (depending on the presence or absence of each covariate). Similarly, in a time series model such as the AR and MA models (Chap. 7), the value of the lag dependence can be left open, depending on the data at hand. Other instances are the determination of the order in a hidden Markov model (Chap. 7), as in DNA sequences where the dependence of the past bases may go back for one, two, or more steps, or even in a capture–recapture experiment (Chap. 5) when one estimates the number of species from the observed species.

While we opt here for a testing perspective, a more generic simulation technique called *reversible jump* has been developed by Green (1995). While it was exposed in the earlier edition (Marin and Robert, 2007), it requires both a high degree of formalization and a very sensitive calibration. In the specific case of mixtures, the number of models under comparison (i.e., the range of k) is usually small enough to prefer an enumeration of all models and hence an approximation of all marginal likelihoods.

Similarly, *Dirichlet processes* are often advanced as alternative to the estimation of the number of components for mixtures because they naturally embed a clustering mechanism. A Dirichlet process is a nonparametric object that formally involves a countably infinite number of components. Nonetheless, inference on Dirichlet processes for a finite sample size produces a random number of clusters, which can be used as an estimate of the number of components. Since the technical complexity of those objects is too high for this book, we refer to Hjort et al. (2010) for detail.

Once testing is adopted as the setting of reference, the implementation of the principle boils down to study some proposals regarding approximations of the Bayes factor oriented towards the direct exploitation of outputs from single model MCMC runs.

In fact, the major difference between approximations of Bayes factors based on those outputs and approximations based on the output from the reversible jump chains is that the latter requires a sufficiently efficient choice of proposals to move around models, which can be difficult. If we can instead concentrate the simulation effort on single models, the complexity of the algorithm decreases (a lot) and there exist ways to evaluate the performance of the corresponding MCMC samples. In addition, it is often the case that few models are in competition when estimating k and it is therefore possible to visit the whole range of potentials models in an exhaustive manner.

We have

$$f_J(\mathbf{x}|\boldsymbol{\lambda}_J) = \prod_{i=1}^{n} \sum_{j=1}^{J} p_j f(x_i|\theta_j)$$

where $\boldsymbol{\lambda}_J = (\boldsymbol{\theta}, \mathbf{p}) = (\theta_1, \ldots, \theta_J, p_1, \ldots, p_J)$. Most solutions (see, e.g. Frühwirth-Schnatter, 2006, Sect. 5.4) revolve around an importance sampling approximation to the marginal likelihood integral

$$m_J(x) = \int f_J(\mathbf{x}|\boldsymbol{\lambda}_J)\, \pi_J(\boldsymbol{\lambda}_J)\, \mathrm{d}\boldsymbol{\lambda}_J$$

where J denotes the model index (that is the number of components in the present case). A different possibility is to use Gelfand and Dey (1994) representation: starting from an arbitrary density g_J, the equality

$$
\begin{aligned}
1 = \int g_J(\boldsymbol{\lambda}_J)\,\mathrm{d}\boldsymbol{\lambda}_J &= \int \frac{g_J(\boldsymbol{\lambda}_J)}{f_J(\mathbf{x}|\boldsymbol{\lambda}_J)\,\pi_J(\boldsymbol{\lambda}_J)} f_J(\mathbf{x}|\boldsymbol{\lambda}_J)\,\pi_J(\boldsymbol{\lambda}_J)\,\mathrm{d}\boldsymbol{\lambda}_J \\
&= m_J(\mathbf{x}) \int \frac{g_J(\boldsymbol{\lambda}_J)}{f_J(\mathbf{x}|\boldsymbol{\lambda}_J)\,\pi_J(\boldsymbol{\lambda}_J)} \pi_J(\boldsymbol{\lambda}_J|\mathbf{x})\,\mathrm{d}\boldsymbol{\lambda}_J
\end{aligned}
$$

implies that a potential estimate of $m_J(\mathbf{x})$ is

$$
\hat{m}_J(\mathbf{x}) = 1 \Big/ \frac{1}{T}\sum_{t=1}^{T} \frac{g_J(\boldsymbol{\lambda}_J^{(t)})}{f_J(\mathbf{x}|\boldsymbol{\lambda}_J^{(t)})\,\pi_J(\boldsymbol{\lambda}_J^{(t)})}
$$

when the $\boldsymbol{\lambda}_J^{(t)}$'s are produced by a Monte Carlo or an MCMC sampler targeted at $\pi_J(\boldsymbol{\lambda}_J|\mathbf{x})$.

While this solution can be easily implemented in low dimensional settings, calibrating the auxiliary density g_k is always an issue. The auxiliary density could be selected as a non-parametric estimate of $\pi_k(\boldsymbol{\lambda}_J|x)$ based on the sample itself but this is very costly. Another difficulty is that the estimate may have an infinite variance and thus be too variable to be trustworthy.

Yet another approximation to the integral $m_J(\mathbf{x})$ is to consider it as the expectation of $f_J(\mathbf{x}|\boldsymbol{\lambda}_J)$, when $\boldsymbol{\lambda}_J$ is distributed from the prior. While a brute force approach simulating $\boldsymbol{\lambda}_J$ from the prior distribution requires a huge number of simulations!

We consider here a further solution, first proposed by Chib (1995), that is straightforward to implement in the setting of mixtures. Although this method may fail because of the lack of label switching, we show below how the difficulty can easily be removed. Chib's method is directly based on the expression of the marginal distribution (loosely called *marginal likelihood* in this section) in Bayes' theorem:

$$
m_J(\mathbf{x}) = \frac{f_J(\mathbf{x}|\boldsymbol{\lambda}_J)\,\pi_J(\boldsymbol{\lambda}_J)}{\pi_J(\boldsymbol{\lambda}_J|\mathbf{x})}
$$

and on the property that the rhs of this equation is constant in $\boldsymbol{\lambda}_J$. Therefore, if an arbitrary value of $\boldsymbol{\lambda}_J$, $\boldsymbol{\lambda}_J^*$ say, is selected and if a good approximation to $\pi_J(\boldsymbol{\lambda}_J|\mathbf{x})$ can be constructed, $\hat{\pi}_J(\boldsymbol{\lambda}_J|\mathbf{x})$, Chib's approximation to the marginal likelihood is

$$
\hat{m}_J(\mathbf{x}) = \frac{f_J(\mathbf{x}|\boldsymbol{\lambda}_J^*)\,\pi_J(\boldsymbol{\lambda}_J^*)}{\hat{\pi}_J(\boldsymbol{\lambda}_J^*|\mathbf{x})}\,. \tag{6.8}
$$

In the case of mixtures, a natural approximation to $\pi_J(\boldsymbol{\lambda}_J|\mathbf{x})$ is the Rao–Blackwell estimate

$$
\hat{\pi}_J(\boldsymbol{\lambda}_J^*|\mathbf{x}) = \frac{1}{T}\sum_{t=1}^{T} \pi_J(\boldsymbol{\lambda}_J^*|\mathbf{x},\mathbf{z}^{(t)})\,, \tag{6.9}
$$

where the $\mathbf{z}^{(t)}$'s are the latent variables simulated by the MCMC sampler. To be efficient, this method requires

(a) a good choice of $\boldsymbol{\lambda}_J^*$ but, since in the case of mixtures, the likelihood is computable, $\boldsymbol{\lambda}_J^*$ can be chosen as the MCMC approximation to the MAP estimator (see Algorithm 6.12) and,
(b) a good approximation to $\pi_J(\boldsymbol{\lambda}_J|\mathbf{x})$.

This latter requirement is paramount: while, at a formal level, $\hat{\pi}_J(\boldsymbol{\lambda}_J^*|\mathbf{x})$ is a converging (parametric) approximation to $\pi_J(\boldsymbol{\lambda}_J|\mathbf{x})$ by virtue of the ergodic theorem, this obviously requires the chain $(\mathbf{z}^{(t)})$ to converge to its stationarity distribution. Unfortunately, as discussed previously, in the case of mixtures, the Gibbs sampler rarely converges because of the label switching phenomenon, so the approximation $\hat{\pi}_J(\boldsymbol{\lambda}_J^*|\mathbf{x})$ is untrustworthy. It is easily seen via a numerical experiment that (6.8) is *significantly different from the true value* $m_J(\mathbf{x})$ when label switching *does not occur.* There is, however, a fix to this problem which is to recover the label switching symmetry a posteriori, replacing $\hat{\pi}_J(\boldsymbol{\lambda}_J^*|\mathbf{x})$ in (6.9) above with

$$\hat{\pi}_J(\boldsymbol{\lambda}_J^*|\mathbf{x}) = \frac{1}{T\,J!}\sum_{\sigma\in\mathfrak{S}_J}\sum_{t=1}^{T}\pi_J(\sigma(\boldsymbol{\lambda}_J^*)|\mathbf{x},\mathbf{z}^{(t)}),$$

where $\mathfrak{S}_J$ denotes the set of all permutations of $\{1,\ldots,J\}$ and $\sigma(\boldsymbol{\lambda}_J^*)$ denotes the transform of $\boldsymbol{\lambda}_J^*$ where components are switched according to the permutation σ. Note that the permutation can equally be applied to $\boldsymbol{\lambda}_J^*$ or to the $\mathbf{z}^{(t)}$'s but that the former is usually more efficient from a computational point of view given that the sufficient statistics only have to be computed once. The justification for this modification stems from a Rao–Blackwellization argument, namely that the permutations are ancillary for the problem and should be integrated out.

Example 6.3. In the case of the normal mixture case and a benchmark called the "galaxy dataset" (Robert and Casella, 2004, Chap. 11, Table 11.1) Gibbs sampling does not produce any label switching. If we compute $\log\hat{m}_J(\mathbf{x})$ using Chib's original estimate (6.8), the [logarithm of the] estimated marginal likelihood is

$$\hat{\rho}_J(\mathbf{x}) = -105.1396$$

for $J = 3$ (based on 10^3 simulations), while introducing the permutations leads to

$$\hat{\rho}_J(\mathbf{x}) = -103.3479\,.$$

As noted by Frühwirth-Schnatter (2006), the difference between the original Chib's approximation and the true marginal likelihood is close to $\log(J!)$ (only) when the Gibbs sampler remains concentrated around a single mode of the posterior distribution. In the current case, we have that

$$-116.3747 + \log(2!) = -115.6816$$

exactly! (We also checked this numerical value of the marginal likelihood against a brute-force estimate obtained by simulating from the prior and averaging the likelihood, up to a fourth digit agreement.) A similar result holds for $J = 3$, with

$$-105.1396 + \log(3!) = -103.3479\,.$$

For $J = 4$, we get for instance that the original Chib's approximation is -104.1936, while the average over permutations gives -102.6642. Similarly, for $J = 5$, the difference between -103.91 and -101.93 is less than $\log(5!)$. The $\log(J!)$ difference cannot therefore be used as a direct correction for Chib's approximation because of this difficulty in controlling the amount of overlap. However, it is unnecessary since using the permutation average resolves the difficulty. Table 6.1 shows that the preferred value of J for the **Galaxy** dataset and the current choice of prior distribution is $J = 5$.

J	2	3	4	5	6	7	8
$\hat\rho_J(\mathbf{x})$	-115.68	-103.35	-102.66	-101.93	-102.88	-105.48	-108.44

Table 6.1. Dataset **Galaxy**: estimations of the marginal log-likelihoods by the symmetrized Chib's approximation

When the number of components J grows too large for all permutations in $\mathfrak{S}_J$ to be considered in the average, a (random) subsample of permutations can be simulated to keep the computing time to a reasonable level (obviously keeping the identity as one of the selected permutations!), as in Table 6.1 for $J = 6, 7$. Note also that the discrepancy between the original Chib's (1995) approximation and the average over permutations is a good indicator of the mixing properties of the Markov chain, if a further convergence indicator is requested.

We implemented Chib's method for the **License** dataset in the function `gibbsnorm(niter,mix)`. The code relies on the combinatorial package `combinat` in order to store all possible permutations:

```
lolik=rep(0,niter)
library(combinat)
perms=matrix(unlist(permn(k)),ncol=k,byrow=T)
nperms=dim(perms)[1]
```

The marginal likelihood is then averaged over iterations and permutations

```
chibdeno=0
for (j in 1:nperms)
chibdeno=chibdeno+exp(sum(dnorm(mug[i+1,perms[j,]],
    mean=(mean(datha)+nxj)/(1+ssiz),
```

```
    sd=sqrt(sigg[i+1,perms[j,]])/sqrt((1+ssiz)),log=TRUE))+
  sum(dgamma(1/sigg[i+1,perms[j,]],shape=.5*(20+ssiz),
    rate=var(datha)+.5*ssum+.5*ssiz/
(ssiz+1)*(mean(datha)-nxj/ssiz)^2,log=TRUE)
  -2*log(sigg[i+1,perms[j,]]))+
  sum((ssiz-0.5)*log(prog[i+1,perms[j,]]))+
  lgamma(sum(ssiz+0.5))-sum(lgamma(ssiz+0.5)))
```

the function returning a list `list(...,lolik=lolik,deno=chibdeno)`. Using the code,

```
> simu=gibbsnorm(1000,mix)
> lopos=order(simu$lopost)[1000]
> lnum1=simu$lolik[lopos]
> lnum2=sum(dnorm(simu$mu[lopos,],
+ mean=mean(datha),sd=simu$sig[lopos,],log=TRUE)+
+ dgamma(1/simu$sig[lopos,],10,var(datha),log=TRUE)-
+ 2*log(simu$sig[lopos,]))+
+ sum((rep(0.5,k)-1)*log(simu$p[lopos,]))+
+ lgamma(sum(rep(0.5,k)))-sum(lgamma(rep(0.5,k)))
> lchibapprox2=lnum1+lnum2-log(simu$deno)
```

we obtain Table 6.2 which gives the approximations of the marginal likelihoods from $k = 2$ to $k = 8$. For the **License** dataset, the favored number of components is thus $k = 4$.

k	2	3	4	5	6
$\hat{\rho}_k(\mathbf{x})$	-5373.445	-5315.351	-5308.79	-5336.23	-5341.524

Table 6.2. Dataset **License**: estimations of the marginal log-likelihoods by the symmetrized Chib's approximation

6.9 Exercises

6.1 Show that a mixture of Bernoulli distributions is again a Bernoulli distribution. Extend this to the case of multinomial distributions.

6.2 Show that the number of nonnegative integer solutions of the decomposition of n into k parts such that $n_1 + \ldots + n_k$ is equal to

$$\mathfrak{r} = \binom{n+k-1}{n}.$$

Deduce that the number of partition sets is of order $\mathrm{O}(n^{k-1})$. (*Hint:* This is a classical combinatoric problem.)

6.3 For a mixture of two normal distributions with all parameters unknown,

$$p\mathscr{N}(\mu_1,\sigma_1^2)+(1-p)\mathscr{N}(\mu_2,\sigma_2^2),$$

and for the prior distribution ($j=1,2$)

$$\mu_j|\sigma_j \sim \mathscr{N}(\xi_j,\sigma_j^2/n_j), \quad \sigma_j^2 \sim \mathscr{IG}(\nu_j/2,s_j^2/2), \quad p\sim \mathscr{B}e(\alpha,\beta),$$

show that

$$p|\mathbf{x},\mathbf{z} \sim \mathscr{B}e(\alpha+\ell_1,\beta+\ell_2),$$

$$\mu_j|\sigma_j,\mathbf{x},\mathbf{z} \sim \mathscr{N}\left(\xi_1(\mathbf{z}),\frac{\sigma_j^2}{n_j+\ell_j}\right), \quad \sigma_j^2|\mathbf{x},\mathbf{z}\sim \mathscr{IG}((\nu_j+\ell_j)/2,s_j(\mathbf{z})/2),$$

where ℓ_j is the number of z_i equal to j, $\bar{x}_j(\mathbf{z})$ and $\hat{s}_j^2(\mathbf{z})$ are the empirical mean and variance for the subsample with z_i equal to j, and

$$\xi_j(\mathbf{z})=\frac{n_j\xi_j+\ell_j\bar{x}_j(\mathbf{z})}{n_j+\ell_j}, \quad s_j(\mathbf{z})=s_j^2+\ell_j\hat{s}_j^2(\mathbf{z})+\frac{n_j\ell_j}{n_j+\ell_j}(\xi_j-\bar{x}_j(\mathbf{z}))^2.$$

Compute the corresponding weight $\omega(\mathbf{z})$.

6.4 For the normal mixture model of Exercise 6.3, compute the function $Q(\theta_0,\theta)$ and derive both steps of the EM algorithm. Apply this algorithm to a simulated dataset and test the influence of the starting point θ_0.

6.5 In the mixture model with independent priors on the θ_j's, show that the θ_j's are dependent on each other given (only) $\mathbf{x}$ by summing out the $\mathbf{z}$'s.

6.6 Construct and test the Gibbs sampler associated with the (ξ,μ_0) parameterization of (6.3), when $\mu_1=\mu_0-\xi$ and $\mu_2=\mu_0+\xi$.

6.7 Show that, if an exchangeable prior π is used on the vector of weights $(p_1,\ldots,p_k)$, then, necessarily, $\mathbb{E}^\pi[p_j]=1/k$ and, if the prior on the other parameters $(\theta_1,\ldots,\theta_k)$ is also exchangeable, then $\mathbb{E}^\pi[p_j|x_1,\ldots,x_n]=1/k$ for all j's.

6.8 Show that running an MCMC algorithm with target $\pi(\theta|\mathbf{x})^\gamma$ will increase the proximity to the MAP estimate when $\gamma>1$ is large. (*Note*: This is a crude version of the *simulated annealing* algorithm. See also Chap. 8.) Discuss the modifications required in Algorithm 6.11 to achieve simulation from $\pi(\theta|\mathbf{x})^\gamma$ when $\gamma\in\mathbb{N}^*$ is an integer.

6.9 Show that the ratio (6.7) goes to 1 when α goes to 0 when the proposal q is a random walk. Describe the average behavior of this ratio in the case of an independent proposal.

6.10 If one needs to use importance sampling weights, show that the simultaneous choice of several powers α requires the computation of the normalizing constant of π_α.

6.11 In the setting of the mean mixture (6.3), run an MCMC simulation experiment to compare the influence of a $\mathscr{N}(0,100)$ and of a $\mathscr{N}(0,10000)$ prior on (μ_1,μ_2) on a sample of 500 observations.

6.12 Show that, for a normal mixture $0.5\,\mathscr{N}(0,1)+0.5\,\mathscr{N}(\mu,\sigma^2)$, the likelihood is unbounded. Exhibit this feature by plotting the likelihood of a simulated sample using the R image procedure.

7

Time Series

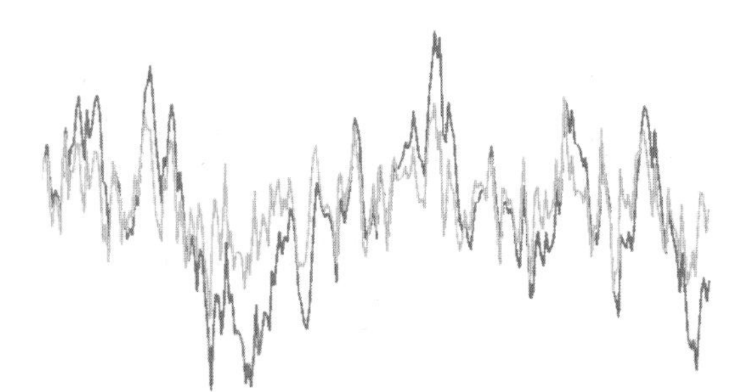

Rebus was intrigued by the long gaps
in the chronology.
—**Ian Rankin, *The Falls*.**—

Roadmap

At one point or another, everyone has to face modeling time series datasets, by which we mean series of dependent observations that are indexed by time (like both series in the picture above!). As in the previous chapters, the difficulty in modeling such datasets is to balance the complexity of the representation of the dependence structure against the estimation of the corresponding model—and thus the modeling most often involves model choice or model comparison. We cover here the Bayesian processing of some of the most standard time series models, namely the autoregressive and moving average models, as well as extensions that are more complex to handle like stochastic volatility models used in finance.

This chapter also covers the more complex dependence structure found in hidden Markov models, while spatial dependence in considered in Chap. 8. The reader should be aware that, due to mathematical constraints related to the long-term stability of the series, this chapter contains more advanced material, although we restrained from introducing complex simulation procedures on variable-dimension spaces.

J.-M. Marin and C.P. Robert, *Bayesian Essentials with R*, Springer Texts in Statistics, DOI 10.1007/978-1-4614-8687-9_7,

7.1 Time-Indexed Data

While we started with independent (and even iid) observations, for the obvious reason that they are easier to process, we soon departed from this setup, gathering more complexity either through heterogeneity as in the linear and generalized linear models (Chaps. 3 and 4), or through some dependence structure as in the open capture–recapture models of Chap. 5 that pertain to the generic notion of hidden Markov models covered in Sect. 7.5.

7.1.1 Setting

This chapter concentrates on time-series (or *dynamic*) models, which somehow appear to be simpler because they are unidimensional in their dependence, being indexed only by time. Their mathematical validation and estimation are however not so simple, while they are some of the most commonly used models in applications, ranging from finance and economics to reliability, to medical experiments, and ecology. This is the case, for instance, for series of pollution data, such as ozone concentration levels, or stock market prices, whose value at time t depends at least on the previous value at time $t-1$.

The dataset we use in this chapter is a collection of four time series connected with the stock market. Figure 7.1 plots the successive values from January 1, 1998, to November 9, 2003, of those four stocks[1] which are the first ones (in alphabetical order) to appear in the financial index Eurostoxx50, a financial reference for the euro zone[2] made of 50 major stocks. These four series constitute the **Eurostoxx50** dataset. A perusal of these graphs is sufficient for rejecting the assumption of independence of these series: High values are followed by high values and small values by small values, even though the variability (or *volatility*) of the stocks varies from share to share.

The simplest mathematical structure for a time series is when the series (x_t) is Markov. We recall that a *stochastic process* $(x_t)_{t\in\mathcal{T}}$, that is, a sequence of random variables indexed by the t's in $\mathcal{T}$ (where, here, $\mathcal{T}$ is equal to $\mathbb{N}$ or $\mathbb{Z}$) is a Markov chain when the distribution of x_t conditional on the past values (for instance, $\mathbf{x}_{0:(t-1)} = (x_0 \ldots, x_{t-1})$ when $\mathcal{T} = \mathbb{N}$) only depends on x_{t-1}. This process is *homogeneous* if the distribution of x_t conditional on the past

[1] The four stocks are as follows. ABN Amro is an international bank from Holland. Aegon is a Dutch insurance company. Ahold Kon., namely Koninklijke Ahold N.V., is also a Dutch company, dealing in retail and food-service businesses. Air Liquide is a French company specializing in industrial and medical gases.

[2] At the present time, the euro zone is made up of the following countries: Austria, Belgium, Finland, France, Germany, Greece, Holland, Ireland, Italy, Portugal, and Spain.

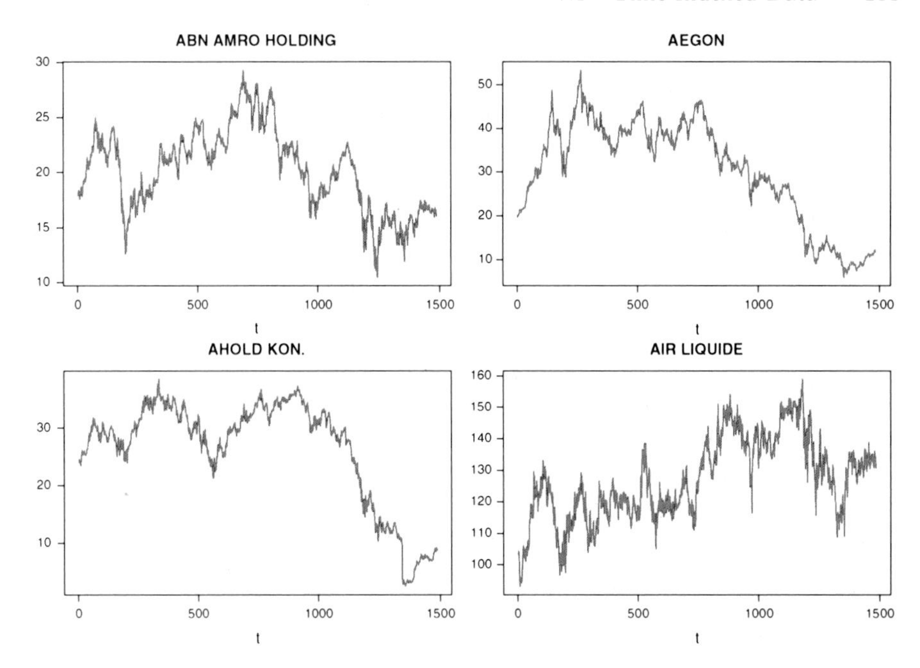

Fig. 7.1. Dataset **Eurostoxx50**: Evolution of the first four stocks over the period January 1, 1998 to November 9, 2003

is constant in $t \in \mathcal{T}$. Thus, given an observed sequence $\mathbf{x}_{0:T} = (x_0, \ldots, x_T)$ from a homogeneous Markov chain, the associated likelihood is given by

$$\ell(\boldsymbol{\theta}|\mathbf{x}_{0:T}) = f_0(x_0|\boldsymbol{\theta}) \prod_{t=1}^{T} f(x_t|x_{t-1}, \boldsymbol{\theta}) ,$$

where f_0 is the distribution of the starting value x_0. From a Bayesian point of view, this likelihood can be processed almost as in an iid model once a prior distribution on $\boldsymbol{\theta}$ is chosen.

However, a generic time series may be represented in formally the same way, namely through the full conditionals as in

$$\ell(\boldsymbol{\theta}|\mathbf{x}_{0:T}) = f_0(x_0|\boldsymbol{\theta}) \prod_{t=1}^{T} f_t(x_t|\mathbf{x}_{0:(t-1)}, \boldsymbol{\theta}) . \tag{7.1}$$

When this function can be obtained in a closed form, a Bayesian analysis is equally possible.

Note that general time-series models can often be represented as Markov models via the inclusion of missing variables and an increase in the dimension of the model. This is called a *state-space representation.*

7.1.2 Stability of Time Series

While we pointed out above that, once the likelihood function is written down, the Bayesian processing of the model is the same as in the iid case,[3] there exists a major difference that leads to a more delicate determination of the corresponding prior distributions in that the new properties of *stationarity* and *causality* constraints must often be accounted for. We cannot embark here on a mathematically rigorous coverage of stationarity for stochastic processes or even for time series (see Brockwell and Davis, 1996), thus simply mention (and motivate) below the constraints found in the time series literature.

A stochastic process $(x_t)_{t\in\mathcal{T}}$ is stationary[4] if the joint distributions of $(x_1,\ldots,x_k)$ and $(x_{1+h},\ldots,x_{k+h})$ are the same for all indices k and $k+h$ in $\mathcal{T}$. Formally, this property is called *strict stationarity* because there exists an alternative version of stationarity, called *second-order stationarity.* This alternative imposes invariance in time only on first and second moments of the process. If we define the autocovariance function $\gamma_x(\cdot,\cdot)$ of the process $(x_t)_{t\in\mathcal{T}}$ by

$$\gamma_x(r,s)=\mathbb{E}[\{x_r-\mathbb{E}(x_r)\}\{x_s-\mathbb{E}(x_s)\}],\quad r,s\in\mathcal{T},$$

namely the covariance between x_r and x_s, $\text{cov}(x_r,x_s)$, assuming that the variance $\mathbb{V}(x_t)$ is finite, a process $(x_t)_{t\in\mathcal{T}}$ with finite second moments is *second-order stationary* if

$$\mathbb{E}(x_t)=\mu\quad\text{and}\quad\gamma_x(r,s)=\gamma_x(r+t,s+t)$$

for all $r,s,t\in\mathcal{T}$.

If $(x_t)_{t\in\mathcal{T}}$ is second-order stationary, then $\gamma_x(r,s)=\gamma_x(|r-s|,0)$ for all $r,s\in\mathcal{T}$. It is therefore convenient to redefine the autocovariance function of a second-order stationary process as a function of just one variable; i.e., with a slight abuse of notation,

$$\gamma_x(h)\equiv\gamma_x(h,0),\quad h\in\mathcal{T}.$$

---

[3]In the sense that, once a closed form of the posterior is available as in (7.1), there exist generic simulation techniques that do not take into account the dynamic structure of the model.

[4]The connection with the stationarity requirement of MCMC methods is that these methods produce a Markov kernel such that, when the Markov chain is started at time $t=0$ from the target distribution π, the whole sequence $(x_t)_{t\in\mathbb{N}}$ is stationary with marginal distribution π.

The function $\gamma_x(\cdot)$ is called the *autocovariance function* of $(x_t)_{t\in\mathcal{T}}$, and $\gamma_x(h)$ is said to be the autocovariance "at lag" h.

The autocorrelation function is implemented in R as `acf()`, already used in Chaps. 4 and 5 for computing the effective sample size of an MCMC sample. By default, the function `acf()` returns $10\log_{10}(m)$ autocorrelations when applied to a series (vector) of size m, the autocovariances being obtained with the option `type="covariance"`, and it also produces a graph of those autocorrelations unless the option `plot=FALSE` is activated.

An illustration of `acf()` for the ABN Amro stock series is given by

```
> data(Eurostoxx50)
> abnamro=Eurostoxx50[,2]
> abnamro=ts(abnamro,freq=365-55*2,start=1998)
> par(mfrow=c(2,2),mar=c(4,4,1,1))
> plot.ts(abnamro,col="steelblue")
> acf(abnamro,lag=365-55*2)
> plot.ts(diff(abnamro),col="steelblue")
> acf(diff(abnamro))
```

whose graphical output is given in Fig. 7.2. The `ts` function turns the vector of ABN Amro stocks into a time series, which explains for the years on the first axis in `plot.ts` and the relative values on the first axis in `acf`, where `1.0` corresponds to a whole year. (The range of a year is computed by adding six bank holidays per year to the weekend breaks.) The second row corresponds to the time-series representation of the first difference $(x_{t+1} - x_t)$, a standard approach used to remove the clear lack of stationarity of the original series. The difference in the autocorrelation graphs is striking: in particular, the complete lack of significant autocorrelation in the first difference is indicative of a random walk behavior for the original series.

Obviously, strict stationarity is stronger than second-order stationarity, and this feature somehow seems more logical from a Bayesian viewpoint as it is a property of the whole model.[5] For a process $(x_t)_{t\in\mathbb{N}}$, this property relates to the distribution f_0 of the starting values.

From a Bayesian point of view, to impose the *stationarity* condition on a model (or rather on its parameters) is however objectionable on the grounds that the data themselves should indicate whether or not the underlying model is stationary. In addition, since the datasets we consider are always finite, the stationarity requirement is at best artificial in practice. For instance, the series in Fig. 7.1 are clearly not stationary on the temporal scale against which they are plotted. However, for reasons ranging from asymptotics (Bayes estimators are not necessarily convergent in nonstationary settings) to causality,

[5]Nonetheless, there exists a huge amount of literature on the study of time series based only on second-moment assumptions.

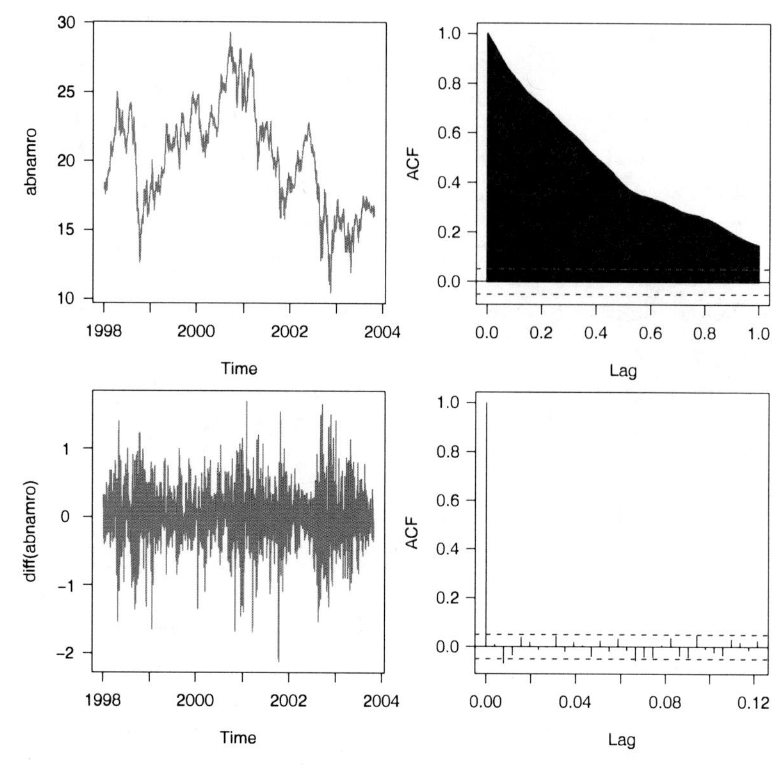

Fig. 7.2. Representations (*left*) of the time series ABN Amro (*top*) and of its first difference (*bottom*), along with the corresponding `acf` graphs (*right*)

to identifiability (see below), and to common practice, it is customary to impose stationarity constraints, possibly on transformed data, even though a Bayesian inference on a nonstationary process could be conducted in principle. The practical difficulty is that, for complex models, the stationarity constraints may get quite involved and may even be unknown in some cases, as for some threshold or changepoint models. We will expose (and solve) this difficulty in the following sections.

7.2 Autoregressive (AR) Models

In this section, we consider one of the most common (linear) time series models, the AR(p) model, along with its Bayesian analyses and its Markov connections (which can be exploited in some MCMC implementations).

7.2.1 The Models

An AR(1) process $(x_t)_{t\in\mathbb{Z}}$ (where AR stands for autoregressive) is defined by the conditional relation ($t \in \mathbb{Z}$),

$$x_t = \mu + \varrho(x_{t-1} - \mu) + \epsilon_t \,, \tag{7.2}$$

where $(\epsilon_t)_{t\in\mathbb{Z}}$ is an iid sequence of random variables with mean 0 and variance σ^2 (that is, a so-called *white noise*). Unless otherwise specified, we will only consider the ϵ_t's to be iid $\mathscr{N}(0,\sigma^2)$ variables.[6]

If $|\varrho| < 1$, $(x_t)_{t\in\mathbb{Z}}$ can be written as

$$x_t = \mu + \sum_{j=0}^{\infty} \varrho^j \epsilon_{t-j} \,, \tag{7.3}$$

and it is easy to see that this is a unique second-order stationary representation. More surprisingly, if $|\varrho| > 1$, the unique second-order stationary representation of (7.2) is

$$x_t = \mu - \sum_{j=1}^{\infty} \varrho^{-j} \epsilon_{t+j} \,.$$

This stationary solution is frequently criticized as being artificial because it implies that x_t is correlated with the *future* white noises $(\epsilon_t)_{s>t}$, a property not shared by (7.3) when $|\varrho| < 1$. While mathematically correct, the fact that x_t appears as a weighted sum of random variables that are generated after time t is indeed quite peculiar, and it is thus customary to restrict the definition of AR(1) processes to the case $|\varrho| < 1$ so that x_t has a representation in terms of the past realizations $(\epsilon_t)_{s\leq t}$. Formally, this restriction corresponds to so-called *causal* or future-independent autoregressive processes.[7] Notice that the causal constraint for the AR(1) model can be naturally associated with a uniform prior on $(-1, 1)$.

Note that, when we replace the above normal sequence (ϵ_t) with another white noise sequence, it is possible to express an AR(1) process with $|\varrho| > 1$ as an AR(1) process with $|\varrho| < 1$. However, this modification is not helpful from a Bayesian point of view because of the complex distribution of the transformed white noise.

[6]Once again, there exists a statistical approach that leaves the distribution of the ϵ_t's unspecified and only works with first and second moments. But this perspective is clearly inappropriate within the Bayesian framework, which cannot really work with half-specified models.

[7]Both stationary solutions above exclude the case $|\varrho| = 1$. This is because the process (7.2) is then a random walk with no stationary solution.

A natural generalization of the AR(1) model is obtained by increasing the lag dependence on the past values. An AR(p) process is thus defined by the conditional (against the past) representation ($t \in \mathbb{Z}$),

$$x_t = \mu + \sum_{i=1}^{p} \varrho_i(x_{t-i} - \mu) + \epsilon_t \,, \tag{7.4}$$

where $(\epsilon_t)_{t\in\mathbb{Z}}$ is a white noise. As above, we will assume implicitly that the white noise is normally distributed. This natural generalization assumes that the p most recent values of the process influence (linearly) the current value of the process. As for the AR(1) model, stationarity and causality constraints can be imposed on this model.

A lack of stationarity of a time series theoretically implies that the series ultimately diverges to $\pm\infty$. An illustration of this property is provided by the following R code, which produces four AR(10) series of 260 points based on the same ϵ_t's when the coefficients ϱ_i are uniform over $(-.5, .5)$. The first and the last series either have coefficients that satisfy the stationarity conditions or have not yet exhibited a divergent trend. Both remaining series clearly exhibit divergence.

```
> p=10
> T=260
> dat=seqz=rnorm(T)
> par(mfrow=c(2,2),mar=c(2,2,1,1))
> for (i in 1:4){
+   coef=runif(p,min=-.5,max=.5)
+   for (t in ((p+1):T))
+     seqz[t]=sum(coef*seqz[(t-p):(t-1)])+dat[t]
+ plot(seqz,ty="l",col="sienna",lwd=2,ylab="")
+ }
```

As shown in Brockwell and Davis (1996, Theorem 3.1.1), the AR(p) process (7.4) is both causal and second-order stationary if and only if the roots of the polynomial

$$\mathcal{P}(u) = 1 - \sum_{i=1}^{p} \varrho_i u^i \tag{7.5}$$

are all outside the unit circle in the complex plane. (Remember that polynomials of degree p always have p roots, but that some of those roots may be complex numbers.) While this necessary and sufficient condition on the parameters ϱ_i is clearly defined, it also imposes an *implicit* constraint on the vector $\boldsymbol{\varrho} = (\varrho_1, \ldots, \varrho_p)$. Indeed, in order to verify that a given vector $\boldsymbol{\varrho}$ satisfies this condition, one needs first to find the roots of the pth degree polynomial $\mathcal{P}$ *and then* to check that these roots all are of modulus larger than 1. In other words, there is no clearly defined boundary on the parameter space to define

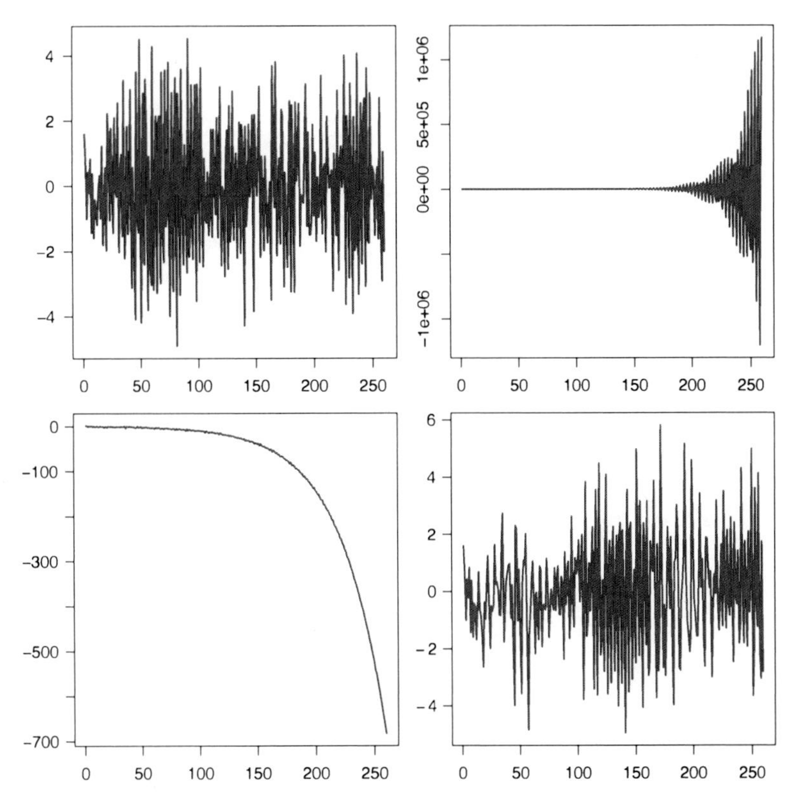

Fig. 7.3. Four simulation of an AR(10) series of 260 points when based on the same standard normal perturbations ϵ_t and when the coefficients ϱ_i are uniform over $(-.5, .5)$

the $\boldsymbol{\varrho}$'s that satisfy (or do not satisfy) this constraint, and this creates a major difficulty for simulation applications, given that simulated values of $\boldsymbol{\varrho}$ need to be tested one at a time. For instance, the R code

```
> maxi=0
> for (i in (1:10^6)) maxi=maxi+
+     (max(Mod(polyroot(c(1,runif(10,-.5,.5)))))>1)
> maxi/10^6
[1] 1
```

shows that no simulation out of one million simulated coefficients for the AR(10) model that satisfy the constraint. It is therefore very likely that all series in Fig. 7.3 are non-stationary.

Note that the general AR(p) model is Markov, just like the AR(1) model, because the distribution of x_{t+1} only depends on a fixed number of past values. It can thus be expressed as a regular Markov chain when considering the vector, for $t \geq p-1$,

$$\mathbf{z}_t = (x_t, x_{t-1}, \ldots, x_{t+1-p})^{\mathsf{T}} = \mathbf{x}_{t:(t+1-p)} \,.$$

Indeed, we can write

$$\mathbf{z}_{t+1} = \mu \mathbf{1}_p + B(\mathbf{z}_t - \mu \mathbf{1}_p) + \varepsilon_{t+1} \,, \tag{7.6}$$

where

$$\mathbf{1}_p = (1, \ldots, 1)^{\mathsf{T}} \in \mathbb{R}^p \,, \quad B = \begin{pmatrix} \varrho_1 & \varrho_2 & \varrho_3 & \ldots & \varrho_{p-2} & \varrho_{p-1} & \varrho_p \\ 1 & 0 & & \ldots & & & 0 \\ 0 & 1 & 0 & \ldots & 0 & 0 & 0 \\ \vdots & & & & & & \vdots \\ 0 & & & \ldots & 0 & 1 & 0 \end{pmatrix} ,$$

and $\varepsilon_t = (\epsilon_t, 0, \ldots, 0)^{\mathsf{T}}$.

If we now consider the likelihood associated with a series $\mathbf{x}_{0:T}$ of observations from a Gaussian AR(p) process, it depends on the unobserved values $x_{-p}, \ldots, x_{-1}$ since

$$\ell(\mu, \varrho_1, \ldots, \varrho_p, \sigma | \mathbf{x}_{0:T}, \mathbf{x}_{-p:-1}) \propto \\ \sigma^{-T-1} \prod_{t=0}^{T} \exp \left\{ - \left[x_t - \mu - \sum_{i=1}^{p} \varrho_i (x_{t-i} - \mu) \right]^2 / 2\sigma^2 \right\} .$$

These unobserved initial values can be processed in various ways that we now describe. First, they can all be set equal to μ, but this is a purely computational convenience with no justification. Second, if the stationarity and causality constraints hold, the process $(x_t)_{t \in \mathbb{Z}}$ has a stationary distribution and one can assume that $\mathbf{x}_{-p:-1}$ is distributed from the corresponding stationary distribution, namely a $\mathscr{N}_p(\mu \mathbf{1}_p, \mathbf{A})$ distribution. We can then integrate those initial values out to obtain the marginal likelihood

$$\int \sigma^{-T-1} \prod_{t=0}^{T} \exp \left\{ \frac{-1}{2\sigma^2} \left(x_t - \mu - \sum_{i=1}^{p} \varrho_i (x_{t-i} - \mu) \right)^2 \right\} f(\mathbf{x}_{-p:-1} | \mu, \mathbf{A}) \, \mathrm{d}\mathbf{x}_{-p:-1} \,,$$

based on the argument that they are not directly observed. This likelihood can be dealt with analytically but is more easily processed via a Gibbs sampler that simulates the initial values. An alternative and equally coherent approach is to consider instead the likelihood conditional on the initial *observed* values $\mathbf{x}_{0:(p-1)}$; that is,

$$\ell^c(\mu, \varrho_1, \ldots, \varrho_p, \sigma|\mathbf{x}_{p:T}, \mathbf{x}_{0:(p-1)}) \propto$$
$$\sigma^{-T+p-1} \prod_{t=p}^{T} \exp\left\{ - \left[x_t - \mu - \sum_{i=1}^{p} \varrho_i (x_{t-i} - \mu) \right]^2 \Big/ 2\sigma^2 \right\} . \quad (7.7)$$

Unless specified otherwise, we will adopt this approach. In this case, if we do not restrict the parameter space through stationarity conditions, a natural conjugate prior can be found for the parameter $\boldsymbol{\theta} = (\mu, \boldsymbol{\varrho}, \sigma^2)$, made up of a normal distribution on $(\mu, \boldsymbol{\varrho})$ and an inverse gamma distribution on σ^2. Instead of the Jeffreys prior, which is controversial in this setting (see Robert, 2007, Note 4.7.2), we can also propose a more traditional noninformative prior such as $\pi(\boldsymbol{\theta}) = 1/\sigma^2$.

7.2.2 Exploring the Parameter Space by MCMC Algorithms

If we do impose the causal stationarity constraint on $\boldsymbol{\varrho}$ that all the roots of $\mathcal{P}$ in (7.5) be outside the unit circle, the set of acceptable $\boldsymbol{\varrho}$'s becomes quite involved and we cannot, for instance, use as prior distribution a normal distribution restricted to this set, if only because we lack a simple algorithm to properly describe the set. While a feasible solution is based on the partial autocorrelations of the AR(p) process (see Robert, 2007, Sect. 4.5.2), we cover here a different and somehow simpler reparameterization approach using the inverses of the real *and* complex roots of the polynomial $\mathcal{P}$, which are within the unit interval $(-1, 1)$ and the unit sphere, respectively. Because of this unusual structure of the parameter space, involving two subsets of completely different natures, we introduce an MCMC algorithm that could be related with *birth and death processes* and simulation in variable dimension spaces.

If we represent the polynomial (7.5) in its factorized form

$$\mathcal{P}(x) = \prod_{i=1}^{p} (1 - \lambda_i x) ,$$

the inverse roots, λ_i $(i = 1, \ldots, p)$, are either real numbers or complex conjugates.[8] Under the causal stationarity constraint, a natural prior is then to use uniform priors for these roots, taking a uniform distribution on the number r_p of conjugate complex roots and uniform distributions on $[-1, 1]$ and on the unit sphere $\mathscr{S} = \{\lambda \in \mathbb{C}; |\lambda| \leq 1\}$ for the real and nonconjugate complex roots, respectively. In other words,

$$\pi(\boldsymbol{\lambda}) = \frac{1}{\lfloor p/2 \rfloor + 1} \prod_{\lambda_i \in \mathbb{R}} \frac{1}{2} \mathbb{I}_{|\lambda_i| < 1} \prod_{\lambda_i \notin \mathbb{R}} \frac{1}{\pi} \mathbb{I}_{|\lambda_i| < 1} , \quad (7.8)$$

[8] The term *conjugate* is to be understood here in the complex calculus sense that if $\iota^2 = -1$ defines the standard root of -1, $\lambda = \mathfrak{r}\, e^{\iota\theta}$ is a (complex) root of $\mathcal{P}$, then $\bar{\lambda} = \mathfrak{r}\, e^{-\iota\theta}$ is also a (complex) root of $\mathcal{P}$.

where $\lfloor p/2 \rfloor + 1$ is the number of different values of r_p and the second product is restricted to the nonconjugate roots of $\mathcal{P}$. (Note that the quantity π in the denominator is in fact the surface of the unit sphere of $\mathbb{C}$.)

Note that this $\lfloor p/2 \rfloor + 1$ factor, while unimportant for a fixed p setting, must necessarily be included within the posterior distribution when using a birth-and-death MCMC algorithm to estimate the lag order p since it does not vanish in the acceptance probability of a move between an AR(p) model and an AR(q) model.

While the connection between the inverse roots and the coefficients of the polynomial $\mathcal{P}$ is straightforward (Exercise 7.10), there is no closed-form expression of the posterior distribution either on the roots or on the coefficients. Therefore, a numerical approach is once again compulsory to approximate the posterior distribution. However, *any* Metropolis–Hastings scheme can work here, given that the likelihood function can be easily computed in every point.

The derivation of the coefficients ϱ_i of the autoregressive model from the roots follows from a recursive linear procedure explained in Exercise 7.10. If `pr` and `pc` denote the number of real and complex roots and if `lr` and `lc` are the real and (non-conjugate) complex roots, respectively, the former being a vector and the latter a two-column matrix, we have

```
Psi=matrix(0,ncol=p,nrow=p+1)
Psi[1,]=1
if (pr>0){
  Psi[2,1]=-lr[1]
  if (pr>1){
    for (i in 2:pr)
      Psi[2:(i+1),i]=Psi[2:(i+1),i-1]-lr[i]*Psi[1:i,i-1]
    }
  }
if (pc>0){
  if (pr>0){
    Psi[2,pr+2]=-2*lc[1]+Psi[2,pr]
    Psi[3:(pr+3),pr+2]=(lc[1]^2+lc[2]^2)*Psi[1:(pr+1),pr]
             -2*lc[1]*Psi[2:(pr+2),pr]+Psi[3:(pr+3),pr]
         }else{
    Psi[2,2]=-2*lc[1];
    Psi[3,2]=(lc[1]^2+lc[2]^2);
    }
  if (pc>2){
  for (i in seq(4,pc,2)){
     pri=pr+i
     prim=pri-2
     Psi[2,pri]=-2*lc[i-1]+Psi[2,prim]
```

```
      Psi[3:(pri+1),pri]=(lc[i-1]^2+lc[i]^2)*Psi[1:(pri-1),
           prim]-2*lc[i-1]*Psi[2:pri,prim]+Psi[3:(pri+1),prim]
      }
      }
   }
Rho=Psi[1:(p+1),p]
```

where the ρ_i's are the opposites of the components of `Rho[2:p]`. The log-likelihood (7.7) is then derived in a straightforward manner:

```
x=x-mu
loglike=0
for (i in (p+1):T)
 loglike=loglike-(t(Rho)%*%x[i:(i-p)])^2
loglike=(loglike/sig2-(T-p)*log(sig2))/2
x=x+mu
```

Since the conditional likelihood function (7.7) is a standard Gaussian likelihood in both μ and σ, we can directly use a Gibbs sampler on those parameters and opt for a Metropolis-within-Gibbs step on the remaining (inverse) roots of $\mathcal{P}$, $\boldsymbol{\lambda} = (\lambda_1, \ldots, \lambda_p)$. A potentially inefficient[9] if straightforward Metropolis–Hastings implementation is to use the prior distribution $\pi(\boldsymbol{\lambda})$ itself as a proposal on $\boldsymbol{\lambda}$. This means selecting first one or several roots of $\mathcal{P}$, $\lambda_1, \ldots, \lambda_q$ $(1 \le q \le p)$, and then proposing new values for these roots that are simulated from the prior, $\lambda'_1, \ldots, \lambda'_q \sim \pi(\boldsymbol{\lambda})$. (Reordering the roots so that the modified values are the first ones is not restrictive since both the prior and the likelihood are permutation invariant.) The acceptance ratio then simplifies into the likelihood ratio by virtue of Bayes' theorem:

$$\frac{\ell(\mu, \boldsymbol{\lambda}', \sigma|\mathbf{x}_{p:T}, \mathbf{x}_{0:(p-1)})\, \pi(\mu, \boldsymbol{\lambda}', \sigma)}{\ell(\mu, \boldsymbol{\lambda}, \sigma|\mathbf{x}_{p:T}, \mathbf{x}_{0:(p-1)})\, \pi(\mu, \boldsymbol{\lambda}, \sigma)} \frac{\pi(\boldsymbol{\lambda})}{\pi(\boldsymbol{\lambda}')} = \frac{\ell(\mu, \boldsymbol{\lambda}', \sigma|\mathbf{x}_{p:T}, \mathbf{x}_{0:(p-1)})}{\ell(\mu, \boldsymbol{\lambda}, \sigma|\mathbf{x}_{p:T}, \mathbf{x}_{0:(p-1)})}$$

The main difficulty with this scheme is that one must take care to modify complex roots by (conjugate) pairs. This means, for instance, that to create a complex root (and its conjugate) either another complex root (and its conjugate) or two real roots must be chosen and modified. Formally, this is automatically satisfied by simulations from the prior (7.8).

One possible algorithmic representation is therefore:

[9]Simulating from the prior distribution when aiming at the posterior distribution is inevitably leading to a waste of simulations if the data is informative about the parameters. The solution is of course unavailable when the prior is improper.

Algorithm 7.13 Metropolis–Hastings AR(p) Sampler

Initialization: Choose $\boldsymbol{\lambda}^{(0)}$, $\mu^{(0)}$, and $\sigma^{(0)}$.
Iteration t ($t \geq 1$):

1. Select one root at random.
   If the root is real, generate a new real root from the prior distribution.
   Otherwise, generate a new complex root from the prior distribution and update the conjugate root.
   Replace $\boldsymbol{\lambda}^{(t-1)}$ with $\boldsymbol{\lambda}^{\star}$ using these new values.
   Calculate the corresponding $\boldsymbol{\varrho}^{\star} = (\varrho_1^{\star}, \ldots, \varrho_p^{\star})$.
   Take $\boldsymbol{\xi} = \boldsymbol{\lambda}^{\star}$ with probability
   $$\frac{\ell^c(\mu^{(t-1)}, \boldsymbol{\varrho}^{\star}, \sigma^{(t-1)} | \mathbf{x}_{p:T}, \mathbf{x}_{0:(p-1)})}{\ell^c(\mu^{(t-1)}, \boldsymbol{\varrho}^{(t-1)}, \sigma^{(t-1)} | \mathbf{x}_{p:T}, \mathbf{x}_{0:(p-1)})} \wedge 1,$$
   and $\boldsymbol{\xi} = \boldsymbol{\lambda}^{(t-1)}$ otherwise.
2. Select two real roots or two complex conjugate roots at random.
   If the roots are real, generate a new complex root from the prior distribution and compute the conjugate root.
   Otherwise, generate two new real roots from the prior distribution.
   Replace $\boldsymbol{\xi}$ with $\boldsymbol{\lambda}^{\star}$ using these new values.
   Calculate the corresponding $\boldsymbol{\varrho}^{\star} = (\varrho_1^{\star}, \ldots, \varrho_p^{\star})$.
   Accept $\boldsymbol{\lambda}^{(t)} = \boldsymbol{\lambda}^{\star}$ with probability
   $$\frac{\ell^c(\mu^{(t-1)}, \boldsymbol{\varrho}^{\star}, \sigma^{(t-1)} | \mathbf{x}_{p:T}, \mathbf{x}_{0:(p-1)})}{\ell^c(\mu^{(t-1)}, \boldsymbol{\varrho}^{(t-1)}, \sigma^{(t-1)} | \mathbf{x}_{p:T}, \mathbf{x}_{0:(p-1)})} \wedge 1,$$
   and set $\boldsymbol{\lambda}^{(t)} = \boldsymbol{\xi}$ otherwise.
3. Generate $\mu^{\star}$ by a random walk proposal.
   Accept $\mu^{(t)} = \mu^{\star}$ with probability
   $$\frac{\ell^c(\mu^{\star}, \boldsymbol{\varrho}^{(t)}, \sigma^{(t-1)} | \mathbf{x}_{p:T}, \mathbf{x}_{0:(p-1)})}{\ell^c(\mu^{(t-1)}, \boldsymbol{\varrho}^{(t)}, \sigma^{(t-1)} | \mathbf{x}_{p:T}, \mathbf{x}_{0:(p-1)})} \wedge 1,$$
   and set $\mu^{(t)} = \mu^{(t-1)}$ otherwise.
4. Generate $\sigma^{\star}$ by a log-random walk proposal.
   Accept $\sigma^{(t)} = \sigma^{\star}$ with probability
   $$\frac{\ell^c(\mu^{(t)}, \boldsymbol{\varrho}^{(t)}, \sigma^{\star} | \mathbf{x}_{p:T}, \mathbf{x}_{0:(p-1)})}{\ell^c(\mu^{(t)}, \boldsymbol{\varrho}^{(t)}, \sigma^{(t-1)} | \mathbf{x}_{p:T}, \mathbf{x}_{0:(p-1)})} \wedge 1,$$
   and set $\sigma^{(t)} = \sigma^{(t-1)}$ otherwise.

While the whole R code is too long (300 lines) to be reproduced here, the core part about the modification of the roots can be implemented as follows: "down" moves removing one pair of complex roots are chosen with probability 0.1 while "up" moves creating one pair of complex roots are chosen with probability 0.9, in order to compensate for the inherently higher difficulty in accepting complex proposals from the prior. Those uneven weights must then be accounted for in the acceptance probability, along with the changes in the masses of the uniform priors on the real and complex roots in (7.8).

```
if (runif(1)<.1){         #down
  ppropcomp=pcomp-2; ppropreal=preal+2
  ind=sample(1:pcomp,1) #indices of removed complex root
  ind=ind-(ind%%2==0)

  if (ppropcomp>0){
    lambpropcomp=lambdacomp[((1:pcomp)[-(ind:(ind+1))]
    }else{ #no complex root
      lambpropcomp=0 #dummy necessary for ARllog function
      }
  lambpropreal=c(lambdareal,2*runif(2)-1)

  coef=9*(1+(preal<2))*(pi/4)  #if new case is boundary

  }else{                 #up
   ppropreal=preal-2; ppropcomp=pcomp+2
   ind=sample(1:preal,2)  #indices of removed real roots

   if (ppropreal>0){
      lambpropreal=lambdareal[(1:preal)[-ind]]
      }else{
        lambpropreal=0 #dummy necessary for ARllog function
        }
   theta=2*pi*runif(1); rho=sqrt(runif(1))
   lambpropcomp=c(lambdacomp,rho*cos(theta),rho*sin(theta))

   coef=(4/pi)*(1+(ppropcomp<p-1))/9   #if new case is
   boundary
   }
```

the boundary cases with no complex root or less than two real roots requiring a special processing (not reproduced here). The Metropolis–Hastings acceptance step is then simple:

```
lloprop=ARllog(pr=ppropreal,pc=ppropcomp,
        lr=lambpropreal,lc=lambpropcomp,mu,sig2)

if (log(runif(1))<log(coef)+lloprop-llo){
```

```
llo=llloprop
preal=ppropreal; pcomp=ppropcomp
lambdacomp=lambpropcomp; lambdareal=lambpropreal
}
```

illustrating the role of the `coef` correction.

As an application of the above, we processed the Ahold Kon. series of **Eurostoxx50**. We ran the algorithm for the whole series with $p = 5$, with satisfactory jump behavior between the different numbers of complex roots. The same behavior can be observed with larger values of p. Note that a call to the non-Bayesian R ar( ) procedure gives an order of 1 for this series, as

```
> ar(x = Eurostoxx50[, 4])
Coefficients:
     1
0.9968
Order selected 1 sigma^2 estimated as  0.5399
```

This standard analysis is very unstable, for instance, using the following alternative produces a very different order estimate!

```
> ar(x = Eurostoxx50[, 4], method = "ml")
Coefficients:
      1       2       3      4       5       6       7       8
  1.042  -0.080  -0.038  0.080  -0.049   0.006   0.080  -0.043
Order selected 8 sigma^2 estimated as  0.3228
```

Figure 7.4 summarizes the MCMC output for 50,000 iterations. The top left graph shows that jumps between 2 and 0 complex roots occur with high frequency and therefore that the MCMC algorithm mixes well between both (sub)models. Both following graphs on the first row relate to the hyperparameters μ and σ, which are updated outside the reversible jump steps. The parameter μ appears to be mixing better than σ, which is certainly due to the choice of the same scaling factor in both cases. The middle rows correspond to the first three coefficients of the autoregressive model, $\varrho_1, \varrho_2, \varrho_3$. Their stability is a good indicator of the convergence of the reversible jump algorithm. Note also that, except for ϱ_1, the other coefficients are close to 0 (since their posterior means are approximately 0.052, -0.0001, 2.99×10^{-5}, and -2.66×10^{-7}, respectively). The final row is an assessment of the fit of the model and the convergence of the MCMC algorithm. The first graph provides the sequence of corresponding log-likelihoods, which remain stable almost from the start, the second the distribution of the complex (inverse) roots, and the last one the connection between the actual series and its one-step-ahead prediction $\mathbb{E}[X_{t+1}|x_t, x_{t-1}, \ldots]$: On this scale, both series are well-related.

While the above algorithm is a regular Metropolis–Hastings algorithm on a parameter space with a fixed number of parameters, $\varrho_1, \ldots, \varrho_p$, the real–complex dichotomy gives us the opportunity to mention a new class of MCMC algorithms, variable dimension MCMC algorithms. The class of *variable dimension models* is made of models characterized by a collection of submodels, $\mathfrak{M}_k$, often nested, that are considered simultaneously and associated with different parameter spaces. The number of submodels can be infinite, and the "parameter" is defined conditionally on the index of the submodel, $\boldsymbol{\theta} = (k, \theta_k)$, with a dimension that generally depends on k. It naturally occurs in settings like Bayesian model choice and Bayesian model assessment.

Inference on such structures is obviously more complicated than on single models, especially when there are an infinite number of submodels, and it can

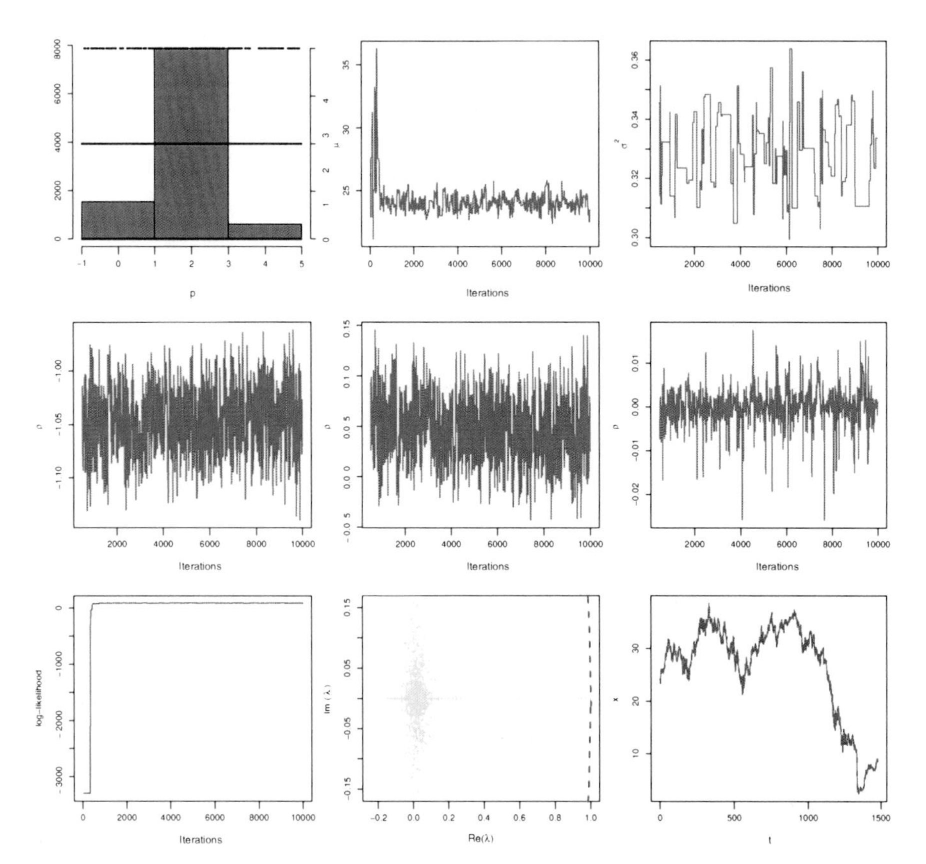

Fig. 7.4. Dataset **Eurostoxx50**: Output of the MCMC algorithm for the Ahold Kon. series and an AR(5) model: (*top row, left*) histogram and sequence of numbers of complex roots (ranging from 0 to 4), (*top row, middle and right*) sequence of μ and σ^2, (*middle row*) sequences of ϱ_i ($i = 1, 2, 3$), (*bottom row, left*) sequence of observed log-likelihood, (*bottom row, middle*) representation of the cloud of complex roots, with a part of the boundary of the unit circle on the right, (*bottom row, right*) comparison of the series and the one-step-ahead prediction

be tackled from two different (or even opposite) perspectives. The first approach is to consider the variable dimension model as a whole and to estimate quantities that are meaningful for the whole model (such as moments or predictives) as well as quantities that only make sense for submodels (such as posterior probabilities of submodels and posterior moments of θ_k). From a Bayesian perspective, once a prior is defined on $\boldsymbol{\theta}$, the only difficulty is in finding an efficient way to explore the complex parameter space in order to produce these estimators. The second perspective on variable dimension models is to resort to *testing*, rather than estimation, by adopting a *model choice* stance. This requires choosing among all possible submodels the "best one" in terms of an appropriate criterion. The drawbacks of this second approach are far from benign. The computational burden may be overwhelming when the number of models is infinite, the interpretation of the selected model is delicate and the variability of the resulting inference is underestimated since it is impossible to include the effect of the selection of the model in the assessment of the variability of the estimators built in later stages. Nonetheless, this is an approach often used in linear and generalized linear models (Chaps. 3 and 4) where subgroups of covariates are compared against a given dataset. It is obviously the recommended approach when the number of models is small, as in the mixture case (Chap. 6) or in the selection of the order p of an AR(p) model, provided the Bayes factors can be approximated.

MCMC algorithms that can handle such variable-dimension structures are facing measure theoretic difficulties and, while a universal and elegant solution through reversible jump algorithms exists (Green, 1995), we have made the choice of not covering these in this book. An introductory coverage can be found in the earlier edition (Marin and Robert, 2007, Sect. 6.7), as well as in Robert and Casella (2004). Nonetheless, we want to point out that the above MCMC algorithm happens to be a special case of the birth-and-death MCMC algorithm (and of its generalization, the *reversible jump algorithm*) where, in nested models, additional components are generated from the prior distribution and the move to a larger model is accepted with a probability equal to the ratio of the likelihoods (with the proper reweighting to account for the multiplicity of possible moves). For instance, extending the above algorithm to the case of the unknown order p is straightforward.

7.3 Moving Average (MA) Models

A second type of time series model that still enjoys linear dependence and closed-form expression is the MA(q) model, where MA stands for moving average. It appears as a dual version of the AR(p) model.

An MA(1) process $(x_t)_{t\in\mathbb{Z}}$ is such that, conditionally on the past

$$x_t = \mu + \epsilon_t - \vartheta\epsilon_{t-1}\,, \qquad t \in \mathcal{T}\,, \tag{7.9}$$

where $(\epsilon_t)_{t\in\mathcal{T}}$ is a white noise sequence. For the same reasons as above, we will assume the white noise is normally distributed unless otherwise specified. Thus,

$$\mathbb{E}[x_t] = \mu\,,\ \mathbb{V}(x_t) = (1+\vartheta^2)\sigma^2\,,\ \gamma_x(1) = \vartheta\sigma^2\,, \quad \text{and} \quad \gamma_x(h) = 0 \quad (h > 1)\,.$$

An important feature of (7.9) is that the model is not identifiable per se. Indeed, we can also rewrite x_t as

$$x_t = \mu + \tilde{\epsilon}_{t-1} - \frac{1}{\vartheta}\tilde{\epsilon}_t, \qquad \tilde{\epsilon} \sim \mathcal{N}(0, \vartheta^2\sigma^2)\,.$$

Therefore, both pairs (ϑ, σ) and $(1/\vartheta, \vartheta\sigma)$ are equivalent representations of the *same* model. To achieve identifiability, it is therefore customary in (non-Bayesian environments) to restrict the parameter space of MA(1) processes by

$$|\vartheta| < 1\,,$$

and we will follow suit. Such processes are called invertible. As with causality, the property of inversibility is not a property of the sole process $(x_t)_{t\in\mathbb{Z}}$ but of the connection between the two processes $(x_t)_{t\in\mathcal{T}}$ and $(\epsilon_t)_{t\in\mathcal{T}}$.

A natural extension of the MA(1) model is to increase the dependence on the past innovations, namely to introduce the MA(q) process as the process $(x_t)_{t\in\mathcal{T}}$ defined by

$$x_t = \mu + \epsilon_t - \sum_{i=1}^{q} \vartheta_i \epsilon_{t-i}\,, \tag{7.10}$$

where $(\epsilon_t)_{t\in\mathcal{T}}$ is a white noise (once again assumed to be normal unless otherwise specified). The corresponding identifiability condition in this model is that the roots of the polynomial

$$\mathcal{Q}(u) = 1 - \sum_{i=1}^{q} \vartheta_i u^i$$

are all outside the unit circle in the complex plane (see Brockwell and Davis, 1996, Theorem 3.1.2, for a proof). Thus, we end up with exactly the same parameter space as in the AR(q) case!

The intuition behind the MA(q) representation is however less straightforward than the regression structure underlying the AR(p) model. This representation assumes that the dependence between observables stems from a dependence between the (unobserved) noises rather than directly through the observables. Furthermore, in contrast with the AR(p) models, where the covariance between the terms of the series is exponentially decreasing to zero but always different from 0, the autocovariance function for the MA(q) model is such that $\gamma_x(s)$ is equal to 0 for $|s| > q$, meaning that x_{t+s} and x_t are independent. In addition, the MA(q) process is obviously (second-order and strictly) *stationary*, whatever the vector $(\vartheta_1, \ldots, \vartheta_q)$, since the white noise is iid and the distribution of (7.10) is thus independent of t. A major difference between the MA(q) and the AR(p) models, though, is that the MA(q) dependence structure is not Markov (even though it can be represented as a Markov process through a *state-space representation*, introduced below).

While, in the Gaussian case, the whole (observed) vector $\mathbf{x}_{1:T}$ is a realization of a normal random variable, with constant mean μ and covariance matrix Σ, and thus provides a formally explicit likelihood function, both the computation and the integration (or maximization) of this likelihood are quite costly since they involve inverting the huge matrix Σ.[10]

A more manageable representation of the MA(q) likelihood is to use the likelihood of $\mathbf{x}_{1:T}$ conditional on the past white noises $\epsilon_0, \ldots, \epsilon_{-q+1}$,

$$\ell^c(\mu, \vartheta_1, \ldots, \vartheta_q, \sigma | \mathbf{x}_{1:T}, \epsilon_{(-q+1):0}) \propto$$

$$\sigma^{-T} \prod_{t=1}^{T} \exp\left\{ - \left(x_t - \mu + \sum_{j=1}^{q} \vartheta_j \hat{\epsilon}_{t-j} \right)^2 \Big/ 2\sigma^2 \right\}, \tag{7.11}$$

where $\hat{\epsilon}_0 = \epsilon_0, \ldots, \hat{\epsilon}_{1-q} = \epsilon_{1-q}$ and ($t > 0$)

$$\hat{\epsilon}_t = x_t - \mu + \sum_{j=1}^{q} \vartheta_j \hat{\epsilon}_{t-j} \,.$$

This recursive definition of the likelihood is still costly since it involves T sums of q terms. Nonetheless, even though the problem of handling the conditioning values $\epsilon_{(-q+1):0}$ must be treated separately via an MCMC step, the complexity $\mathrm{O}(Tq)$ of this representation is much more manageable than the normal exact representation mentioned above.

Since the transform of the roots into the coefficients is *exactly the same* as with the AR(q) model, the expression of the log-likelihood function conditional on the past white noises `eps` is quite straightforward. Taking for `Psi` the subvector `Psi[2:(p+1),p]`, the computation goes as follows:

```
x=x-mu
# construction of the epsilonhats
heps=rep(0,T+q)
heps[1:q]=eps   # past noises
for (i in 1:T)
  heps[p+i]=x[i]+sum(rev(Psi)*heps[i:(q+i-1)])
# completed loglikelihood (includes negative epsilons)
loglike=-((sum(heps^2)/sig2)+(T+q)*log(sig2))/2
x=x+mu
```

[10]Obviously, taking advantage of the block diagonal structure of Σ—due to the fact that $\gamma_x(s) = 0$ for $|s| > q$— may reduce the computational cost, but this requires advanced programming abilities!

Given both $\mathbf{x}_{1:T}$ and the past noises $\epsilon_{(-q+1):0}$, the conditional posterior distribution of the parameters $(\mu, \vartheta_1, \ldots, \vartheta_q, \sigma)$ is formally very close to the posterior associated with an AR(q) posterior distribution. This proximity is such that we can recycle the code of Algorithm 7.13 to some extent since the simulation of the (inverse) roots of the polynomial $\mathcal{Q}$ is identical once we modify the likelihood according to the above changes. The past noises ϵ_{-i} ($i = 1, \ldots, q$) are simulated conditional both on the x_t's and on the parameters μ, σ and $\boldsymbol{\vartheta} = (\vartheta_1, \ldots, \vartheta_q)$. While the exact distribution

$$f(\epsilon_{(-q+1):0}|\mathbf{x}_{1:T}, \mu, \sigma, \boldsymbol{\vartheta}) \propto \prod_{i=-q+1}^{0} e^{-\epsilon_i^2/2\sigma^2} \prod_{t=1}^{T} e^{-\hat{\epsilon}_t^2/2\sigma^2}, \tag{7.12}$$

where the $\hat{\epsilon}_t$'s are defined as above, is exactly a normal distribution on the vector $\epsilon_{(-q+1):0}$ (Exercise 7.13), its computation is too costly to be available for realistic values of T. We therefore implement a hybrid Gibbs algorithm where the missing noise $\epsilon_{(-q+1):0}$ is simulated from a proposal based either on the previous simulated value of $\epsilon_{(-q+1):0}$ (in which case we use a simple termwise random walk) or on the first part of (7.12) (in which case we can use normal proposals).[11] More specifically, one can express $\hat{\epsilon}_t$ ($1 \leq t \leq q$) in terms of the ϵ_{-t}'s and derive the corresponding (conditional) normal distribution on either each ϵ_{-t} or on the whole vector $\boldsymbol{\epsilon}$[12] (see Exercise 7.14).

The additional step, when compared with the AR(p) function, is the conditional simulation of the past noises $\epsilon_{(-q+1):0}$. For $1 \leq i \leq q$ (the indices are drifted to start at 1 rather than $-q$), the corresponding part of our R code is as follows. Unfortunately, the derivation of the Metropolis–Hastings acceptance probability does require computing the inverse $\hat{\epsilon}_{-t}$'s as they are functions of the proposed noises.

```
x=x-mu
heps[1:q]=eps       # simulated ones
for (j in (q+1):(2*p+1)) # epsilon hat
   heps[j]=x[j]+sum(rev(Psi)*heps[(j-q):(j-1)])

heps[i]=0
for (j in 1:(q-i+1))
   keps[j]=x[j]+sum(rev(Psi)*heps[j:(j+q-1)])
x=x+mu
```

[11] In the following output analysis, we actually used a more hybrid proposal with the innovations $\hat{\epsilon}_t$'s ($1 \leq t \leq q$) fixed at their previous values. This approximation remains valid when accounted for in the Metropolis–Hastings acceptance ratio, which requires computing the $\hat{\epsilon}_t$'s associated with the proposed ϵ_{-i}.

[12] Using the horizon $t = q$ is perfectly sensible in this setting given that $x_1, \ldots, x_q$ are the only observations correlated with the ϵ_{-t}'s, even though (7.11) gives the impression of the opposite, since all $\hat{\epsilon}_t$'s depend on the ϵ_{-t}'s.

```
epsvar=1/sum(c(1,Psi[i:q]^2))
epsmean=sum(Psi[i:q]*keps[1:(q-i+1)])*epsvar
epsmean=epsmean/epsvar
epsvar=sig2*epsvar

propeps=rnorm(1,mean=epsmean,sd=sqrt(epsvar))
epspr=eps
epspr[i]=propeps
lloprop=MAllog(pr=preal,pc=pcomp,lr=lambdareal,
    lc=lambdacomp,mu=mu,sig2=sig2,compsi=FALSE,pepsi=Psi,
    eps=epspr)
propsal1=dnorm(propeps,mean=epsmean,sd=sqrt(epsvar),log=TRUE)

x=x-mu
heps[i]=propeps
for (j in (q+1):(2*q+1))
   heps[j]=x[j]+sum(rev(Psi)*heps[(j-q):(j-1)])

heps[i]=0
for (j in 1:(q-i+1))
   keps[j]=x[j]+sum(rev(Psi)*heps[j:(j+q-1)])
x=x+mu

epsvar=1/sum(c(1,Psi[i:q]^2))
epsmean=sum(Psi[i:q]*keps[1:(q-i+1)])
epsmean=epsmean*epsvar
epsvar=sig2*epsvar
propsal0=dnorm(eps[i],mean=epsmean,sd=sqrt(epsvar),log=TRUE)

if (log(runif(1))<lloprop-llo-propsal1+propsal0){
   eps[i]=propeps;
   llo=lloprop
   }
```

The complete R code also includes an additional random walk perturbation of the ϵ_i, centered on the proposal

```
propeps = rnorm(1,mean=eps[i],sd=0.1*sqrt(sig2))
```

in order to increase the mixing properties of the chain. Apart from those changes, the R code is identical to the code used for the AR(p) model.

Algorithm 7.14 MCMC MA(q) SAMPLER

Initialization: Choose $\boldsymbol{\lambda}^{(0)}$, $\epsilon^{(0)}$, $\mu^{(0)}$, and $\sigma^{(0)}$ arbitrarily.
Iteration t ($t \geq 1$):

1. Run steps 1–4 of Algorithm 7.13 conditional on $\epsilon^{(t-1)}$ with the correct corresponding conditional likelihood.
2. Simulate $\epsilon^{(t)}$ by a Metropolis–Hastings step.

To illustrate the behavior of this algorithm, we considered the first 350 points of the Air Liquide series in **Eurostoxx50**. The output is represented on Fig. 7.5 for $q = 9$ and 10,000 iterations of Algorithm 7.14, with the same conventions as in Fig. 7.4, except that the lower right graph represents the series of the simulated ϵ_{-t}'s rather than the predictive behavior.

Interestingly, the likelihood found by the algorithm as the iteration proceeds is (numerically) much higher than the one found by the classical R arima procedure since it differs by a factor of 450 on the log scale (assuming we are talking of the same quantity since R arima computes the log-likelihood associated with the observations without the ϵ_{-i}'s!). The details of the call to arima are as follows:

```
> arima(x = Eurostoxx50[1:350, 5], order = c(0, 0, 9))
Coefficients:
         ma1     ma2     ma3     ma4     ma5     ma6     ma7
      1.0605  0.9949  0.9652  0.8542  0.8148  0.7486  0.5574
s.e.  0.0531  0.0760  0.0881  0.0930  0.0886  0.0827  0.0774
         ma8     ma9  intercept
      0.3386  0.1300  114.3146
s.e.  0.0664  0.0516  1.1281
sigma^2 estimated as 8.15:  log likelihood = -864.97
```

The favored number of complex roots is 6, and the smaller values 0 and 2 are not visited after the initial warmup. The mixing over the σ parameter is again lower than over the mean μ, despite the use of three different proposals. The first one is based on the inverted gamma distribution associated with $\widehat{\epsilon}_{-(q-1):q}$, the second one is based on a (log) random walk with scale $0.1\hat{\sigma}_x$, and the third one is an independent inverted gamma distribution with scale $\hat{\sigma}_x/(1 + \vartheta_1^2 + \ldots + \vartheta_q^2)^{1/2}$. Note also that, except for ϑ_9, the other coefficients ϑ_i are quite different from 0 (since their posterior means are approximately 1.0206, 0.8403, 0.8149, 0.6869, 0.6969, 0.5693, 0.2889, and 0.0895, respectively). This is also the case for the estimates above obtained in R arima. The prediction being of little interest for MA models (Exercise 7.15), we represent instead the range of simulated ϵ_t's in the bottom right figure. The range is compatible with the $\mathscr{N}(0, \sigma^2)$ distribution.

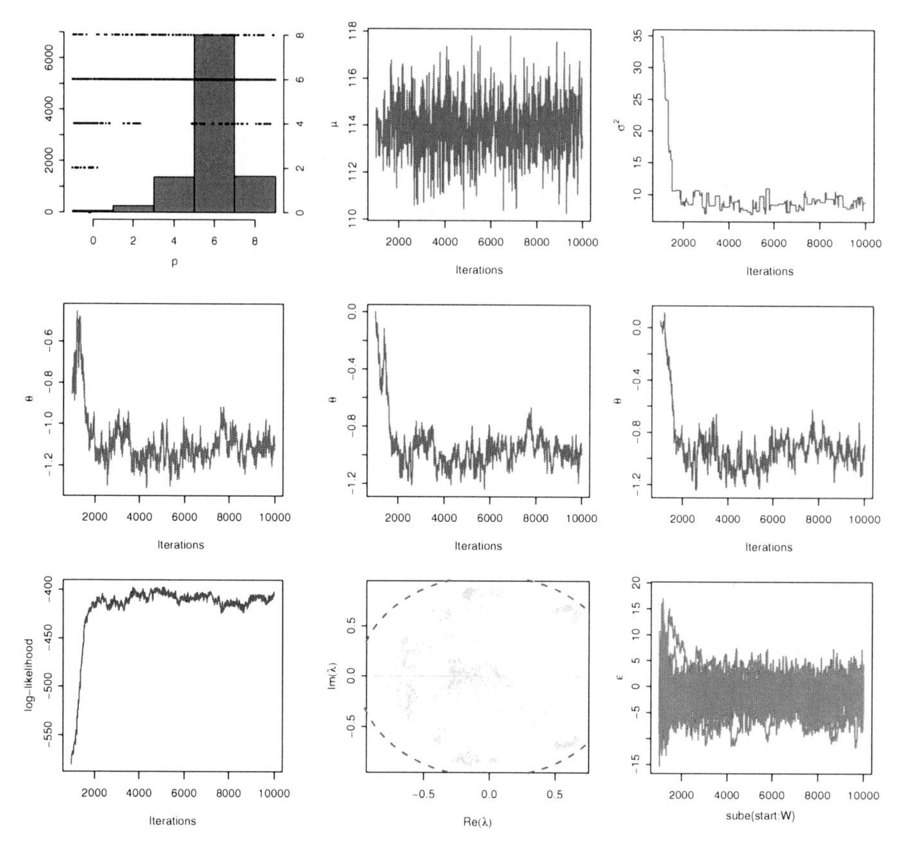

Fig. 7.5. Dataset **Eurostoxx50**: Output of the MCMC algorithm for the Air Liquide series and an MA(9) model: (*top row, left*) histogram and sequence of numbers of complex roots (ranging from 0 to 8); (*top row, middle and right*) sequence of μ and σ^2; (*middle row*) sequences of ϑ_i $(i = 1, 2, 3)$; (*bottom row, left*) sequence of observed likelihood; (*bottom row, middle*) representation of the cloud of complex roots, with the boundary of the unit circle; and (*bottom row, right*) evolution of the simulated ϵ_{-t}'s

7.4 ARMA Models and Other Extensions

An alternative approach that is of considerable interest for the representation and analysis of the MA(q) model and its generalizations is the so-called *state-space representation*, which relies on missing variables to recover both the Markov structure *and* the linear framework.[13]

The general idea is to represent a time series $(\mathbf{x}_t)$ as a system of two equations,

[13]It is also inspired from the *Kalman filter*, ubiquitous for prediction, smoothing, and filtering in time series.

$$\mathbf{x}_t = G\mathbf{y}_t + \boldsymbol{\varepsilon}_t\,, \tag{7.13}$$

$$\mathbf{y}_{t+1} = F\mathbf{y}_t + \boldsymbol{\xi}_t\,, \tag{7.14}$$

where $\boldsymbol{\varepsilon}_t$ and $\boldsymbol{\xi}_t$ are multivariate normal vectors[14] with general covariance matrices that may depend on t and $\mathbb{E}[\varepsilon_u^\mathsf{T}\xi_v] = 0$ for all (u, v)'s. Equation (7.13) is called the *observation equation*, while (7.14) is called the *state equation*. This representation embeds the process of interest $(\mathbf{x}_t)$ into a larger space, the *state space*, where the *missing* process $(\mathbf{y}_t)$ is Markov and linear. For instance, (7.6) is a *state-space representation* of the AR(p) model (see Exercise 7.16).

The MA(q) model can be written that way by defining $\mathbf{y}_t$ as

$$\mathbf{y}_t = (\epsilon_{t-q}, \ldots, \epsilon_{t-1}, \epsilon_t)^\mathsf{T}\,.$$

Then the state equation is

$$\mathbf{y}_{t+1} = \begin{pmatrix} 0\,1\,0 \ldots 0 \\ 0\,0\,1 \ldots 0 \\ \ldots \\ 0\,0\,0 \ldots 1 \\ 0\,0\,0 \ldots 0 \end{pmatrix} \mathbf{y}_t + \epsilon_{t+1} \begin{pmatrix} 0 \\ 0 \\ \vdots \\ 0 \\ 1 \end{pmatrix}, \tag{7.15}$$

while the observation equation is

$$\mathbf{x}_t = x_t = \mu - \left(\vartheta_q\ \vartheta_{q-1}\ \ldots\ \vartheta_1\ -1\right)\mathbf{y}_t\,,$$

with no perturbation ε_t.

The state-space decomposition of the MA(q) model thus involves no vector $\boldsymbol{\varepsilon}_t$ in the observation equation, while $\boldsymbol{\xi}_t$ is degenerate in the state equation. The degeneracy phenomenon is quite common in state-space representations, but this is not a hindrance in conditional uses of the model, as in MCMC implementations. Notice also that the state-space representation of a model is not unique, again a harmless feature for MCMC uses. For instance, for the MA(1) model, the observation equation can also be chosen as $x_t = \mu + (1\ \ 0)\mathbf{y}_t$ with $\mathbf{y}_t = (y_{1t}, y_{2t})^\mathsf{T}$ directed by the state equation

$$\mathbf{y}_{t+1} = \begin{pmatrix} 0\,1 \\ 0\,0 \end{pmatrix} \mathbf{y}_t + \begin{pmatrix} 1 \\ -\vartheta_1 \end{pmatrix} \epsilon_{t+1}\,.$$

Note that, while the state-space representation is wide-ranging and convenient, it does not mean that the derived MCMC strategies are necessarily efficient. In particular, when the hidden state $\boldsymbol{x}_t$ is too large, a naïve completion may prove itself disastrous. Alternative solutions based in sequential importance sampling (SMC) have been shown to be usually more efficient. (See Del Moral et al., 2006.)

[14] Notice the different fonts that distinguish the $\boldsymbol{\varepsilon}_t$'s used in the state-space representation from the ϵ_t's used in the AR and MA models.

A straightforward extension of both previous AR and MA models are the (normal) ARMA(p,q) models, where x_t $(t \in \mathbb{Z})$ is conditionally defined by

$$x_t = \mu - \sum_{i=1}^{p} \varrho_i (x_{t-i} - \mu) + \epsilon_t - \sum_{j=1}^{q} \vartheta_j \epsilon_{t-j} \,, \quad \epsilon_t \sim \mathscr{N}(0, \sigma^2) \,, \tag{7.16}$$

the (ϵ_t)'s being independent. The role of such models, as compared with both AR and MA models, is to aim toward parsimony; that is, to resort to much smaller values of p and q than in a pure AR(p) or a pure MA(q) modeling.

The causality and inversibility conditions on the parameters of (7.16) still correspond to the roots of both polynomials $\mathcal{P}$ and $\mathcal{Q}$ being outside the unit circle, respectively, with a further condition that both polynomials have no common root. (But this almost surely never happens under a continuous prior on the parameters.) The root reparameterization can therefore be implemented for both the ϑ_j's and the ϱ_i's, still calling for MCMC techniques owing to the complexity of the posterior distribution.

State-space representations also exist for ARMA(p,q) models, one possibility being

$$\mathbf{x}_t = x_t = \mu - \begin{pmatrix} \vartheta_{r-1} & \vartheta_{r-2} & \ldots & \vartheta_1 & -1 \end{pmatrix} \mathbf{y}_t$$

for the observation equation and

$$\mathbf{y}_{t+1} = \begin{pmatrix} 0 & 1 & 0 & \ldots & 0 \\ 0 & 0 & 1 & \ldots & 0 \\ & & & \ldots & \\ 0 & 0 & 0 & \ldots & 1 \\ \varrho_r & \varrho_{r-1} & \varrho_{r-2} & \ldots & \varrho_1 \end{pmatrix} \mathbf{y}_t + \epsilon_{t+1} \begin{pmatrix} 0 \\ 0 \\ \vdots \\ 0 \\ 1 \end{pmatrix} \tag{7.17}$$

for the state equation, with $r = \max(p, q+1)$ and the convention that $\varrho_m = 0$ if $m > p$ and $\vartheta_m = 0$ if $m > q$.

Similarly to the MA(q) case, this state-space representation is handy in devising MCMC algorithms that converge to the posterior distribution of the parameters of the ARMA(p,q) model.

A straightforward MCMC processing of the ARMA model is to take advantage of the AR and MA algorithms that have been constructed above by using both algorithms sequentially. Indeed, conditionally on the AR parameters, the ARMA model can be expressed as an MA model and, conversely, conditionally on the MA parameters, the ARMA model can be expressed almost as an AR model. This is quite obvious for the MA part since, if we define $(t > p)$

$$\tilde{x}_t = x_t - \mu + \sum_{i=1}^{p} \varrho_i (x_{t-i} - \mu) \,,$$

the likelihood is formally equal to a standard MA(q) likelihood on the $\tilde{x}_t$'s. The reconstitution of the AR(p) likelihood is more involved: If we now define the residuals $\tilde{\epsilon}_t = \sum_{j=1}^{q} \vartheta_j \epsilon_{t-j}$, the log-likelihood conditional on $\mathbf{x}_{0:(p-1)}$ is

$$-\sum_{t=p}^{T}\left(x_t-\mu-\sum_{j=1}^{p}\varrho_j[x_{t-j}-\mu]-\tilde{\epsilon}_t\right)^2 \Big/ 2\sigma^2\,,$$

which is obviously close to an AR(p) log-likelihood, except for the $\tilde{\epsilon}_t$'s. The original AR(p) MCMC code can then be recycled modulo this modification in the likelihood.

Another extension of the AR model is the ARCH model, used to represent processes, particularly in finance, with independent errors but time-dependent variances, as in the ARCH(p) process[15] ($t \in \mathbb{Z}$)

$$x_t = \sigma_t \epsilon_t\,, \qquad \epsilon_t \overset{\text{iid}}{\sim} \mathcal{N}(0,1)\,, \qquad \sigma_t^2 = \alpha + \sum_{i=1}^{p} \beta_i x_{t-i}^2\,.$$

The ARCH(p) process defines a Markov chain since x_t only depends on $\mathbf{x}_{t-p:t-1}$. It can be shown that a stationarity condition for the ARCH(1) model is that $\mathbb{E}[\log(\beta_1\epsilon_t^2)] < 0$, which is equivalent to $\beta_1 < 3.4$. This condition becomes much more involved for larger values of p. Contrary to the stochastic volatility model defined below, the ARCH(p) model enjoys a closed-form likelihood when conditioning on the initial values $x_1, \ldots, x_p$. However, because of the nonlinearities in the variance terms, approximate methods based on MCMC algorithms must be used for their analysis.

State-space models are special cases of hidden Markov models (detailed below in Sect. 7.5) in the sense that (7.13) and (7.14) are a special occurrence of the generic representation

$$\begin{aligned} \mathbf{x}_t &= G(\mathbf{y}_t, \epsilon_t)\,, \\ \mathbf{y}_t &= F(\mathbf{y}_{t-1}, \zeta_t)\,. \end{aligned} \tag{7.18}$$

Note, however, that it is not necessarily appealing to resort to this hidden Markov representation, in comparison with state-space models, because the complexity of the functions F or G may hinder the processing of this representation to unbearable levels (while, for state-space models, the linearity of the relations always allows for a generic if not necessarily efficient processing based on, e.g., Gibbs sampling steps).

Stochastic volatility models are quite popular in financial applications, especially in describing series with sudden and correlated changes in the magnitude of variation of the observed values. These models use a hidden chain $(y_t)_{t\in\mathbb{N}}$, called the *stochastic volatility*, to model the variance of the observables $(x_t)_{t\in\mathbb{N}}$ in the following way: Let $y_0 \sim \mathcal{N}(0,\sigma^2)$ and, for $t = 1, \ldots, T$, define

[15] The acronym ARCH stands for *autoregressive conditional heteroscedasticity*, heteroscedasticity being a term favored by econometricians to describe heterogeneous variances. Gouriéroux (1996) provides a general reference on these models, as well as classical inferential methods of estimation.

$$\begin{cases} y_t = \varphi y_{t-1} + \sigma \epsilon^*_{t-1}\,, \\ x_t = \beta e^{y_t/2} \epsilon_t\,, \end{cases} \tag{7.19}$$

where both ϵ_t and ϵ^*_t are iid $\mathscr{N}(0,1)$ random variables. In this simple version, the observable is thus a white noise, except that the variance of this noise enjoys a particular AR(1) structure on the logarithmic scale. Quite obviously, this structure makes the computation of the (observed) likelihood a formidable challenge!

Figure 7.6 gives the sequence $\{\log(x_t) - \log(x_{t-1})\}$ when (x_t) is the Aegon stock sequence plotted in Fig. 7.1. While this real-life sequence is not necessarily a stochastic volatility process, it presents some features that are common with those processes, including an overall stationary structure and periods in the magnitude of the variation of the sequence.

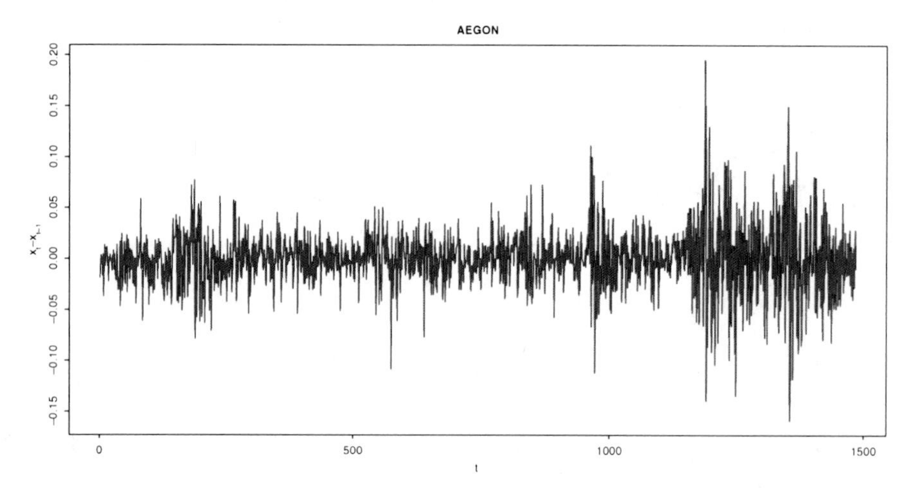

Fig. 7.6. Dataset **Eurostoxx50**: First-order difference $\{\log(x_t) - \log(x_{t-1})\}$ of the Aegon stock sequence regarded as a potential stochastic volatility process (7.19)

When comparing ARMA with the hidden Markov models of the following Section, it may appear that the former are more general in the sense that they allow a different dependence on the past values. Resorting to the state-space representation (7.18) shows that this is not the case. Different horizons p of dependence can also be included for hidden Markov models simply by (a) using a vector $\mathbf{x}_t = (x_{t-p+1}, \ldots, x_t)$ for the observables or by (b) using a vector $\mathbf{y}_t = (y_{t-q+1}, \ldots, y_t)$ for the latent process in (7.18).

7.5 Hidden Markov Models

Hidden Markov models are a generalization of the mixture models of Chap. 6. Their appeal within this chapter is that they constitute an interesting case of non-Markov time series, besides being extremely useful in modeling, e.g., for financial, telecommunication, and genetic data. We refer the reader to McDonald and Zucchini (1997) for a deeper introduction to these models and to Cappé et al. (2004) and Frühwirth-Schnatter (2006) for a complete coverage of their statistical processing.

7.5.1 Basics

The family of *hidden Markov models* (abbreviated to HMM) consists of a bivariate process $(x_t, y_t)_{t\in\mathbb{N}}$, where the *unobserved* subprocess $(y_t)_{t\in\mathbb{N}}$ is a homogeneous Markov chain on a state space $\mathscr{Y}$ and, conditional on $(y_t)_{t\in\mathbb{N}}$, $(x_t)_{t\in\mathbb{N}}$ is a series of random variables on $\mathscr{X}$ such that the conditional distribution of x_t given y_t and the past $(x_j, y_j)_{j<t}$ *only depends on* y_t, as represented by the DAG in Fig. 7.7. When $\mathscr{Y} = \{1, \ldots, \kappa\}$, i.e. when the hidden Markov chain takes a finite number of possible values, we have, in particular,

$$x_t|y_t \sim f(x|\xi_{y_t})$$

where $(y_t)_{t\in\mathbb{N}}$ thus is a finite state-space Markov chain, meaning that $y_t|y_{t-1}$ is distributed from

$$\mathbb{P}(y_t = i|y_{t-1} = j) = p_{ji}\,, \quad 1 \leq i \leq \kappa\,,$$

and the ξ_i's are the different parameters indexing the conditional distribution. In the general case, the joint distribution of (x_t, y_t) given the past values $\mathbf{x}_{0:(t-1)} = (x_0, \ldots, x_{t-1})$ and $\mathbf{y}_{0:(t-1)} = (y_0, \ldots, y_{t-1})$ factorizes as

$$(x_t, y_t)|\mathbf{x}_{0:(t-1)}, \mathbf{y}_{0:(t-1)} \sim f(y_t|y_{t-1})\, f(x_t|y_t)\,,$$

in agreement with Fig. 7.7. The process $(y_t)_{t\in\mathbb{N}}$ is usually referred to as the *state* of the model and, again, is *not observable* (hence, *hidden*). Inference thus has to be carried out only in terms of the observable process $(x_t)_{t\in\mathbb{N}}$.

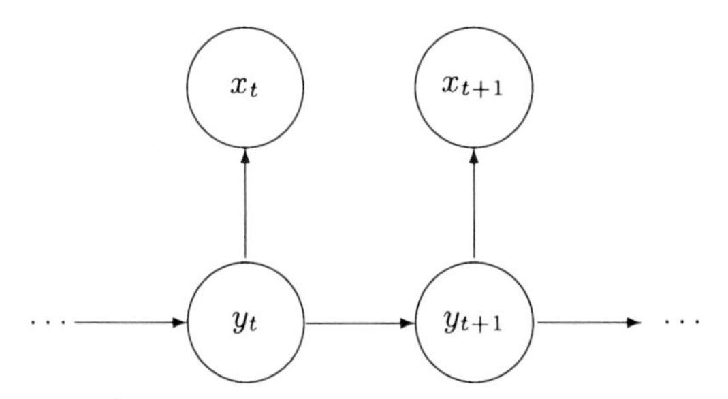

Fig. 7.7. Directed acyclic graph (DAG) representation of the dependence structure of a hidden Markov model, where $(x_t)_{t\in\mathbb{N}}$ is the observable process and $(y_t)_{t\in\mathbb{N}}$ the hidden process

Simulating a hidden Markov chain is then straightforward: we start with the simulation of the hidden layer, i.e. of the process $(y_t)_{t=1,\ldots,T}$ and proceed

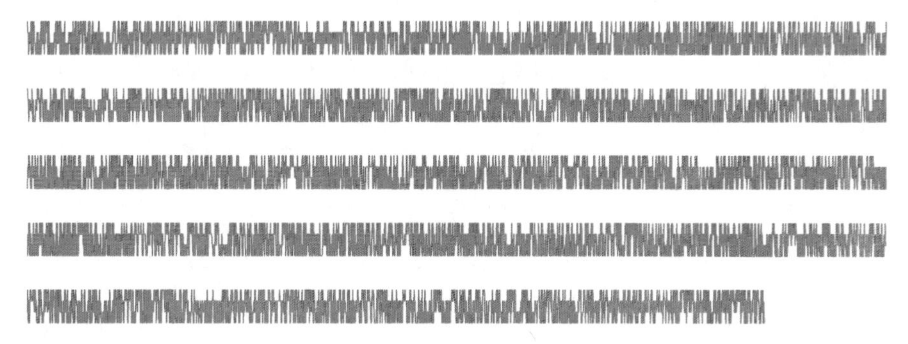

Fig. 7.8. Dataset Dnadataset: Sequence of 9718 amine bases for an HIV genome. The four bases A, C, G, and T have been recoded as $1, \ldots, 4$

to simulating each x_t conditional on the corresponding y_t ($t = 1, \ldots, T$). The corresponding computing time is linear in T (Exercise 7.18).

Hidden Markov models have been used in genetics since the early 1990s for the modeling of DNA sequences. In short (and with no ambition at completeness!), DNA, which stands for deoxyribonucleic acid, is a molecule that carries the genetic information about a living organism and is replicated in each of its cells. This molecule is made up of a sequence of amine bases—adenine, cytosine, guanine, and thymine—abbreviated as A, C, G, and T. The particular arrangement of bases in different parts of the sequence is thought to be related to different characteristics of the living organism to which it corresponds. **Dnadataset** is a particular sequence corresponding to a complete HIV (which stands for Human Immunodeficiency Virus) genome where A, C, G, and T have been recoded as $1, \ldots, 4$. Figure 7.8 represents this sequence of 9718 bases by decomposing it into five blocks. The simplest modeling of this sequence is to assume a two-state hidden Markov model with $\mathscr{Y} = \{1, 2\}$ and $\mathscr{X} = \{1, 2, 3, 4\}$, the assumption being that one hidden state corresponds to noncoding regions and the other hidden state to coding regions.[16]

For statistical purposes, the distributions of both x_t and y_t are usually parameterized, that is, (7.18) looks like

$$\begin{aligned} x_t &= G(y_t, \epsilon_t | \theta)\,, \\ y_t &= F(y_{t-1}, \zeta_t | \delta)\,, \end{aligned} \tag{7.20}$$

where ϵ_t and ζ_t are independent perturbations (*white noise*) and where θ and δ are finite-dimensional parameters.

To draw inference on either the parameters of the HMM or on the hidden chain, it is generally necessary to take advantage of the missing-variable nature of HMMs and to use simultaneous simulation both of $(y_t)_{t \in \mathbb{N}}$ and

[16] There obviously is no reason why the data should fit this formalized model.

of the parameters of the model. There is, however, one exception to that requirement, which is revealed in Sect. 7.5.2, and that is when the state space $\mathcal{Y}$ of the hidden chain $(y_t)_{t\in\mathbb{N}}$ is finite.

In the event that both the hidden and the observed chains are on finite state-spaces, with $\mathscr{Y} = \{1, \ldots, \kappa\}$ and $\mathscr{X} = \{1, \ldots, k\}$, as in **Dnadataset**, the parameter θ is made up of p probability vectors

$$\mathbf{q}^1 = (q_1^1, \ldots, q_k^1), \ldots, \mathbf{q}^\kappa = (q_1^\kappa, \ldots, q_k^\kappa)$$

and the parameter δ is the $\kappa \times \kappa$ Markov transition matrix $\mathbb{P} = (p_{ij})$ on $\mathscr{Y}$. Given that the joint distribution of $(x_t, y_t)_{0\leq t\leq T}$ is

$$\varrho_{y_0} q_{x_0}^{y_0} \prod_{t=1}^{T} \{p_{y_{t-1}y_t} q_{x_t}^{y_t}\} ,$$

where $\boldsymbol{\varrho} = (\varrho_1, \ldots, \varrho_\kappa)$ is the stationary distribution of $\mathbb{P}$ (i.e., such that $\boldsymbol{\varrho}\mathbb{P} = \boldsymbol{\varrho}$), the posterior distribution of (θ, δ) given $(x_t, y_t)_t$ factorizes as

$$\pi(\theta, \delta)\, \varrho_{y_0} \prod_{i=1}^{\kappa} \prod_{j=1}^{k} (q_j^i)^{n_{ij}} \times \prod_{i=1}^{\kappa} \prod_{j=1}^{p} p_{ij}^{m_{ij}} ,$$

where the n_{ij}'s and the m_{ij}'s are sufficient statistics representing

- the number of visits to state j by the x_t's when the corresponding y_t's are equal to i

and

- the number of transitions from state i to state j on the hidden chain $(y_t)_{t\in\mathbb{N}}$,

respectively. If we condition on the starting value y_0, set equal to 1 for (partial) identifiability reasons, and thus omit ϱ_{y_0} in the expression above and if we use a flat prior on the p_{ij}'s and q_j^i's, the posterior distributions are Dirichlet. Similarly to the ARMA case processed in Chap. 7, if we include the starting values in the posterior distribution, this introduces a non-conjugate structure in the simulation of the p_{ij}'s, but this can be handled with a Metropolis–Hastings substitute that uses the Dirichlet distribution as the proposal. Note that, in the non-conditional case, we need to simulate y_0.

Conditional on the parameters, the simulation of the chain $(y_t)_{0\leq t\leq T}$ can be processed Gibbs-wise (i.e., one term at a time), using the fully conditional distributions

$$\mathbb{P}(y_t = i | x_t, y_{t-1}, y_{t+1}) \propto p_{y_{t-1}i}\, p_{iy_{t+1}}\, q_{x_t}^i .$$

Therefore, the overall algorithm looks as follows:

Algorithm 7.15 Finite–State HMM Gibbs Sampler

Initialization:

1. Generate random values (or pick arbitrary estimators) of the p_{ij}'s and the q^i_j's.
2. Generate the hidden Markov chain $(y_t)_{0\leq t\leq T}$ by $(i=1,2)$

$$\mathbb{P}(y_t=i) \propto \begin{cases} p_{ii}\, q^i_{x_0} & \text{if } t=0\,, \\ p_{y_{t-1}i}\, q^i_{x_t} & \text{if } t>0\,, \end{cases}$$

and compute the corresponding sufficient statistics.

Iteration m $(m\geq 1)$:

1. Generate

$$(p_{i1},\ldots,p_{i\kappa}) \sim \mathscr{D}(1+n_{i1},\ldots,1+n_{i\kappa})\,,$$
$$(q^i_1,\ldots,q^i_k) \sim \mathscr{D}(1+m_{i1},\ldots,1+m_{ik})\,,$$

and correct for the missing initial probability by a Metropolis–Hastings step with acceptance probability $\varrho'_{y_0}/\varrho_{y_0}$.

2. Generate successively each y_t $(0\leq t\leq T)$ by

$$\mathbb{P}(y_t=i|x_t,y_{t-1},y_{t+1}) \propto \begin{cases} p_{ii}\, q^i_{x_1}\, p_{iy_1} & \text{if } t=0\,, \\ p_{y_{t-1}i}\, q^i_{x_t}\, p_{iy_{t+1}} & \text{if } t>0\,, \end{cases}$$

and compute the corresponding sufficient statistics.

In the initialization step of Algorithm 7.15, any distribution on $(y_t)_{t\in\mathbb{N}}$ is obviously valid, but this particular choice is of interest since it is related to the true conditional distribution, simply omitting the dependence on the next value.

The main loop in the Gibbs sampler is then of the form (for $\kappa=2$ and $k=4$ as in **Dnadataset**)

```
# Beta/Dirichlet simulations for P
a=1/(1+rgamma(1,nab+1)/rgamma(1,naa+1))
b=1/(1+rgamma(1,nba+1)/rgamma(1,nbb+1))
P=matrix(c(a,1-a,1-b,b),ncol=2,byrow=T)

q1=rgamma(4,ma+1) # and Q
q2=rgamma(4,mb+1)
q1=q1/sum(q1); q2=q2/sum(q2)

# (hidden) Markov conditioning
x[1]=sample(1:2,1,prob=c(a*P[1,x[2]]*q1[y[1]],b*P[2,x[2]]
```

```
        *q2[y[1]]))
for (m in 2:(T-1))
  x[m]=sample(1:2,1,prob=c(P[x[m-1],1]*P[1,x[m+1]]*q1[y[m]],
         P[x[m-1],2]*P[2,x[m+1]]*q2[y[m]]))
x[T]=sample(1:2,1,prob=c(P[x[T-1],1]*q1[y[T]],P[x[T-1],2]
        *q2[y[T]]))

# Sufficient statistics for next iteration
naa=sum((x[1:(T-1)]==1)*(x[2:T]==1))
nab=sum((x[1:(T-1)]==1)*(x[2:T]==2))
nba=sum((x[1:(T-1)]==2)*(x[2:T]==1))
nbb=sum((x[1:(T-1)]==2)*(x[2:T]==2))
ya=y[x==1]
ma=c(sum(ya==1),sum(ya==2),sum(ya==3),sum(ya==4))
yb=y[x==2]
mb=c(sum(yb==1),sum(yb==2),sum(yb==3),sum(yb==4))
```

We ran several Gibbs samplers for 1,000 iterations, starting from small, medium and high values for p_{11} and p_{22}, and got very similar results in both first and both last cases for the approximations to the Bayes posterior means, as shown by Table 7.1. The raw output also gives a sense of stability, as shown by Fig. 7.9.

For the third case, started at small values of both p_{11} and p_{22}, the simulated chain had not visited the same region of the posterior distribution after those 1,000 iterations, and it produced an estimate with a smaller log-likelihood[17] value of $-13,160$. However, running the Gibbs sampler longer (for 4,000 more iterations) did produce a similar estimate, as shown by the third replication in Table 7.1. This phenomenon is slightly related to the phenomenon, discussed in the context of Figs. 6.4 and 6.3, that the Gibbs sampler tends to "stick" to lower modes for lack of sufficient energy. In the current situation, the energy required to leave the lower mode appears to be available. Note that we have reordered the output to compensate for a possible switch between hidden states 1 and 2 among experiments. This is quite natural, given the lack of identifiability of the hidden states (Exercise 7.17). Flipping the indices 1 and 2 does not modify the likelihood, and thus all these experiments explore the same mode of the posterior.

7.5.2 Forward–Backward Representation

When the state space of the hidden Markov chain $\mathcal{Y}$ is finite, that is, when

$$\mathcal{Y} = \{1, \ldots, \kappa\} ,$$

[17] The log-posterior is proportional to the log-likelihood in that special case, and the log-likelihood is computed using a technique described below in Sect. 7.5.2.

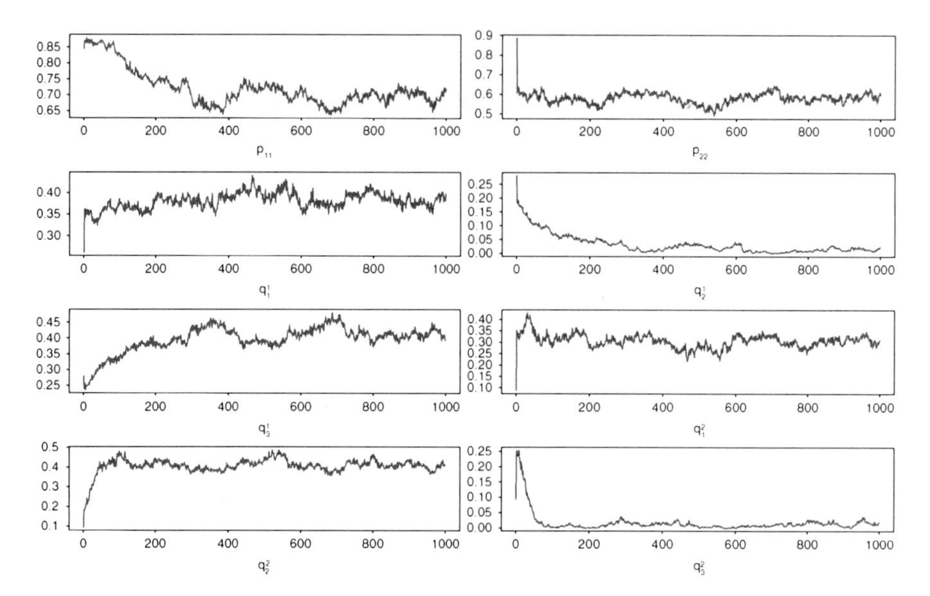

Fig. 7.9. Dataset Dnadataset: Convergence of a Gibbs sequence to the region of interest on the posterior surface for the hidden Markov model (this is replication 2 in Table 7.1). The row-wise order of the parameters is the same as in Table 7.1

Table 7.1. Dataset Dnadataset: Five runs of the Gibbs sampling approximations to the Bayes estimates of the parameters for the hidden Markov model along with final log-likelihood (starting values are indicated on the line below in parentheses) based on $M = 1000$ iterations (except for replication 3, based on 5,000 iterations)

Run	p_{11}	p_{22}	q_1^1	q_2^1	q_3^1	q_1^2	q_2^2	q_3^2	Log-like.
1	0.720	0.581	0.381	0.032	0.396	0.306	0.406	0.018	−13,121
	(0.844)	(0.885)	(0.260)	(0.281)	(0.279)	(0.087)	(0.094)	(0.0937)	
2	0.662	0.620	0.374	0.016	0.423	0.317	0.381	0.034	−13,123
	(0.628)	(0.621)	(0.203)	(0.352)	(0.199)	(0.066)	(0.114)	(0.0645)	
3	0.696	0.609	0.376	0.023	0.401	0.318	0.389	0.030	−13,118
	(0.055)	(0.150)	(0.293)	(0.200)	(0.232)	(0.150)	(0.102)	(0.119)	
4	0.704	0.580	0.377	0.024	0.407	0.313	0.403	0.020	−13,121
	(0.915)	(0.610)	(0.237)	(0.219)	(0.228)	(0.079)	(0.073)	(0.076)	
5	0.694	0.585	0.376	0.0218	0.410	0.315	0.395	0.0245	−13,119
	(0.600)	(0.516)	(0.296)	(0.255)	(0.288)	(0.110)	(0.095)	(0.107)	

the likelihood function[18] of the observed process $(x_t)_{1\le t\le T}$ can be computed in a manageable $\mathrm{O}(T \times \kappa^2)$ time by a recurrence relation called the *forward–*

[18]To lighten notation, we will not use the parameters appearing in the various distributions of the HMM, even though they are obviously of central interest.

backward or *Baum–Welch* formulas.[19] We now explain how those formulas are derived.

As illustrated in Fig. 7.7, a generic feature of HMMs is that ($t = 2, \ldots, T$)

$$p(y_t|y_{t-1}, \mathbf{x}_{0:T}) = p(y_t|y_{t-1}, \mathbf{x}_{t:T}) \, .$$

In other words, knowledge of the past observations is redundant for the distribution of the hidden Markov chain when we condition on its previous value. Therefore, when $\mathcal{Y}$ is finite, we can write that

$$p(y_T|y_{T-1}, \mathbf{x}_{0:T}) \propto p_{y_{T-1}y_T} f(x_T|y_T) \equiv p^\star_T(y_T|y_{T-1}, \mathbf{x}_{0:T}) \, ,$$

meaning that we define $p^\star_T(y_T|y_{T-1}, \mathbf{x}_{0:T})$ as the unnormalized version of the density $p(y_T|y_{T-1}, \mathbf{x}_{0:T})$. Then we can process backward the definition of the previous conditionals, so that ($1 < t < T$)

$$\begin{aligned}
p(y_t|y_{t-1}, \mathbf{x}_{0:T}) &= \sum_{i=1}^{\kappa} p(y_t, y_{t+1} = i|y_{t-1}, \mathbf{x}_{t:T}) \\
&\propto \sum_{i=1}^{\kappa} p(y_t, y_{t+1} = i, \mathbf{x}_{t:T}|y_{t-1}) \\
&= \sum_{i=1}^{\kappa} p(y_t|y_{t-1}) f(x_t|y_t) p(y_{t+1} = i, \mathbf{x}_{(t+1):T}|y_t) \\
&\propto p_{y_{t-1}y_t} f(x_t|y_t) \sum_{i=1}^{\kappa} p(y_{t+1} = i|y_t, \mathbf{x}_{(t+1):T}) \\
&\propto p_{y_{t-1}y_t} f(x_t|y_t) \sum_{i=1}^{\kappa} p^\star_{t+1}(i|y_t, \mathbf{x}_{1:T}) \equiv p^\star_t(y_t|y_{t-1}, \mathbf{x}_{1:T}) \, .
\end{aligned}$$

At last, the conditional distribution of the first hidden value y_0 is

$$p(y_0|\mathbf{x}_{0:T}) \propto \varrho_{y_0} f(x_0|y_0) \sum_{i=1}^{\kappa} p^\star_1(i|y_0, \mathbf{x}_{0:t}) \equiv p^\star_0(y_0|\mathbf{x}_{0:T}) \, ,$$

where $(\varrho_k)_k$ is the stationary distribution associated with the Markov transition matrix $\mathbb{P}$. (This is unless the first hidden value y_0 is automatically set equal to 1 for identifiability reasons.)

While this construction amounts to a straightforward conditioning argument, the use of the unnormalized functions $p^\star_{t+1}(y_{t+1} = i|y_t, x_{(1:T)})$ is crucial for deriving the joint conditional distribution of $y_{1:T}$ since resorting to the normalized conditionals instead would result in a useless identity.

---

[19] This recurrence relation has been known for quite a while in the signal processing literature and is also used in the corresponding EM algorithm; see Cappé et al. (2004) for details.

Notice that, as stated above, the derivation of the $p_t^\star$'s indeed has a cost of $O(T\times\kappa^2)$ since, for each t and each of the κ values of y_t, a sum of κ terms has to be computed. So, in terms of raw computational time, computing the observed likelihood does not take less time than simulating the sequence $(y_t)_{t\in\mathbb{N}}$ in the Gibbs sampler. However, the gain in using this forward–backward formula may impact in subtler ways a resulting Metropolis–Hastings algorithm, such as a better mixing of the chain of the parameters, given that we are simulating the whole vector at once.

Once we have all the conditioning functions (or *backward* equations), it is possible to simulate sequentially the hidden sequence $\mathbf{y}_{0:T}$ given $\mathbf{x}_{0:T}$ by generating first y_0 from $p(y_0|\mathbf{x}_{0:T})$, second y_1 from $p(y_1|y_0,\mathbf{x}_{0:T})$ and so on. However, there is (much) more to be done. Indeed, when considering the joint conditional distribution of $\mathbf{y}_{0:T}$ given $\mathbf{x}_{0:T}$, we have

$$
\begin{aligned}
p(\mathbf{y}_{0:T}|\mathbf{x}_{0:T}) &= p(y_0|\mathbf{x}_{0:T})\prod_{t=1}^{T} p(y_t|y_{t-1},\mathbf{x}_{0:T}) \\
&= \frac{\pi(y_1)f(x_0|y_0)}{\sum_{i=1}^{\kappa} p_0^\star(i|\mathbf{x}_{0:T})}\prod_{t=1}^{T}\frac{p_{y_{t-1}y_t}f(x_t|y_t)\sum_{i=1}^{\kappa}p_{t+1}^\star(i|y_t,\mathbf{x}_{1:T})}{\sum_{i=1}^{\kappa}p_t^\star(i|y_{t-1},\mathbf{x}_{(1:T)})} \\
&= \pi(y_0)f(x_0|y_0)\prod_{t=1}^{T}p_{y_{t-1}y_t}f(x_t|y_t)\Big/\sum_{i=1}^{\kappa}p_1^\star(i|\mathbf{x}_{0:T})
\end{aligned}
$$

since all the other sums cancel. This joint conditional distribution immediately leads to the derivation of the observed likelihood since, by Bayes' formula,

$$
f(\mathbf{x}_{0:T}) = \frac{f(\mathbf{x}_{0:T}|\mathbf{y}_{1:T})\,p(\mathbf{y}_{0:T})}{p(\mathbf{y}_{0:T}|\mathbf{x}_{0:T})} = \sum_{i=1}^{\kappa} p_1^\star(i|\mathbf{x}_{0:T})\,,
$$

which is the normalizing constant of the initial conditional distribution! Therefore, working with the unnormalized densities has this supplementary advantage to provide an approximation to the observed likelihood. (Keep in mind that all the expressions above implicitly depend on the model parameters.)

A forward derivation of the likelihood can similarly be constructed. Besides the obvious construction that is symmetrical to the previous one, consider the so-called *prediction filter*

$$
\varphi_t(i) = \mathbb{P}(y_t = i|x_{1:t-1})\,,
$$

with $\varphi_1(j) = \pi(j)$ (where the term *prediction* refers to the conditioning on the observations prior to time t). The *forward* equations are then given by $(t = 1,\ldots,T)$

$$
\varphi_{t+1}(j) = \frac{1}{c_t}\sum_{i=1}^{\kappa} f(x_t|y_t = i)\varphi_t(i)p_{ij}\,,
$$

where

$$c_t = \sum_{k=1}^{\kappa} f(x_t|y_t = k)\varphi_t(k)$$

is the normalizing constant. (This formula uses exactly the same principle as the backward equations.) Exploiting the Markov nature of the joint process $(x_t, y_t)_t$, we can then derive the log-likelihood as

$$\begin{aligned}\log p(\mathbf{x}_{1:t}) &= \sum_{r=1}^{t} \log \left[\sum_{i=1}^{\kappa} p(x_t, y_t = i | x_{1:(r-1)}) \right] \\ &= \sum_{r=1}^{t} \log \left[\sum_{i=1}^{\kappa} f(x_r|y_t = i)\varphi_r(i) \right],\end{aligned}$$

which also requires a $O(T \times \kappa^2)$ computational time.

The resulting R function for computing the (observed) likelihood is therefore (for $\kappa = 2$ and $k = 4$ as in **Dnadataset**)

```
likej=function(vec,log=TRUE){
# vec is the aggregated parameter vector
  P=matrix(c(vec[1],1-vec[1],1-vec[2],vec[2]),ncol=2,byrow=
     TRUE)
  Q1=vec[3:6]; Q2=vec[7:10]

  pxy=c(P[1,1],P[2,2])
  pxy=pxy/sum(pxy) # stationary distribution of P

  pyy=rep(1,T)
  pyy[1]=pxy[1]*Q1[y[1]]+pxy[2]*Q2[y[1]]

  for (t in 2:T){
    pxy=pxy[1]*Q1[y[t-1]]*P[1,]+pxy[2]*Q2[y[t-1]]*P[2,]
    pxy=pxy/sum(pxy)
    pyy[t]=(pxy[1]*Q1[y[t]]+pxy[2]*Q2[y[t]])
    }

  if (log){
   ute=sum(log(pyy))
   }
   else{
     ute=prod(pyy)
     }
  ute
}
```

Obviously, to be able to handle directly the observed likelihood when T is reasonable opens new avenues for simulation methods. For instance, the completion step (of simulating the hidden Markov chain) is no longer necessary, and Metropolis–Hastings alternatives such as random-walk proposals can be used.

Returning to **Dnadataset**, we can compute the log-likelihood (and hence the posterior up to a normalizing constant) associated with a given parameter using, for instance, the prediction filter. In that case,

$$\log p(\mathbf{x}_{1:T}) = \sum_{t=1}^{T} \log \left[\sum_{i=1}^{k} q^i_{x_t} \varphi_t(i) \right] ,$$

where $\varphi_t(j) \propto \sum_{i=1}^{2} q^i_{x_t} \varphi_t(i) p_{ij}$. This representation of the log-likelihood is used in the computation given above for the Gibbs sampler.

Furthermore, given that all parameters to be simulated are probabilities, using a normal random walk proposal in the Metropolis–Hastings algorithm is not adequate. Instead, a more appropriate proposal is based on Dirichlet distributions centered at the current value, with scale factor $\alpha > 0$; that is $(j = 1, 2)$,

$$\tilde{p}_{jj} \sim \mathscr{B}e(\alpha p_{jj}, \alpha(1 - p_{jj})) \quad \tilde{q^j} \sim \mathscr{D}(\alpha q^j_1, \ldots, \alpha q^j_4) .$$

The Metropolis–Hastings acceptance probability is then the ratio of the likelihoods over the ratio of the proposals, $f(\theta|\theta')/f(\theta'|\theta)$. Since larger values of α produce more local moves, we could test a range of values to determine the "proper" scale. However, this requires a long calibration step. Instead, the algorithm can take advantage of the different scales by picking at random for each iteration a value of α from among 1, 10, 100, 10,000 or 100,000. (The randomness in α can then be either ignored in the computation of the proposal density f or integrated by a Rao–Blackwell argument.) For **Dnadataset**, this range of α's was wide enough since the average probability of acceptance is 0.25 and a chain $(\theta_m)_m$ started at random does converge to the same values as the Gibbs chains simulated above, as shown by Fig. 7.10, which also indicates that more iterations would be necessary to achieve complete stability. We can note in particular that the maximum log-posterior value found along the iterations of the Metropolis–Hastings algorithm is $-13{,}116$, which is larger than the values found in Table 7.1 for the Gibbs sampler, for parameter values of $(0.70, 0.58, 0.37, 0.011, 0.42, 0.19, 0.32, 0.42, 0.003, 0.26)$.

When the state space $\mathcal{Y}$ is finite, it may be of interest to estimate the order of the hidden Markov chain. For instance, in the case of **Dnadataset**, it is relevant to infer on how many hidden coding states there are. A possible approach, not covered here, is to use a reversible jump MCMC algorithm that resemble very much the reversible jump algorithm for the mixture model. The reference in this direction is Cappé et al. (2004, Chap. 16) where the authors construct a reversible jump algorithm in this setting. However, the availability

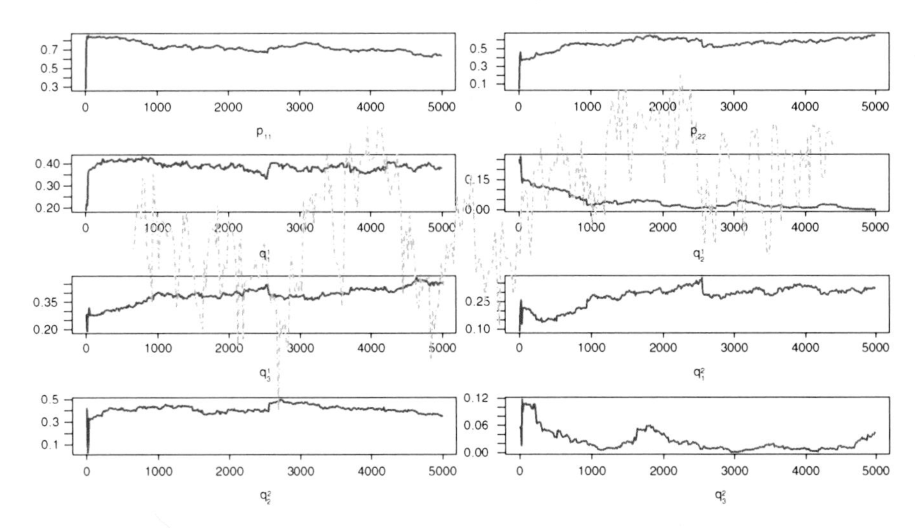

Fig. 7.10. Dataset Dnadataset: Convergence of a Metropolis–Hastings sequence for the hidden Markov model based on 5,000 iterations. The overlayed *curve in the background* is the sequence of log-posterior values

of the (observed) likelihood means that the marginal solution of Chib (1995), exposed in Chap. 6 (Sect. 6.8) for the mixtures of distributions also applies in the current setting (Exercise 7.19).

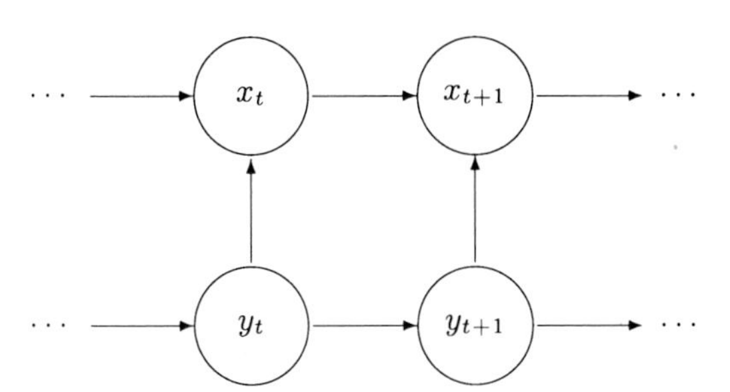

Fig. 7.11. DAG representation of the dependence structure of a Markov-switching model where $(x_t)_t$ is the observable process and $(y_t)_t$ is the hidden chain

The model first introduced for **Dnadataset** is overly simplistic in that, at least within the coding regime, the x_t's are not independent. A more realistic modeling thus assumes that the x_t's constitute a Markov chain within each state of the hidden chain, resulting in the dependence graph of Fig. 7.11. To distinguish this case from the earlier one, it is often called *Markov-switching*.

This extension is much more versatile than the model of Fig. 7.7, and we can hope to capture the time dependence better. However, it is far from parsimonious, as the use of different Markov transition matrices for *each* hidden state induces an explosion in the number of parameters. For instance, if there are two hidden states, the number of parameters is 26; if there are four hidden states, the number jumps to 60.

7.6 Exercises

7.1 Consider the process $(x_t)_{t\in\mathbb{Z}}$ defined by

$$x_t = a + bt + y_t\,,$$

where $(y_t)_{t\in\mathbb{Z}}$ is an iid sequence of random variables with mean 0 and variance σ^2, and where a and b are constants. Define

$$w_t = (2q+1)^{-1}\textstyle\sum_{j=-q}^{q} x_{t+j}\,.$$

Compute the mean and the autocovariance function of $(w_t)_{t\in\mathbb{Z}}$. Show that $(w_t)_{t\in\mathbb{Z}}$ is not stationary but that its autocovariance function $\gamma_w(t+h,t)$ does not depend on t.

7.2 Suppose that the process $(x_t)_{t\in\mathbb{N}}$ is such that $x_0 \sim \mathscr{N}(0,\tau^2)$ and, for all $t \in \mathbb{N}$,

$$x_{t+1}|\mathbf{x}_{0:t} \sim \mathscr{N}(x_t/2, \sigma^2)\,, \qquad \sigma > 0\,.$$

Give a necessary condition on τ^2 for $(x_t)_{t\in\mathbb{N}}$ to be a (strictly) stationary process.

7.3 Suppose that $(x_t)_{t\in\mathbb{N}}$ is a *Gaussian random walk* on $\mathbb{R}$: $x_0 \sim \mathscr{N}(0,\tau^2)$ and, for all $t \in \mathbb{N}$,

$$x_{t+1}|\mathbf{x}_{0:t} \sim \mathscr{N}(x_t, \sigma^2)\,, \qquad \sigma > 0\,.$$

Show that, whatever the value of τ^2 is, $(x_t)_{t\in\mathbb{N}}$ is not a (strictly) stationary process.

7.4 Give the necessary and sufficient condition under which an AR(2) process with autoregressive polynomial $\mathcal{P}(u) = 1 - \varrho_1 u - \varrho_2 u^2$ (with $\varrho_2 \neq 0$) is causal.

7.5 Consider the process $(x_t)_{t\in\mathbb{N}}$ such that $x_0 = 0$ and, for all $t \in \mathbb{N}$,

$$x_{t+1}|\mathbf{x}_{0:t} \sim \mathscr{N}(\varrho\, x_t, \sigma^2)\,.$$

Suppose that $\pi(\varrho,\sigma) = 1/\sigma$ and that there is no constraint on ϱ. Show that the conditional posterior distribution of ϱ, conditional on the observations $\mathbf{x}_{0:T}$ and on σ^2, is a $\mathscr{N}(\mu_T, \omega_T^2)$ distribution with

$$\mu_T = \sum_{t=1}^{T} x_{t-1}x_t \Big/ \sum_{t=1}^{T} x_{t-1}^2 \quad \text{and} \quad \omega_T^2 = \sigma^2 \Big/ \sum_{t=1}^{T} x_{t-1}^2\,.$$

Show that the marginal posterior distribution of ϱ is a Student $\mathscr{T}(T-1, \mu_T, \nu_T^2)$ distribution with

$$\nu_T^2 = \frac{1}{T-1}\left(\sum_{t=1}^{T} x_t^2 \Big/ \sum_{t=0}^{T-1} x_t^2 - \mu_T^2\right).$$

Apply this modeling to the Aegon series in **Eurostoxx50** and evaluate its predictive abilities.

7.6 For Algorithm 7.13, show that, if the proposal on σ^2 is a log-normal distribution $\mathscr{LN}(\log(\sigma_{t-1}^2), \tau^2)$ and if the prior distribution on σ^2 is the noninformative prior $\pi(\sigma^2) = 1/\sigma^2$, the acceptance ratio also reduces to the likelihood ratio because of the Jacobian.

7.7 Write down the joint distribution of $(y_t, x_t)_{t\in\mathbb{N}}$ in (7.19) and deduce that the (observed) likelihood is not available in closed form.

7.8 Show that the stationary distribution of $\mathbf{x}_{-p:-1}$ in an AR(p) model is a $\mathscr{N}_p(\mu\mathbf{1}_p, \mathbf{A})$ distribution, and give a fixed point equation satisfied by the covariance matrix $\mathbf{A}$.

7.9 Show that the posterior distribution on $\boldsymbol{\theta}$ associated with the prior $\pi(\boldsymbol{\theta}) = 1/\sigma^2$ and an AR(p) model is well-defined for $T > p$ observations.

7.10 Show that the coefficients of the polynomial $\mathcal{P}$ in (7.5) associated with an AR(p) model can be derived in $O(p^2)$ time from the inverse roots λ_i using the recurrence relations ($i = 1, \ldots, p, j = 0, \ldots, p$)

$$\psi_0^i = 1\,, \qquad \psi_j^i = \psi_j^{i-1} - \lambda_i \psi_{j-1}^{i-1}\,,$$

where $\psi_0^0 = 1$ and $\psi_j^i = 0$ for $j > i$, and setting $\varrho_j = -\psi_j^p$ ($j = 1, \ldots, p$).

7.11 Given the polynomial $\mathcal{P}$ in (7.5), the fact that all the roots are outside the unit circle can be determined without deriving the roots, thanks to the Schur–Cohn test. If $\mathcal{A}_p = \mathcal{P}$, a recursive definition of decreasing degree polynomials is ($k = p, \ldots, 1$)

$$u\mathcal{A}_{k-1}(u) = \mathcal{A}_{k-1}(u) - \varphi_k \mathcal{A}_k^\star(u)\,,$$

where $\mathcal{A}_k^\star$ denotes the reciprocal polynomial $\mathcal{A}_k^\star(u) = u^k \mathcal{A}_{k-1}(1/u)$.

1. Given the expression of φ_k in terms of the coefficients of $\mathcal{A}_k$.
2. Show that the degree of $\mathcal{A}_k$ is at most k.
3. If $a_{m,k}$ denotes the m-th degree coefficient in $\mathcal{A}_k$, show that $a_{k,k} \neq 0$ for $k = 0, \ldots, p$ if, and only if, $a_{0,k} \neq a_{k,k}$ for all k's.
4. Check by simulation that, in cases when $a_{k,k} \neq 0$ for $k = 0, \ldots, p$, the roots are outside the unit circle if, and only if, all the coefficients $a_{k,k}$ are positive.

7.12 For an MA(q) process, show that ($s \leq q$)

$$\gamma_x(s) = \sigma^2 \sum_{i=0}^{q-|s|} \vartheta_i \vartheta_{i+|s|}\,.$$

7.13 Show that the conditional distribution of $(\epsilon_0, \ldots, \epsilon_{-q+1})$ given both $\mathbf{x}_{1:T}$ and the parameters is a normal distribution. Evaluate the complexity of computing the mean and covariance matrix of this distribution.

7.14 Give the conditional distribution of ϵ_{-t} given the other ϵ_{-i}'s, $\mathbf{x}_{1:T}$, and the $\hat{\epsilon}_i$'s. Show that this distribution only depends on the other ϵ_{-i}'s, $\mathbf{x}_{1:q-t+1}$, and $\hat{\epsilon}_{1:q-t+1}$.

7.15 Show that the (useful) predictive horizon for the MA(q) model is restricted to the first q future observations x_{t+i}.

7.16 Show that the system of equations given by (7.13) and (7.14) induces a Markov chain on the completed variable $(\mathbf{x}_t, \mathbf{y}_t)$. Deduce that state-space models are special cases of hidden Markov models.

7.17 Show that, for a hidden Markov model, when the support $\mathcal{Y}$ is finite and when $(y_t)_{t\in\mathbb{N}}$ is stationary, the marginal distribution of x_t is the same mixture distribution for all t's. Deduce that the same identifiability problem as in mixture models occurs in this setting.

7.18 Given a hidden Markov chain (x_t, y_t) with both x_t and y_t taking a finite number of possible values, k and κ, show that the time required for the simulation of T consecutive observations is in $\mathrm{O}(k\kappa T)$.

7.19 Implement Chib's method of Sect. 6.8 in the case of a doubly finite hidden Markov chain. First, show that an equivalent to the approximation (6.9)) is available for the denominator of (6.8). Second, discuss whether or not the label switching issue also rises in this framework. Third, apply this approximation to **Dnadataset**.

7.20 Show that the counterpart of the prediction filter in the Markov-switching case is given by

$$\log p(\mathbf{x}_{1:t}) = \sum_{r=1}^{t} \log \left[\sum_{i=1}^{\kappa} f(x_r|x_{r-1}, y_r = i)\varphi_r(i) \right],$$

where $\varphi_r(i) = \mathbb{P}(y_r = i|\mathbf{x}_{1:r-1})$ is given by the recursive formula

$$\varphi_r(i) \propto \sum_{j=1}^{\kappa} p_{ji} f(x_{r-1}|x_{r-2}, y_{r-1} = j)\varphi_{r-1}(j).$$

8

Image Analysis

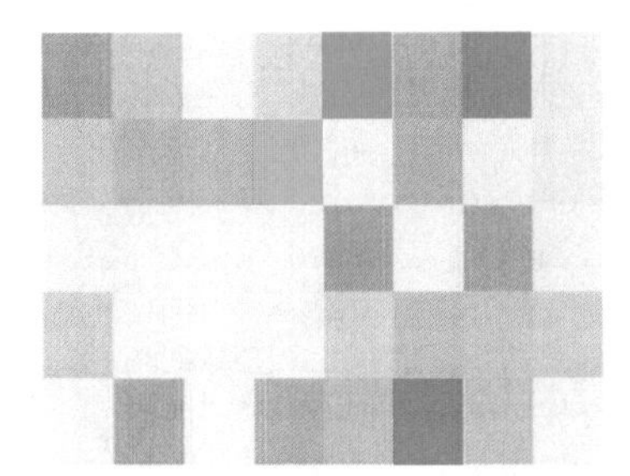

"Reduce it to binary, Siobhan," she told herself.
—**Ian Rankin**, ***Resurrection Men***.—

Roadmap

This final chapter covers the analysis of pixelized images through Markov random field models, towards pattern detection and image correction. We start with the statistical analysis of Markov random fields, which are extensions of Markov chains to the spatial domain, as they are instrumental in this chapter. This is also the perfect opportunity to cover the ABC method, as these models do not allow for a closed form likelihood. Image analysis has been a very active area for both Bayesian statistics and computational methods in the past 30 years, so we feel it well deserves a chapter of its own for its specific features.

J.-M. Marin and C.P. Robert, *Bayesian Essentials with R*, Springer Texts in Statistics, DOI 10.1007/978-1-4614-8687-9_8,

8.1 Image Analysis as a Statistical Problem

If we think of a computer image as a (large) collection of colored pixels disposed on a grid, there does not seem to be any randomness involved nor any need for statistical analysis! Nonetheless, image analysis seen as a statistical analysis is a thriving field that saw the emergence of several major statistical advances, including, for instance, the Gibbs sampler. (Moreover, this field has predominantly adopted a Bayesian perspective both because this was a natural thing to do and because the analytical power of this approach was higher than with other methods.) The reason for this apparent paradox is that, while pixels usually are deterministic objects, the complexity and size of images require one to represent those pixels as the random output of a distribution governed by an object of much smaller dimension. For instance, this is the case in computer vision, where specific objects need to be extracted out of a much richer (or noisier) background.

In this spirit of extracting information from huge dimensional structure, we thus build in Sect. 8.2 a specific family of distributions inspired from particle physics, the Potts model, in order to structure images and other spatial structures in terms of local homogeneity. Unfortunately, this is a mostly theoretical section with very few illustrations. In Sect. 8.3, we address the fundamental issue of handling the missing normalizing constant in these models by introducing a new computational technique called ABC that operates on intractable likelihoods (with the penalty of producing an approximative answer). In Sect. 8.4, we impose a strong spatial dimension on the prior associated with an image in order to gather homogeneous structures out of a complex or blurry image.

8.2 Spatial Dependence

8.2.1 Grids and Lattices

An image (in the sense of a computer generated image) is a special case of a *lattice*, in the sense that it is a random object whose elements are indexed by the location of the pixels and are therefore related by the geographical proximity of those locations. In full generality, a *lattice* is a mathematical multidimensional object on which a neighbourhood relation can be defined. Even though the original analysis of lattice models by Besag (1974) focussed on plant ecology and agricultural experiments, the neighbourhood relation is only constrained to be a symmetric relation and it does not necessarily have a connection with a geographical proximity, nor with an image. For instance, the relation can describe social interactions between Amazon tribes or words in a manuscript sharing a linguistic root. (The neighbourhood relation between two points of the lattice is generally translated in statistical terms into a probabilistic dependence between those points.) The lattice associated with an image is a regular $n \times m$ array made of (i,j)'s $(1 \leq i \leq n, 1 \leq j \leq m)$, whose nearest (but not necessarily only) neighbors are made of the four entries $(i,j-1)$, $(i,j+1)$, $(i-1,j)$ and $(i+1,j)$. In order to properly describe a dependence structure in images or in other spatial objects indexed by a lattice, we need to expand the notion of Markov chain on those structures. Since a lattice is a multidimensional object—as opposed to the unidimensional line corresponding to the times of observation of the Markov chain—, a first requirement for the generalization is to define a proper neighbourhood structure.

In order to illustrate this notion, we consider a small dataset[1] depicting the presence of tufted sedges[2] in a part of a wetland. This dataset, called **Laichedata**, is simply a 25×25 matrix of zeroes and ones. The corresponding lattice is the 25×25 array (Fig. 8.1).

Fig. 8.1. Presence/absence of the tufted sedge plant (*Carex elata*) on a rectangular patch

Given a lattice $\mathcal{I}$ of sites $i \in \mathcal{I}$ on a map or of pixels in an image,[3] a neighbourhood relation on $\mathcal{I}$ is denoted by $\sim$, $i \sim j$ meaning that i and j are *neighbors*. If we associate a probability distribution on a vector $\mathbf{x}$ indexed by the lattice, $\mathbf{x} = (x_i)_{i \in \mathcal{I}}$, with this relation, meaning that two components x_i and x_j are correlated if the sites i and j are neighbors, a fundamental

[1] Taken from Gaetan and Guyon (2010), kindly provided by the authors.

[2] Wikipedia: "*Carex* is a genus of plants in the family *Cyperaceae*, commonly known as *sedges*. Most (but not all) sedges are found in wetlands, where they are often the dominant vegetation." Laîche is the French for sedge.

[3] We will indiscriminately use *site* and *pixel* in the remainder of the chapter.

requirement for the existence of this distribution is that the neighbourhood relation is symmetric (Cressie, 1993): if i is a neighbor of j (written as $i \sim j$), then j is a neighbor of i. (By convention, i is not a neighbor of itself.) Figure 8.2 illustrates this notion for three types of neighborhoods on a regular grid. For instance, **Laichedata** could be associated with a northwest-southeast neighbourhood to account for dominant winds: an entry (i, j) would have as neighbors $(i-1, j-1)$ and $(i+1, j+1)$.

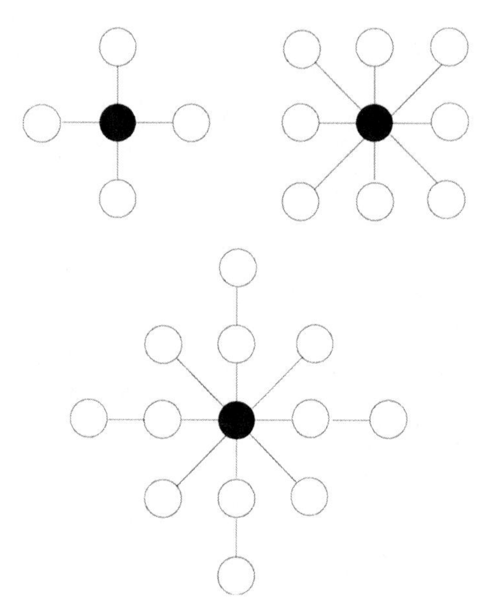

Fig. 8.2. Some common neighbourhood structures used in imaging, with four (*upper left*), eight (*upper right*), or twelve neighbors (*lower*)

8.2.2 Markov Random Fields

A *random field* on $\mathcal{I}$ is a random structure indexed by the lattice $\mathcal{I}$, a collection of random variables $\{x_i; i \in \mathcal{I}\}$ where each x_i takes values in a finite set χ. Obviously, the interesting case is when the x_i's are dependent random variables in relation with the neighbourhood structure on $\mathcal{I}$.

If $n(i)$ is the set of neighbors of $i \in \mathcal{I}$ and if $\mathbf{x}_A = \{x_i; i \in A\}$ denotes the subset of $\mathbf{x}$ for indices in a subset $A \subset \mathcal{I}$, then $\mathbf{x}_{n(i)}$ is the set of values taken by the neighbors of i. The extension from a Markov chain to a Markov random field then assumes only dependence on the neighbors.[4] More precisely, if, as before, we denote by $\mathbf{x}_{-A} = \{x_i; i \notin A\}$ the coordinates that are *not* in

[4]This dependence immediately forces the neighbourhood relation to be symmetric.

a given subset $A \subset \mathcal{I}$, a random field is a *Markov random field* (MRF) if the conditional distribution of any pixel given the other pixels only depends on the values of the neighbors of that pixel; i.e., for $i \in \mathcal{I}$,

$$\pi(x_i|\mathbf{x}_{-i}) = \pi(x_i|\mathbf{x}_{n(i)}) .$$

Markov random fields have been used for quite a while in imaging, not necessarily because images obey Markov laws but rather because these dependence structures offer highly stabilizing properties in modeling. Indeed, constructing the joint prior distribution of an image is a daunting task because there is no immediate way of describing the global properties of an image via a probability distribution. Just as for the directed acyclic graphs (DAG) models at the core of the BUGS software, using the full conditional distributions breaks the problem down to a sequence of *local* problems and this is therefore more manageable in the sense that we may be able to express more clearly how we think x_i behaves when the configuration of its neighbors is known.[5]

Before launching into the use of specific MRFs to describe prior assumptions on a given lattice, we need to worry[6] about the very existence of MRFs! Indeed, defining a set of full conditionals does not guarantee that there is a joint distribution behind them (Exercise 8.1). In our case, this means that general forms of neighborhoods and general types of dependences on the neighbors do not usually correspond to a joint distribution on $\mathbf{x}$.

We first obtain a representation that can be used for testing the existence of a joint distribution. Starting from a complete set of full conditionals on a lattice $\mathcal{I}$, if there indeed exists a corresponding joint distribution, $\pi(\mathbf{x})$, it is completely defined by the ratio $\pi(\mathbf{x})/\pi(\mathbf{x}^*)$ for a given fixed value $\mathbf{x}^*$ since the normalizing constant is automatically determined. Now, if $\mathcal{I} = \{1, \ldots, n\}$, it is simple to exhibit a full conditional density within the joint density by writing the natural decomposition

$$\pi(\mathbf{x}) = \pi(x_1|\mathbf{x}_{-1})\pi(\mathbf{x}_{-1})$$

and then to introduce $\mathbf{x}^*$ by the simple divide-and-multiply trick

$$\pi(\mathbf{x}) = \frac{\pi(x_1|\mathbf{x}_{-1})}{\pi(x_1^*|\mathbf{x}_{-1})}\,\pi(x_1^*, \mathbf{x}_{-1}) .$$

If we iterate this trick for all terms in the lattice (assuming we never divide by 0), we eventually get to the representation

[5]It is no surprise that computational techniques such as the Gibbs sampler stemmed from this area, as the use of conditional distributions is deeply ingrained in the imaging community.

[6]For those that do not want nor do not need to worry, the end of this section can be skipped, it being of a more theoretical nature and not used in the rest of the chapter.

$$\frac{\pi(\mathbf{x})}{\pi(\mathbf{x}^*)} = \prod_{i=0}^{n-1} \frac{\pi(x_{i+1}|x_1^*, \ldots, x_i^*, x_{i+2}, \ldots, x_n)}{\pi(x_{i+1}^*|x_1^*, \ldots, x_i^*, x_{i+2}, \ldots, x_n)} . \tag{8.1}$$

Hence, we can truly write the joint density as a product of ratios of its full conditionals modulo one renormalization.[7]

This result can also be used toward our purpose of checking for compatibility of the full conditional distributions: if there exists a joint density such that the full conditionals never cancel, then (8.1) must hold for every representation of $\mathcal{I} = \{1, \ldots, n\}$; that is, for every ordering of the indices, and for every choice of reference value $\mathbf{x}^*$. Although we cannot provide here the reasoning behind the result, there exists a necessary and sufficient condition for the existence of an MRF. This condition relies on the notion of *clique*: Given a lattice $\mathcal{I}$ and a neighbourhood relation $\sim$, a clique is a maximal subset of $\mathcal{I}$ made of sites that are all neighbors. The corresponding existence result (Cressie, 1993) is that an MRF associated with $\mathcal{I}$ and the neighbourhood relation $\sim$ necessarily is of the form

$$\pi(\mathbf{x}) \propto \exp\left(- \sum_{C \in \mathscr{C}} \Phi_C(\mathbf{x}_C) \right) , \tag{8.2}$$

where $\mathscr{C}$ is the collection of all cliques. This result amounts to saying that the joint distribution must separate in terms of its system of cliques.

We now embark on the description of two specific MRFs that are appropriate for image analysis, namely the *Ising model* used for binary images and its extension, the *Potts model*, used for images with more than two colors.

8.2.3 The Ising Model

If pixels of the image $\mathbf{x}$ under study can only take two colors (black and white, say, as in Fig. 8.1), $\mathbf{x}$ is binary. We typically refer to each pixel x_i as being *foreground* if $x_i = 1$ (black) and *background* if $x_i = 0$ (white). The conditional distribution of a pixel is then Bernoulli, with the corresponding probability parameter depending on the other pixels. A simplification step is to assume that it is a function of the number of black neighboring pixels, using for instance a logit link as ($j = 0, 1$)

$$\pi(x_i = j|\mathbf{x}_{-i}) \propto \exp(\beta n_{i,j}) , \qquad \beta > 0 , \tag{8.3}$$

where $n_{i,j} = \sum_{\ell \in n(i)} \mathbb{I}_{x_\ell = j}$ is the number of neighbors of x_i with color j. The *Ising model* is then defined via these full conditionals

$$\pi(x_i = 1|\mathbf{x}_{-i}) = \frac{\exp(\beta n_{i,1})}{\exp(\beta n_{i,0}) + \exp(\beta n_{i,1})} ,$$

[7] This representation is by no means limited to MRFs: it holds for every joint distribution such that the full conditionals never cancel. It is called the *Hammersley–Clifford theorem*, and a two-dimensional version of it was introduced in Exercise 3.10.

and the joint distribution therefore satisfies

$$\pi(\mathbf{x}) \propto \exp\left(\beta \sum_{j \sim i} \mathbb{I}_{x_j = x_i}\right), \tag{8.4}$$

where the summation is taken over all pairs (i, j) of neighbors (Exercise 8.17).

When inferring on β and thus simulating the posterior distribution β, we will be faced with a major obstacle, namely that the normalizing constant of (8.4), $Z(\beta)$, is intractable except for very small lattices $\mathcal{I}$, while depending on β. Therefore the likelihood function cannot be computed. We will introduce in Sect. 8.3 a computational technique called ABC that is intended to fight this very problem. At this early stage, however, we consider β to be known and focus on the simulation of $\mathbf{x}$ in preparation for the inference on both β and $\mathbf{x}$ given a noisy version of the image, $\mathbf{y}$, as presented in Sect. 8.4.

The computational conundrum of Ising models goes deeper as, due to the convoluted correlation structure of the Ising model, a direct simulation of $\mathbf{x}$ is not possible, expect in very specific cases. Faced with this difficulty, the image community very early developed computational tools which eventually led in 1984 to the proposal of the Gibbs sampler (Sect. 3.5.1).[8] The specification of Markov random fields and in particular of the Ising model implies the full conditional distributions of those models are available in closed form. The local structure of Markov random fields thus provides an immediate site-by-site update for the Gibbs sampler:

Algorithm 8.16 Ising Gibbs Sampler

Initialization: For $i \in \mathcal{I}$, generate independently

$$x_i^{(0)} \sim \mathscr{B}(1/2)\,.$$

Iteration t $(t \geq 1)$:

1. Generate $\mathbf{u} = (u_i)_{i \in \mathcal{I}}$, a random ordering of the elements of $\mathcal{I}$.
2. For $1 \leq \ell \leq |\mathcal{I}|$, update $n_{u_\ell,0}^{(t)}$ and $n_{u_\ell,1}^{(t)}$, and generate

$$x_{u_\ell}^{(t)} \sim \mathscr{B}\left(\frac{\exp(\beta n_{u_\ell,1}^{(t)})}{\exp(\beta n_{u_\ell,0}^{(t)}) + \exp(\beta n_{u_\ell,1}^{(t)})}\right).$$

In this implementation, the order of the updates of the pixels of $\mathcal{I}$ is random in order to overcome possible bottlenecks in the exploration of the distribu-

[8] The very name "Gibbs sampling" was proposed in reference to Gibbs random fields, related to the physicist Willard Gibbs. Interestingly, both of the major MCMC algorithms are thus named after physicists and were originally developed for problems that were beyond the boundaries of (standard) statistical inference.

tion, although this is not a necessary condition for the algorithm to converge. In fact, when considering two pixels x_1 and x_2 that are m pixels apart, the influence of a change in x_1 is not felt in x_2 before at least m iterations of the basic Gibbs sampler. Of course, if m is large, the dependence between x_1 and x_2 is quite moderate, but this slow propagation of changes is indicative of slow mixing in the Markov chain. For instance, to see a change of color of a relatively large homogeneous region is an event of very low probability, even though the distribution of the colors is exchangeable (Exercise 8.18).

⚡ If β is large, the Ising distribution (8.4) is very peaked around both single color configurations. In such settings, the Gibbs sampler will face enormous difficulties to simply change the value of a single pixel.

Running Algorithm 8.16 in R is straightforward: opting for a four-neighbor relation, if we use the following function for the number of neighbors at (a, b),

```
xneig4=function(x,a,b,col){
n=dim(x)[1];m=dim(x)[2]
nei=c(x[a-1,b]==col,x[a,b-1]==col)
if (a!=n)
  nei=c(nei,x[a+1,b]==col)
if (b!=m)
  nei=c(nei,x[a,b+1]==col)
sum(nei)
}
```

the above Gibbs sampler can be written as

```
isingibbs=function(niter,n,m=n,beta){
  # initialization
  x=sample(c(0,1),n*m,prob=c(0.5,0.5),rep=TRUE)
  x=matrix(x,n,m)
  for (i in 1:niter){
    sampl1=sample(1:n)
    sampl2=sample(1:m)
    for (k in 1:n){
    for (l in 1:m){
     n0=xneig4(x,sampl1[k],sampl2[l],0)
     n1=xneig4(x,sampl1[k],sampl2[l],1)
     x[sampl1[k],sampl2[l]]=sample(c(0,1),1,
                  prob=exp(beta*c(n0,n1)))
     }}}
  x
  }
```

where niter is the number of times the whole matrix x is modified. (It should therefore be scaled against n*m, the size of x.) Figure 8.3 presents the output of simulations from Algorithm 8.16

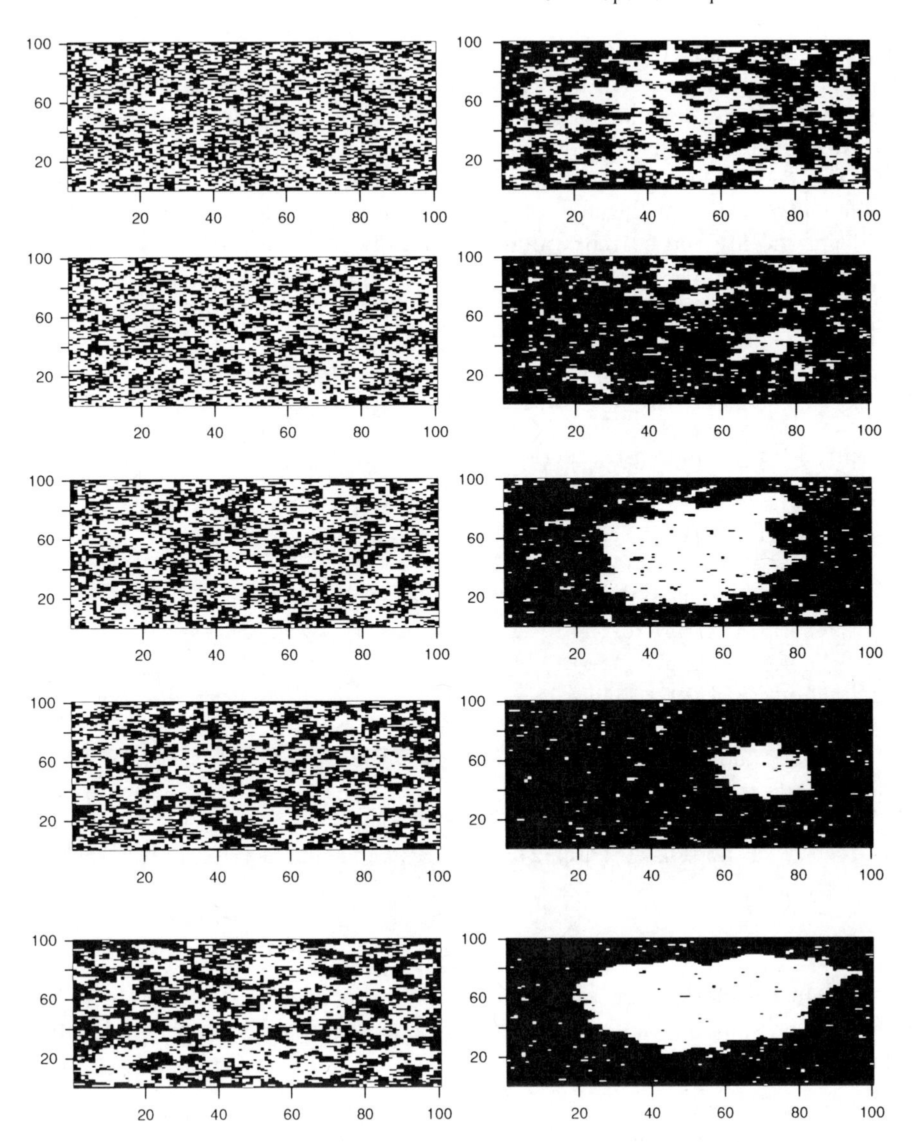

Fig. 8.3. Simulations from the Ising model with a four-neighbor neighbourhood structure on a 100×100 array after 1,000 iterations of the Gibbs sampler: β varies in steps of 0.1 from 0.3 to 1.2 (*first column, then second column*)

```
> image(1:100,1:100,isingibbs(10^3,100,100,beta))
```

for different values of β. Although we cannot discuss here convergence assessment for the Gibbs sampler (see Robert and Casella, 2009, Chap. 8), the images thus produced are representative of the Ising distributions: the larger

β, the more homogeneous the image (and also the slower the Gibbs sampler).[9] When looking at the result associated with the larger values of β, we can start to see the motivations for using such representations to model images like the **Menteith** dataset, discussed in Sect. 8.4.

Along with the slow dynamic induced by the single-site updating, we can point out another inefficiency of this algorithm, namely that many updates will not modify the current value of $\mathbf{x}$ simply because the new value of x_l is equal to its previous value! It is, however, straightforward to modify the algorithm so that it only proposes changes of values. The update of each pixel l is then a Metropolis–Hastings step with acceptance probability

$$\rho = \exp(\beta n_{l,1-x_l}) / \exp(\beta n_{l,x_l}) \wedge 1 \,,$$

with the corresponding R function

```
isinghm=function(niter,n,m=n,beta){
  x=sample(c(0,1),n*m,prob=c(0.5,0.5),rep=TRUE)
  x=matrix(x,n,m)
  for (i in 1:niter){
   sampl1=sample(1:n)
   sampl2=sample(1:m)
   for (k in 1:n){
   for (l in 1:m){
    n0=xneig4(x,sampl1[k],sampl2[l],x[sampl1[k],sampl2[l]])
    n1=xneig4(x,sampl1[k],sampl2[l],1-x[sampl1[k],sampl2[l]])
    if (runif(1)<exp(beta*(n1-n0)))
      x[sampl1[k],sampl2[l]]=1-x[sampl1[k],sampl2[l]]
   }}}
  x
  }
```

Although the details are too involved to be included here, Liu (1996) has shown that this alternative is faster (to converge) than the original Gibbs sampler.

8.2.4 The Potts Model

The generalization of the Ising model to cases when the image has more than two colors, G say, is straightforward. If $n_{i,g}$ denotes the number of neighbors of $i \in \mathcal{I}$ with color g $(1 \le g \le G)$, that is,

[9] In fact, there exists a critical value of β, $\beta_c = 2.269185$ in the case of the four neighbor relation, such that, when $\beta > \beta_c$, the Markov chain converges to one of two different stationary distributions, depending on the starting point. In other words, the chain is no longer irreducible. In particle physics, this phenomenon is called *phase transition*.

$$n_{i,g} = \sum_{j \sim i} \mathbb{I}_{x_j = g} ,$$

the full conditional distribution of x_i is chosen as

$$\pi(x_i = g|\mathbf{x}_{-i}) \propto \exp(\beta n_{i,g}) .$$

This choice corresponds to a (true) joint probability model, the *Potts model*, whose density is given by (Exercise 8.6)

$$\pi(\mathbf{x}) \propto \exp\left(\beta \sum_{j \sim i} \mathbb{I}_{x_j = x_i}\right) . \tag{8.5}$$

This model is a clear generalization of the Ising model and it suffers from the same drawback, namely that the normalizing constant of this density—which is a function of β—is not available in closed form and thus hinders inference and the computation of the likelihood function.

Once again, we face the hindrance that, when simulating $\mathbf{x}$ from a Potts model with a large β, the single-site Gibbs sampler may be quite slow. More efficient alternatives are available, including the Swendsen–Wang algorithm (Exercise 8.7). For instance, Algorithm 8.17 below is again a Metropolis–Hastings algorithm that forces moves on the current values. Note the special feature that, while this Metropolis–Hastings proposal is *not* a random walk, using instead a uniform proposal on the $G-1$ other possible values still leads to an acceptance probability that is equal to the ratio of the target densities.

Algorithm 8.17 POTTS METROPOLIS–HASTINGS SAMPLER

Initialization: For $i \in \mathcal{I}$, generate independently

$$x_i^{(0)} \sim \mathscr{U}(\{1, \ldots, G\}) .$$

Iteration t ($t \geq 1$):

1. Generate $\mathbf{u} = (u_i)_{i \in \mathcal{I}}$ a random ordering of the elements of $\mathcal{I}$.
2. For $1 \leq \ell \leq |\mathcal{I}|$,

   generate

   $$\tilde{x}_{u_\ell} \sim \mathscr{U}(\{1, 2, \ldots, x_{u_\ell}^{(t-1)} - 1, x_{u_\ell}^{(t-1)} + 1, \ldots, G\}) ,$$

   compute the $n_{u_l,g}^{(t)}$ and

   $$\rho_l = \left\{\exp(\beta n_{u_\ell, \tilde{x}_{u_\ell}}) / \exp(\beta n_{u_\ell, x_{u_\ell}}^{(t)})\right\} \wedge 1 ,$$

   and set $x_{u_\ell}^{(t)}$ equal to $\tilde{x}_{u_\ell}$ with probability ρ_l.

Figure 8.4 illustrates the result of a simulation using Algorithm 8.17 in a situation where there are $G = 4$ colors, using the following R function

```
pottshm=function(ncol=2,niter=10^4,n,m=n,beta=0){
x=matrix(sample(1:ncol,n*m,rep=TRUE),n,m)
for (i in 1:niter){
  sampl=sample(1:(n*m))
  for (k in 1:(n*m)){
    xcur=x[sampl[k]]
    a=(sampl[k]-1)%%n+1
    b=(sampl[k]-1)%/%n+1
    xtilde=sample((1:ncol)[-xcur],1)
    acpt=beta*(xneig4(x,a,b,xtilde)-xneig4(x,a,b,xcur))
    if (log(runif(1))<acpt) x[sampl[k]]=xtilde
    }}
return(x)
}
```

for the simulation. (The use of a single vector of indices for rows and columns is a programming trick that removes a loop in the code and thus saves a considerable amount of computing time. This also allows a true uniform distribution in `sampl`. Note the call to the congruential operators `%%` for modulo and `%/%` for integer division) We point out the reinforced influence of large β's on Fig. 8.4: not only is the homogeneity higher, but there is also a larger differentiation in the colors.[10] We stress that, while β in Fig. 8.4 ranges over the same values as in Fig. 8.3, the β's are not directly comparable since the larger number of classes in the Potts model induces a smaller value of the $n_{i,g}$'s for the neighbourhood structure.

8.3 Handling the Normalizing Constant

While simulating random variables distributed *from* a Potts model is required in several settings, one of which we will cover in the next section, a more common *statistical* setting is observing $\mathbf{x}$ distributed as

$$f(\mathbf{x}|\beta) = \frac{1}{Z(\beta)} \exp\left(\beta \sum_{j\sim i} \mathbb{I}_{x_j = x_i}\right), \tag{8.6}$$

where $Z(\beta)$ is the normalizing constant of the density in $\mathbf{x}$, and inferring upon the parameter β, using for instance a uniform prior $\beta \sim \mathscr{U}(0,2)$.[11]

[10] Similar to the Ising model mentioned in Footnote 9, there also exist a phase transition phenomenon and a critical value for β in this model.

[11] The upper bound on β in the above prior is chosen for a very precise reason: As mentioned in the previous footnotes, when $\beta \geq 2$, the Potts model associated with a four-neighbor relation is almost surely concentrated on single-color images. It is thus pointless to consider larger values of β.

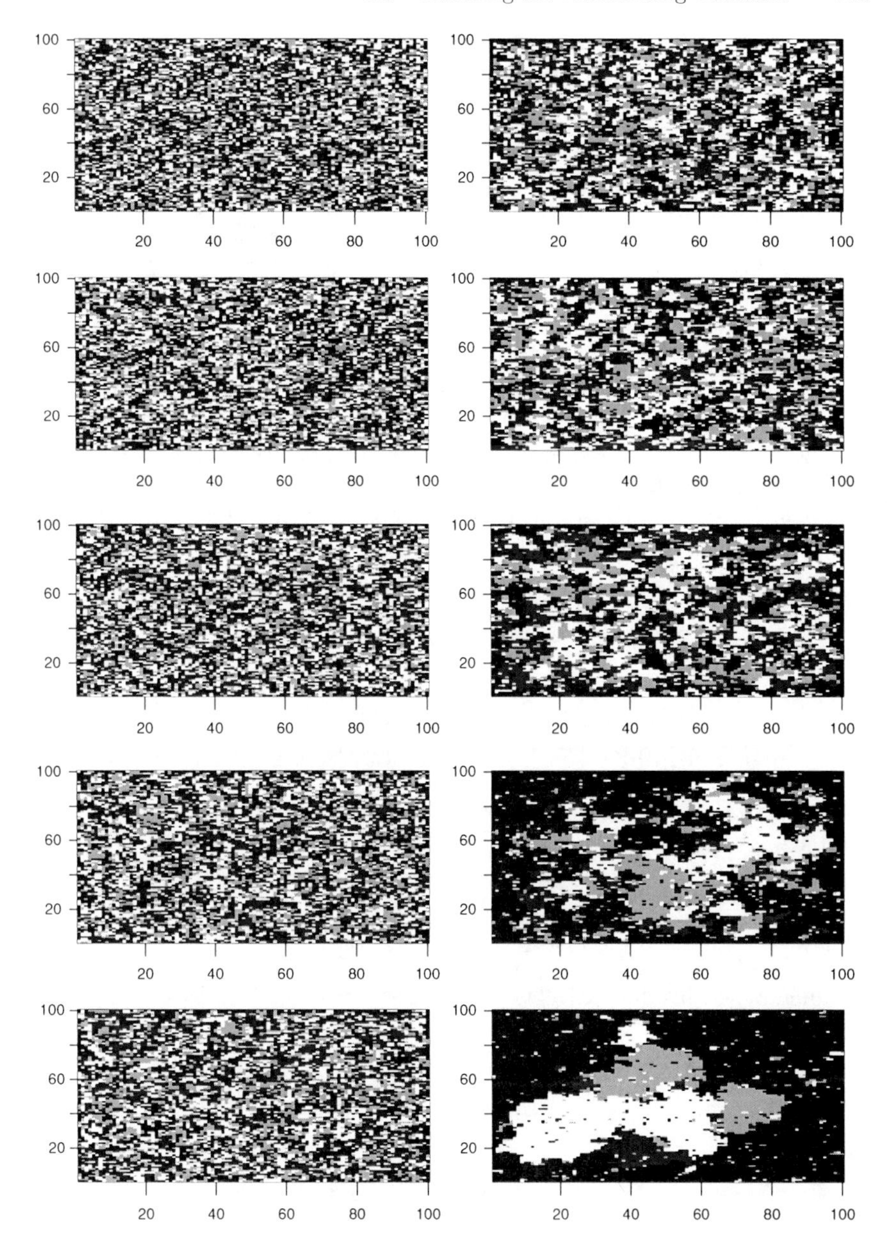

Fig. 8.4. Simulations from the Potts model with four grey levels and a four-neighbor neighbourhood structure based on 1,000 iterations of the Metropolis–Hastings sampler. The parameter β varies in steps of 0.1 from 0.3 to 1.2 (*first column, then second column*)

The primary computational difficulty with this inference is the unavailability of the normalizing constant

$$Z(\beta) = \sum_{\mathbf{x}} \exp\left\{\beta S(\mathbf{x})\right\} ,$$

where $S(\mathbf{x}) = \sum_{i\in\mathcal{I}} \sum_{j\sim i} \mathbb{I}_{x_j=x_i}$. The above summation operates over the $G^{|\mathcal{I}|}$ possible values of $\mathbf{x}$, where $|\mathcal{I}|$ denotes the size of $\mathcal{I}$. It involves too many terms to be manageable. In the case of the Ising model, the number of terms in the sum is for instance 2 to the power the number of points in the lattice. For a small 256×256 black-and-white image, there are therefore 2^{65536} terms in the sum! Furthermore, this is not a setting where a standard MCMC solution would apply because of the same difficulty: a Metropolis–Hastings algorithm also requires the evaluation of the ratio $Z(\tilde{\beta})/Z(\beta)$ in the acceptance probability. Unsurprisingly, addressing the approximation of $Z(\beta)$ has given rise to a huge literature, as shown by Ripley (1988) and Rue and Held (2005), but the solutions are mostly too convoluted for this book (see, e.g., the auxiliary variable method of Møller et al., 2006). We first describe a semi-practical resolution of this difficulty, called *path sampling*, which is costly in computing time for large images, before moving to a more generic if less precise solution.

8.3.1 Path Sampling

The path sampling technique is based on a derivative representation of the normalizing constant. Since

$$\frac{\mathrm{d}Z(\beta)}{\mathrm{d}\beta} = \sum_{\mathbf{x}} S(\mathbf{x}) \exp(\beta S(\mathbf{x})) ,$$

we can express this derivative as an expectation under $\pi(\mathbf{x}|\beta)$,

$$\frac{\mathrm{d}Z(\beta)}{\mathrm{d}\beta} = Z(\beta) \sum_{\mathbf{x}} S(\mathbf{x}) \frac{\exp(\beta S(\mathbf{x}))}{Z(\beta)} = Z(\beta)\, \mathbb{E}_\beta[S(\mathbf{X})] ,$$

that is,

$$\frac{\mathrm{d} \log Z(\beta)}{\mathrm{d}\beta} = \mathbb{E}_\beta[S(\mathbf{X})] .$$

Therefore, the ratio $Z(\beta_1)/Z(\beta_0)$ can be represented as an integral,

$$\log\left\{Z(\beta_1)/Z(\beta_0)\right\} = \int_{\beta_0}^{\beta_1} \mathbb{E}_\beta[S(\mathbf{x})]\mathrm{d}\beta , \tag{8.7}$$

leading to the *path sampling identity* (see Chen et al., 2000, for many more details about this technique.)

Although (8.7) may not look like a considerable improvement, since we now have to compute an expectation in $\mathbf{x}$ plus an integral over β, the representation (8.7) is appealing because we can use standard simulation procedures for its approximation. First, for a given value of β, $\mathbb{E}_\beta[S(\mathbf{X})]$ can be approximated from an MCMC sequence simulated by Algorithm 8.17. Obviously, changing the value of β should involve a new simulation run, however the cost can be attenuated by using instead importance sampling for similar values of β. Second, the integral itself can be approximated by *numerical quadrature*, namely by computing the value of $f(\beta) = \mathbb{E}_\beta[S(\mathbf{X})]$ for a finite number of values of β and approximating $f(\beta)$ by a piecewise-linear function $\hat{f}(\beta)$ for the intermediate values of β. Indeed, for arbitrary β_0 and β_1,

$$\int_{\beta_0}^{\beta_1} f(\beta)\,\mathrm{d}\beta \approx \hat{f}(\beta_0) + \left\{\hat{f}(\beta_1) - \hat{f}(\beta_0)\right\} \frac{(\beta_1-\beta_0)^2}{2}\,,$$

where $\hat{f}(\beta)$ is approximated by the above Monte Carlo method.

The rendering of the above in R for **Laichedata** is as follows for a four-neighbor relation: the expectation $\mathbb{E}_\beta[S(\mathbf{X})]$ is approximated via the following R function

```
sumising=function(niter=10^3,numb,beta){
S=0
x=matrix(sample(c(0,1),numb^2,rep=TRUE),ncol=numb)
for (i in 1:niter){
  s=0
  sampl1=sample(1:numb)
  sampl2=sample(1:numb)
  for (k in 1:numb){
  for (l in 1:numb){
   n0=xneig4(x,sampl1[k],sampl2[l],x[sampl1[k],sampl2[l]])
   n1=xneig4(x,sampl1[k],sampl2[l],1-x[sampl1[k],sampl2[l]])
   if (log(runif(1))<(beta*(n1-n0))){
     x[sampl1[k],sampl2[l]]=1-x[sampl1[k],sampl2[l]]
     n0=n1}
   s=s+n0
   }}
  if (2*i>niter)
    S=S+s
  }
return(2*S/niter)
}
```

for a few selected values of β, while the whole function $f(\beta)$ is then approximated using the R procedure `approxfun` as

```
Z=seq(0,2,by=.1)
for (i in 1:21)
```

```
  Z[i]=sumising(numb=24,beta=Z[i])
lrcst=approxfun(seq(0,2,0.1),Z)
```

This approximation is illustrated by Fig. 8.5. The ratio of the constants, $Z(\tilde{\beta})/Z(\beta)$ is provided by the R numerical integration function, `integrate`, as

```
Zratio=integrate(lrcst,betatilde,beta)$value
```

and can be easily inserted within a random walk Metropolis–Hasting algorithm. Indeed, now that we have painstakingly constructed a satisfactory

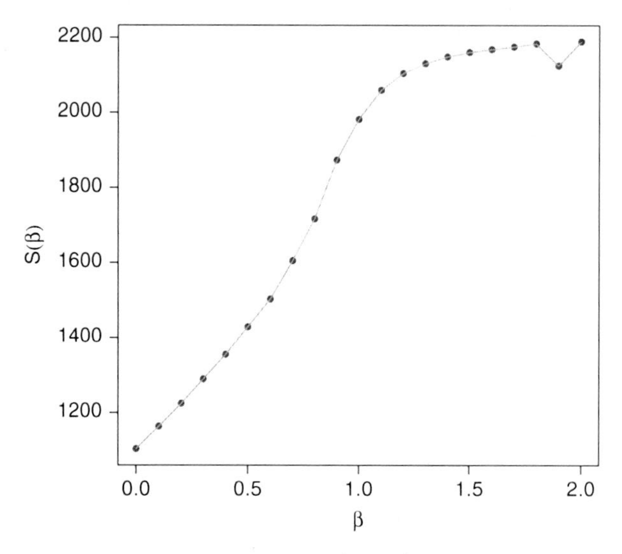

Fig. 8.5. Monte Carlo approximation of $\mathbb{E}_\beta[S(\mathbf{X})]$ for a 24×24 Ising model, based on 10^3 iterations. The irregularity at the penultimate value of β can be attributed to a failed convergence of the Gibbs sampler

approximation of $Z(\beta_1)/Z(\beta_0)$ for any arbitrary pair (β_0, β_1), we can run an MCMC sampler targeting the posterior distribution $\pi(\beta|\mathbf{x})$, where simulation at iteration t is based on the proposal

$$\tilde{\beta} \sim \mathscr{U}([\beta^{(t-1)} - h, \beta^{(t-1)} + h]) ;$$

that is, a uniform move with range $2h$. The acceptance ratio associated with the pair $(\beta^{(t-1)}, \tilde{\beta})$ is thus given by

$$1 \wedge \left(\hat{Z}(\beta^{(t-1)})/\hat{Z}(\tilde{\beta})\right) \exp\left\{(\tilde{\beta} - \beta^{(t-1)})S(\mathbf{x})\right\} ,$$

which translates into the R code

```
betatilde=beta[t-1]+runif(1,-0.05,0.05)
laccept=lvr*(betatilde-beta[t-1])+integrate(lrcst,
        betatilde,beta[t-1])$value
if (runif(1)<exp(laccept)){
  beta[t]=betatilde}else{
  beta[t]=beta[t-1]}
```

The outcome of this MCMC algorithm is represented by the histogram of Fig. 8.6, which exhibits a very regular posterior distribution for β, which is symmetric around 0.47. Thanks to the path sampling approximation to $Z(\beta)$, running 10^5 iterations is almost instantaneous.

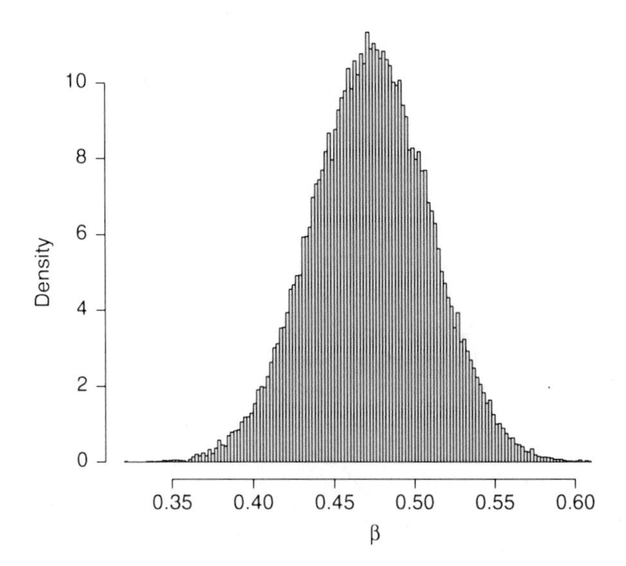

Fig. 8.6. Dataset **Laichedata**: Histogram of the MCMC sample of β's produced using the path sampling approximation to the ratio $Z(\tilde{\beta})/Z(\beta)$, when based on 10^5 iterations

8.3.2 The ABC Method

In a general setting where the likelihood function is not available in a closed form, the trick at the core of the path sampling technique is not always available. (Consider for instance the case of a multivariate β.) We thus need to turn towards faster if more rudimentary approximations and a method of choice is the ABC (approximate Bayesian computation) technique, introduced by Pritchard et al. (1999) in population genetic settings.

The method starts from a valid rejection technique bypassing the computation of the likelihood function. Namely, if we observe $x \sim f(x|\theta)$

and if $\pi(\theta)$ is the prior distribution on the parameter θ, then an algorithm that jointly simulates

$$\theta' \sim \pi(\theta) \quad \text{and} \quad y \sim f(y|\theta')$$

and accepts the simulated θ' if, and only if, the auxiliary variable y is equal to the observed value,

$$x = y\,,$$

is exact in the sense that the accepted θ''s are distributed from the posterior. Obviously, the algorithm is not practical in cases when x is continuous or even takes a large enough number of values.[12] In most standard occurrences, the ABC algorithm starts with an approximation, in the sense that the equality constraint $x = y$ is replaced with a tolerance condition, $\varrho(x, y) \leq \epsilon$, where ϱ is a measure of discrepancy between x and y. We will call $\epsilon > 0$ the tolerance bound and ϱ will be chosen as a distance between summary statistics. The output of the ABC algorithm is then distributed from the distribution with density proportional to

$$\pi(\theta)\,\mathbb{P}_\theta(\varrho(x, y) \leq \epsilon|x)\,,$$

where the probability is associated with $y \sim f(y|\theta)$. This density is denoted by $\pi_\epsilon(\theta \mid x)$.

If the tolerance ϵ is "too large", the approximation is poor; to understand why, consider that, when ϵ goes to ∞, the ABC algorithm amounts to simulating from the prior since all simulations are accepted. If ϵ is sufficiently small, $\pi_\epsilon(\theta|x)$ is a good approximation of $\pi(\theta|x)$, but the acceptance probability may be too low for this value to be practical. Selecting the "right" ϵ is thus crucial. It is customary to pick ϵ as an empirical quantile of $\varrho(x, y)$ when y is simulated from the marginal

$$\int \pi(\theta) f(y|\theta)\mathrm{d}\theta$$

and the choice is often the corresponding 1 % quantile. This quantile is easily approximated by simulation.

In settings when the data x has a large dimension, the ABC algorithm uses instead a distance between summary statistics $\varrho(S(x), S(y))$ rather than a distance between x and y. This choice throws away some information contained in the data about θ, but it also allows to concentrate on important features of the data in order to bring a maximal discrimination between the observed and the simulated statistics. It is thus rarely the case that S is a

[12] Note that, for **Laichedata**, it is possible to wait for the equality $S(x) = S(y)$ with a sufficiently high probability. In that case, since S is a sufficient statistic, we are simulating from the *exact* posterior.

sufficient statistic.[13] In the general case, the output of the ABC algorithm is therefore a simulation from the distribution $\pi_\epsilon(\theta \mid x)$. The ABC algorithm thus reads as follows:

Algorithm 8.18 ABC algorithm For $i = 1, \ldots, N$,

1. Generate θ_i from the prior π.
2. Generate y_i from the model distribution $f(x|\theta_i)$.
3. Compute the distance $\varrho(S(y_i), S(x))$.

Deduce ϵ as the 1% quantile of the distances. Accept the θ_i's such that $\varrho(S(x), S(y_i)) \leq \epsilon$.

To illustrate the ABC method in a simple environment, consider the problem already processed in Chap. 2 about assessing whether a normal $\mathscr{N}(\mu, \sigma^2)$ distribution has a zero mean, $\mu = 0$. As explained in Sect. 2.3.1, the natural Bayesian approach is to include the model index $\mathfrak{M}$ as an extra parameter taking only the values 1 (when $\mu = 0$) and 2 (when $\mu \neq 0$). In other words, Bayesian inference covers the pair $(\mathfrak{M}, \theta)$, conditional on the data $\mathscr{D}_n$. Simulating by ABC from the posterior on $(\mathfrak{M}, \theta)$ given $\mathscr{D}_n$ then follows from Algorithm 8.18:

1. Generate $\mathfrak{M}^i$ uniformly at random on $\{1, 2\}$ $(i = 1, \ldots, n)$.
2. Generate θ_i from the prior $\pi(\theta|\mathfrak{M}^i)$ $(i = 1, \ldots, n)$.
3. Generate $\mathscr{D}_n^i$ from the normal model indexed by $(\mathfrak{M}^i, \theta_i)$ $(i = 1, \ldots, n)$.
4. Compute the distances between the statistics $(\bar{x}(\mathscr{D}_n), s^2(\mathscr{D}_n))$ and $(\bar{x}(\mathscr{D}_n^i), s^2(\mathscr{D}_n^i))$ $(i = 1, \ldots, n)$.
5. Deduce ϵ as the 1 % quantile of the distances.
6. Accept the $\mathfrak{M}^i$'s for which the distances are less than ϵ.

The distance we pick is inspired from the likelihood function, namely

$$\begin{aligned}\varrho\{(\bar{x}(\mathscr{D}_n), s^2(\mathscr{D}_n)), (\bar{x}(\mathscr{D}_n^*), s^2(\mathscr{D}_n^*))\} &= n\{\bar{x}(\mathscr{D}_n) - \bar{x}(\mathscr{D}_n^*)\}^2 \\ &\quad + \{s^2(\mathscr{D}_n)/s^2(\mathscr{D}_n^*)\} - 1 - \log\{s^2(\mathscr{D}_n)/s^2(\mathscr{D}_n^*)\}\,.\end{aligned}$$

The implementation is then straightforward: we select one of the models at random, simulate from the corresponding (necessarily proper) prior on the parameter(s) and create a normal sample $\mathscr{D}_n^*$. The posterior probability of the model associated with $\mu = 0$ is then estimated by the proportion of accepted simulations from the simpler model. Under an $\mathscr{E}(1)$ prior on σ^2 in both models and a $\mathscr{N}(0, \sigma^2)$ on μ under the larger model, with the **normaldata** benchmark, the R code goes as follow:

[13] The setting of Markov random fields like the Ising and the Potts models is an exception in that it allows for a sufficient statistic, while being intractable via classical approaches.

```
> xbar=mean(normaldata)
> s2=(n-1)*var(normaldata)
> Nsim=10^6                      #simulations from the prior
> indem=sample(c(0,1),Nsim,rep=TRUE)
> ssigma=1/rexp(Nsim)
> smu=rnorm(Nsim)*sqrt(ssigma)*(indem==1)
> ss2=s2/(ssigma*rchisq(Nsim,n-1))
> sobs=n*(rnorm(Nsim,smu,sqrt(ssigma/n))-xbar)^2+
+ ss2-1-log(ss2)
> epsi=quantile(sobs,.001)  #bound and selection
> prob=sum(indem[sobs<=epsi]==0)/(0.001*Nsim)
> (1-prob)/prob
[1] 0.1574074
```

producing a numerical value to be compared with the exact Bayes factor

$$(n+1)^{-1/2}\left[\frac{n\bar{x}^2+s^2+2}{n\bar{x}^2/(n+1)+s^2+2}\right]^{n+2/2}$$

(deduced from the derivation on page 45 by modifying for the exponential prior), which is equal to 0.1369 for **normaldata**. Figure 8.7 represents the variability of the ABC approximation compared with the true value.

8.3.3 Inference on Potts Models

If we consider the specific case of the posterior distribution associated with (8.6) and a uniform prior, Algorithm 8.18 simulates values of β uniformly over

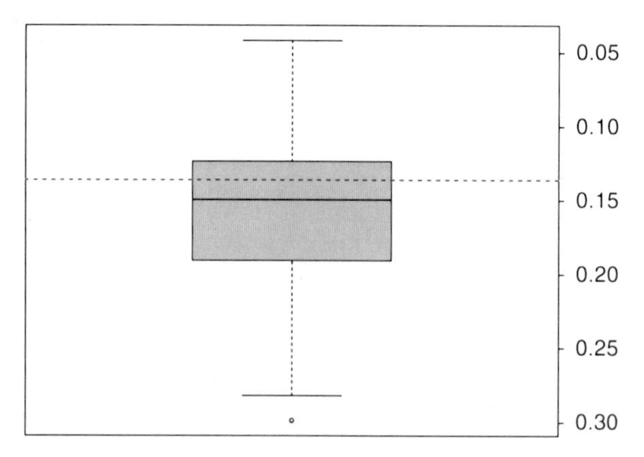

Fig. 8.7. Dataset **normaldata**: Boxplot representation of the ABC approximation to the Bayes factor, which true value is represented by an *horizontal line*, based on 10^5 proposals, a 1 % acceptance rate, and 500 replications

(0, 2) and then values $\mathbf{x}$ from the Potts model (8.6). Simulating a data set $\mathbf{x}$ is unfortunately non-trivial for Markov random fields and in particular for Potts models, as we already discussed. While there exist developments towards this goal in the special case of the Ising model—in the sense that they produce exact simulations, at a high computing cost—, we settle for using a certain number of steps of an MCMC sampler (for instance, Algorithm 8.17) updating one clique at a time conditional on the others. Obviously, this solution brings a further degree of approximation into the picture in that running a fixed number of iterations of the MCMC sampler does not produce an exact simulation from (8.6). There is however little we can do about this if we want to use ABC. (And we can further argue that ABC involves such a significant departure from the exact posterior that an imperfect MCMC simulation does not matter so much!)

Since, for every new value of β, the algorithm runs a full MCMC simulation, we need to discuss the choice of the starting value as well. There are (at least) three natural solutions:

- start completely at random;
- start from the previously simulated $\mathbf{x}$.
- always start from the observed value $\mathbf{x}_0$;

The first one is the closest to the MCMC idea and it produces independent outcomes. The second solution is less compelling as the continuity it creates between draws is not statistically meaningful, given that the simulated β's change (independently or not) from one step to the other. The third solution offers the appealing feature of connecting with the observed value $\mathbf{x}_0$, thus favoring proximity between the simulated and the observed values, but this feature could confuse the issues in that this proximity may be due to a poor mixing of the chain rather than to a proper choice for β. (For instance, in the extreme case the MCMC chain does not move from $\mathbf{x}_0$, $\mathbf{x} = \mathbf{x}_0$ does not mean that the simulated β is at all interesting for $\pi(\beta|\mathbf{x}_0)$...) The distance used in step 3 of Algorithm 8.18 is the (natural) absolute difference between the sufficient statistics $S(\mathbf{x})$ and $S(\mathbf{x}_0)$, with

$$S(\mathbf{x}) = \sum_{i \sim j} \mathbb{I}_{x_i = x_j} \,.$$

For the four-neighbour relation, the statistic can be computed directly without loops as

```
sum(x[-1,]==x[-n,])+sum(x[,-1]==x[,-m])
```

and the whole R code corresponding to a random start of the Metropolis–Hastings algorithm is as follows:

```
> ncol=4; nrow=10; Nsim=2*10^4; Nmc=10^2
> suf0=sum(x0[-1,]==x0[-nrow,])+sum(x0[,-1]==x0[,-nrow])
> outa=dista=rep(0,Nsim)
```

```
> for (tt in 1:Nsim){
+   beta=runif(1,max=2)
+   xprop=pottshm(ncol,nit=Nmc,n=nrow,beta=beta)
+   dista[tt]=abs(suf0-(sum(xprop[-1,]==xprop[-nrow,])+
+    sum(xprop[,-1]==xprop[,-ncol])))
+  outa[tt]=beta
+  }
betas=outa[order(dista)<=.01*Nsim]
```

Note the inequality sign `<=` and the use of `jitter` to get exactly `0.01*Nsim` values in the vector `beta`. This is due to the fact that the statistic S takes integer values.

When applying the above to the **Laichedata** dataset, we obtain the outcome represented in Fig. 8.8. When comparing with Fig. 8.6, we can check that ABC produces an almost exact representation, even though ϵ is not equal to zero. As mentioned above, it would actually be feasible to achieve $\epsilon = 0$ with a larger number of simulations.

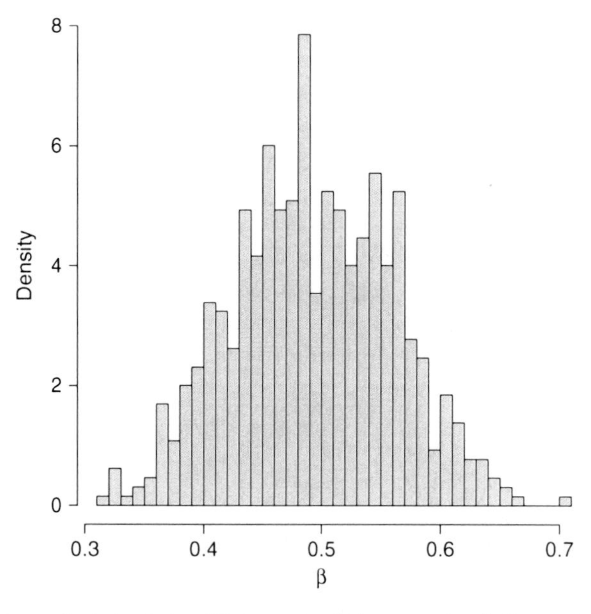

Fig. 8.8. Dataset **Laichedata**: Histogram of the sample of β's produced using an ABC algorithm with 10^4 iterations and a 1 % quantile on the difference between the sufficient statistics as its tolerance bound ϵ

8.4 Image Segmentation

In this section, we still consider images as statistical objects, but they are now "noisy" in the sense that the color or the grey level of a pixel is not observed exactly but with some perturbation (sometimes called *blurring* as in satellite imaging). The purpose of image segmentation is to cluster pixels into homogeneous classes without supervision or preliminary definition of those classes, based only on the spatial coherence of the structure.

This underlying structure of the "true" pixels is denoted by $\mathbf{x}$, while the observed image is denoted by $\mathbf{y}$. Both objects $\mathbf{x}$ and $\mathbf{y}$ are arrays, with each entry of $\mathbf{x}$ taking a finite number of values and each entry of $\mathbf{y}$ taking real values (for modeling convenience rather than reality constraints). We are thus interested in the posterior distribution of $\mathbf{x}$ given $\mathbf{y}$ provided by Bayes' theorem, $\pi(\mathbf{x}|\mathbf{y}) \propto f(\mathbf{y}|\mathbf{x})\pi(\mathbf{x})$. In this posterior distribution, the likelihood, $f(\mathbf{y}|\mathbf{x})$, describes the link between the observed image and the underlying classification; that is, it gives the distribution of the noise, while the prior $\pi(\mathbf{x})$ encodes beliefs about the (possible or desired) properties of the underlying image. Although, as in other chapters, we cannot provide the full story of Bayesian image segmentation, an excellent tutorial on Bayesian image processing based on a summer school course can be found in Hurn et al. (2003).

As indicated above, a proper motivation for image segmentation is satellite processing since images caught by satellites are often blurred, either because of inaccuracies in the instruments or transmission or because of clouds or vegetation cover between the satellite and the area of interest.

The **Menteith** dataset that motivates this section is a 100×100 pixel satellite image of the lake of Menteith, as represented in Fig. 8.9. The lake of Menteith is located in Scotland, near Stirling, and offers the peculiarity of being called "lake" rather than the traditional Scottish "loch." As shown by the image, there are several islands on this lake, one of which houses an ancient abbey. The purpose of analyzing this satellite dataset is to classify all pixels into one of six states in order to detect some homogeneous regions.

The model being introduced, we turn to the central issue, namely how to draw inference on the "true" image, $\mathbf{x}$, given an observed noisy image, $\mathbf{y}$. The prior on $\mathbf{x}$ is a Potts model with G categories,

$$\pi(\mathbf{x}|\beta) = \frac{1}{Z(\beta)} \exp\left(\beta \sum_{j \sim i} \mathbb{I}_{x_j = x_i}\right),$$

where $Z(\beta)$ is the (intractable, see Sect. 8.3) normalizing constant of the Potts model. Given $\mathbf{x}$, we assume that the observations in $\mathbf{y}$ are independent normal random variables,

$$f(\mathbf{y}|\mathbf{x}, \sigma^2, \mu_1, \ldots, \mu_G) = \prod_{i \in \mathcal{I}} \frac{1}{(2\pi\sigma^2)^{1/2}} \exp\left\{-\frac{1}{2\sigma^2}(y_i - \mu_{x_i})^2\right\}.$$

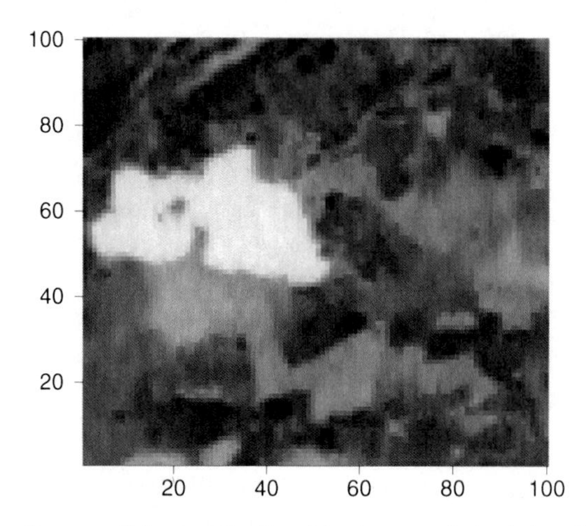

Fig. 8.9. Dataset **Menteith**: Satellite image of the lake of Menteith

This model is not exact in that the y_i's are integer grey levels that vary between 0 and 255, but it is easier to handle than a parameterized distribution on $\{0, \ldots, 255\}$. This setting is clearly reminiscent[14] of the mixture and hidden Markov models of Chaps. 6 and 7 in that a Markov structure, the Markov random field, is only observed through random variables indexed by the states.

In this problem, the parameters $\beta, \sigma^2, \mu_1, \ldots, \mu_G$ are usually considered to be *nuisance* parameters, a point of view that justifies the use of uniform priors like

$$\begin{aligned} \beta &\sim \mathscr{U}([0,2])\,, \\ \boldsymbol{\mu} = (\mu_1, \ldots, \mu_G) &\sim \mathscr{U}(\{\boldsymbol{\mu}\,;\, 0 \leq \mu_1 \leq \ldots \leq \mu_G \leq 255\})\,, \\ \pi(\sigma^2) &\propto \sigma^{-2}\mathbb{I}_{]0,\infty[}(\sigma^2)\,, \end{aligned}$$

the last prior corresponding to a uniform prior on $\log \sigma$.

The upper bound on β has been discussed in the previous section. The ordering of the μ_g's is not necessary, strictly speaking, but it avoids the label switching phenomenon discussed in Sect. 6.5. (The alternative is to use the same uniform prior on all μ_g's and then reorder them once the MCMC simulation is done. While this may avoid slow convergence behaviors in some cases, this strategy also implies more involved bookkeeping and higher storage requirements. In the case of large images, it simply cannot be considered.)

[14] Besides image segmentation, another typical illustration of such structures is *character recognition* where a machine scans handwritten documents, e.g., envelopes, and must infer a sequence of symbols (i.e., numbers or letters) from digitized pictures. Hastie et al. (2001) provide an illustration of this problem.

The corresponding posterior distribution is thus

$$\pi(\mathbf{x},\beta,\sigma^2,\boldsymbol{\mu}|\mathbf{y}) \propto \pi(\beta,\sigma^2,\boldsymbol{\mu}) \times \frac{1}{Z(\beta)} \exp\left(\beta \sum_{j\sim i} \mathbb{I}_{x_j=x_i}\right)$$
$$\times \prod_{i\in\mathcal{I}} \frac{1}{(2\pi\sigma^2)^{1/2}} \exp\left\{\frac{-1}{2\sigma^2}(y_i-\mu_{x_i})^2\right\}.$$

We can therefore construct the various full conditionals of this joint distribution with a view to the derivation of a hybrid Gibbs sampler for this model. First, the full conditional distribution of x_i $(i \in \mathcal{I})$ is $(1 \le g \le G)$

$$\mathbb{P}(x_i = g|\mathbf{y},\beta,\sigma^2,\boldsymbol{\mu}) \propto \exp\left\{\beta \sum_{j\sim i} \mathbb{I}_{x_j=g} - \frac{1}{2\sigma^2}(y_i-\mu_g)^2\right\},$$

which can be simulated directly, even though this is no longer a Potts model. As in the mixture and hidden Markov cases, once $\mathbf{x}$ is known, the groups associated with each category g separate and therefore the μ_g's can be simulated independently conditional on $\mathbf{x}$, $\mathbf{y}$, and σ^2. More precisely, if we denote by

$$n_g = \sum_{i\in\mathcal{I}} \mathbb{I}_{x_i=g} \quad \text{and} \quad s_g = \sum_{i\in\mathcal{I}} \mathbb{I}_{x_i=g} y_i$$

the number of observations and the sum of the observations allocated to category g, respectively, the full conditional distribution of μ_g is a truncated normal distribution on $[\mu_{g-1},\mu_{g+1}]$ (setting $\mu_0 = 0$ and $\mu_{G+1} = 255$) with mean s_g/n_g and variance σ^2/n_g. (Obviously, if no observation is allocated to this group, the conditional distribution turns into a uniform distribution on $[\mu_{g-1},\mu_{g+1}]$.) The full conditional distribution of σ^2 is an inverse gamma distribution with parameters $|\mathcal{I}|^2/2$ and $\sum_{i\in\mathcal{I}}(y_i-\mu_{x_i})^2/2$. Finally, the full conditional distribution of β is such that

$$\pi(\beta|\mathbf{y}) \propto \frac{1}{Z(\beta)} \exp\left(\beta \sum_{j\sim i} \mathbb{I}_{x_j=x_i}\right), \tag{8.8}$$

since β does not depend on σ^2, $\boldsymbol{\mu}$, and $\mathbf{y}$, given $\mathbf{x}$. As discussed in Sect. 8.3.1, a path sampler can provide an approximation for the ratio of normalizing constants.

In the case of the **Menteith** data, we use a four-neighbour neighbourhood and $G = 6$ on a 100×100 image. For β ranging from 0 to 2 by steps of 0.1, the approximation to $f(\beta)$ is based on 15,000 iterations of Algorithm 8.17 (after burn-in), following the same procedure as with Fig. 8.5. The resulting piecewise-linear function is given in Fig. 8.10 and is smooth enough for us to consider the approximation as acceptable. (We use these numerical values in the clustering function `reconstruct` as the vector `dali`.) Note that the increasing nature of the function f in β is intuitive: As β grows, the probability of having more neighbors of the same category increases and so does $S(\mathbf{x})$.

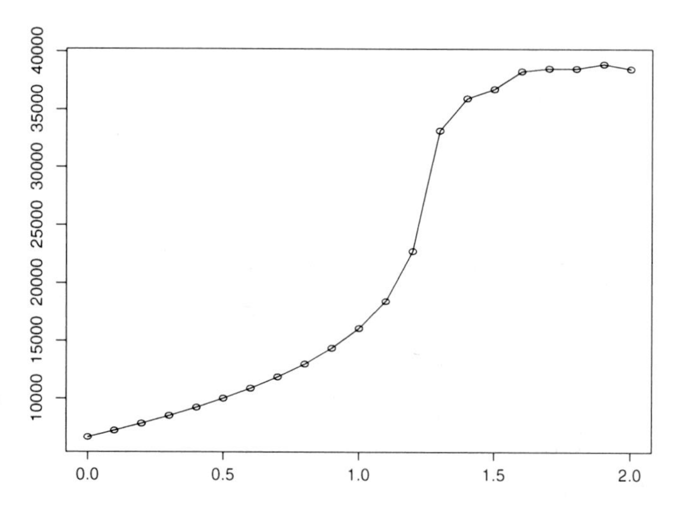

Fig. 8.10. Approximation of $f(\beta)$ for the Potts model on a 100×100 image, a four-neighbour neighbourhood, and $G = 6$, based on 1,500 MCMC iterations after burn-in

The corresponding R code for 6 colors and 4 neighbors (which are the specifications for the **Menteith** dataset) is as follows:

```
reconstruct=function(niter=10^3,y){
numb=dim(y)[1]
x=0*y
mu=matrix(0,niter,6)
sigma2=rep(0,niter)
#prior input
mu[1,]=c(35,50,65,84,92,120)
sigma2[1]=100
beta=rep(1,niter)
xcum=matrix(0,numb^2,6)
n=rep(0,6)
```

```
dali=c(6667.729,7245.159,7856.514,8523.00,9242.127,10025.211,
10896.380,11877.379,12985.344,14360.080,16062.470,18408.592,
22755.124,33163.207,35947.756,36745.675,38286.608,38534.912,
38531.211,38916.662,38495.781)
thefunc=approxfun(seq(0,2,length=21),dali)

for (i in 2:niter){
  lvr=0
  for (k in 1:numb){
    for (l in 1:numb){
      for (co in 1:6)
        n[co]=xneig4(x,k,l,co)
      x[k,l]=sample(1:6,1,prob=exp(beta[i-1]*n)*
             dnorm(y[k,l],mu[i-1,],sqrt(sigma2[i-1])))
      xcum[(k-1)*numb+l,x[k,l]]=xcum[(k-1)*numb+l,x[k,l]]+1
      lvr=lvr+n[x[k,l]]
      }}
  mu[i,1]=truncnorm(1,mean(y[x==1]),sqrt(sigma2[i-1]/
              sum(x==1)),0,mu[i-1,2])
  for (co in 2:5)
     mu[i,co]=truncnorm(1,mean(y[x==co]),sqrt(sigma2[i-1]/
              sum(x==co)),mu[i,co-1],mu[i-1,co+1])
  mu[i,6]=truncnorm(1,mean(y[x==6]),sqrt(sigma2[i-1]/
          sum(x==5)),mu[i,5],255)
  sese=sum((y-mu[i,1])^2*(x==1))
  for (co in 2:6)
    sese=sese+(y-mu[i,co])^2*(x==co))
  sigma2[i]=1/rgamma(1,numb^2/2,sese/2)
  betilde=beta[i-1]+runif(1,-0.05,0.05)
  laccept=vr*(betatilde-beta[i-1])+integrate(thefunc,
   betatilde,beta[i-1])$value
    integrate(lrcst,betilde,beta[i-1])$value
  if (log(runif(1))<laccept){
    beta[i]=betilde}else{beta[i]=beta[i-1]}
  }
list(beta=beta,mu=mu,sigma2=sigma2,xcum=xcum)
}
```

In the above, `truncnorm` is the standard simulator of a truncated normal variate based on the inverse cdf (see Robert and Casella, 2004, Chap. 2, for details).

In the case of the **Menteith** data, we use a four-neighbour neighbourhood and $G = 6$ on a 100×100 image. For β ranging from 0 to 2 by steps of 0.1, the approximation to $f(\beta)$ is based on 1,500 iterations of Algorithm 8.17 (after burn-in), following the same procedure as with Fig. 8.5. The resulting piecewise-linear function is given in Fig. 8.11 and is smooth enough for us to consider the approximation as acceptable. Note that the increasing nature of the function f in β is intuitive: As β grows, the probability of having more neighbors of the same category increases and so does $S(\mathbf{x})$.

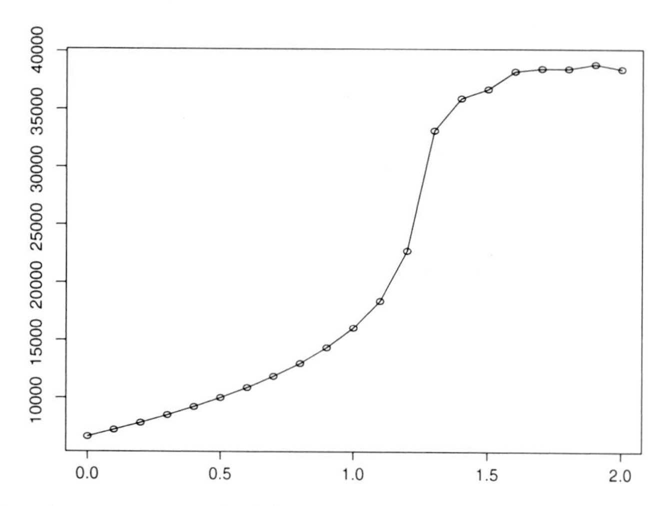

Fig. 8.11. Approximation of $f(\beta)$ for the Potts model on a 100×100 image, a four-neighbour neighbourhood, and $G = 6$, based on 1,500 MCMC iterations after burn-in

Figures 8.12–8.14 illustrate the convergence performances of the hybrid Gibbs sampler for **Menteith**. In that case, using $h = 0.05$ shows that 2,000 MCMC iterations are sufficient for convergence. (Recall, however, that $\mathbf{x}$ is a 100×100 image and thus that a single Gibbs step implies simulating the value of 10^4 pixels. This comes in addition to the cost of approximating the ratio of normalizing constants.)All histograms are smooth and unimodal, even though the moves on β are more difficult than for the other components. (Different values of h were tested for this dataset and none improved this behavior.) Note that large images like **Menteith** often lead to a very concentrated posterior on β. (Other starting values for β were also tested to check for the stability of the stationary region.)

We recall that the primary purpose of this image analysis is to clean (de-noise) and to classify into G categories the pixels of the image. Based on the MCMC output and in particular on the chain $(\mathbf{x}^{(t)})_{1\le t\le T}$ (where T is the number of MCMC iterations), an estimator of $\mathbf{x}$ needs to be derived through an evaluation of the consequences of wrong allocations. Two common ways of running this evaluation are either to count the number of (individual) pixel misclassification,

$$L_1(\mathbf{x},\hat{\mathbf{x}}) = \sum_{i\in\mathcal{I}} \mathbb{I}_{x_i\neq\hat{x}_i},$$

or to use the global "zero–one" loss function (see Sect. 2.3.1),

$$L_2(\mathbf{x},\hat{\mathbf{x}}) = \mathbb{I}_{\mathbf{x}\neq\hat{\mathbf{x}}},$$

which amounts to saying that only a perfect reconstitution of the image is acceptable (and thus sounds rather extreme in its requirements). It is then easy to show that the estimators associated with these loss functions are the marginal posterior mode (MPM), $\hat{\mathbf{x}}^{MPM}$; that is, the image made of the pixels

$$\hat{x}_i^{MPM} = \arg\max_{1\le g\le G} \mathbb{P}^\pi(x_i = g|\mathbf{y}), \quad i\in\mathcal{I},$$

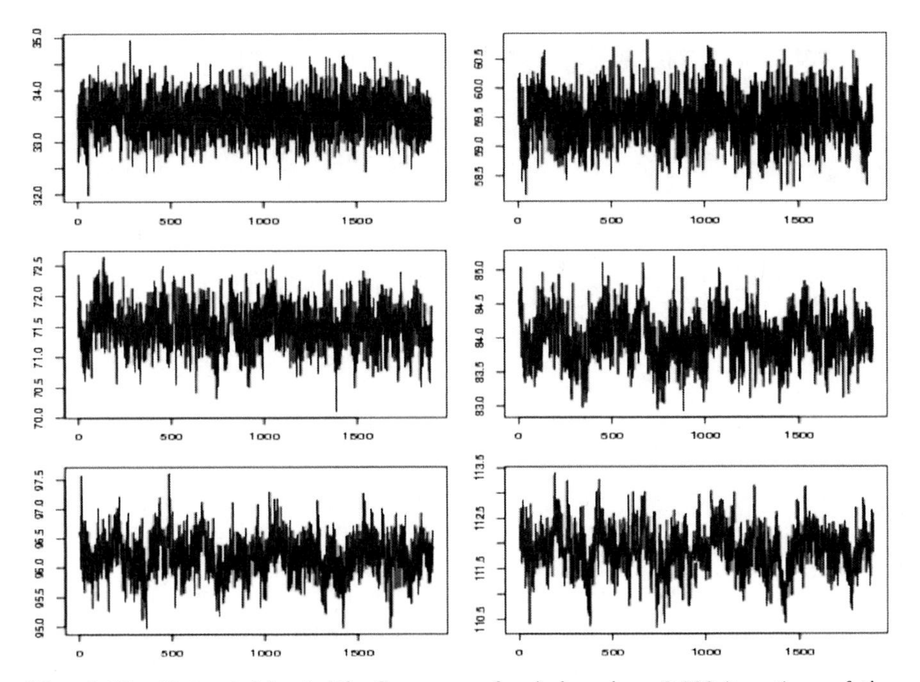

Fig. 8.12. Dataset **Menteith**: Sequence of μ_g's based on 2,000 iterations of the hybrid Gibbs sampler (*read row-wise from* μ_1 *to* μ_6)

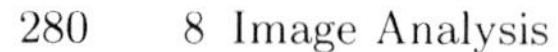

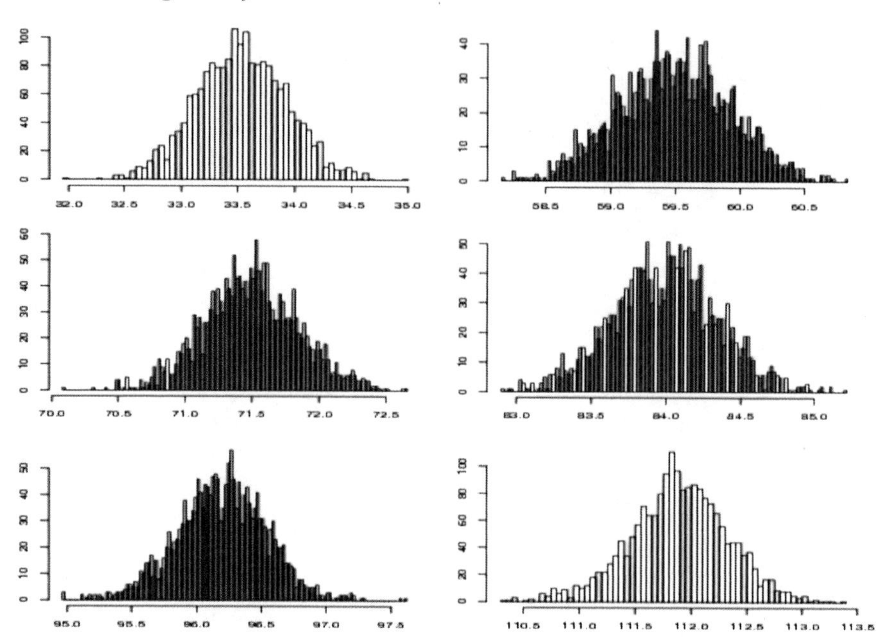

Fig. 8.13. Dataset **Menteith**: Histograms of the μ_g's represented in Fig. 8.12

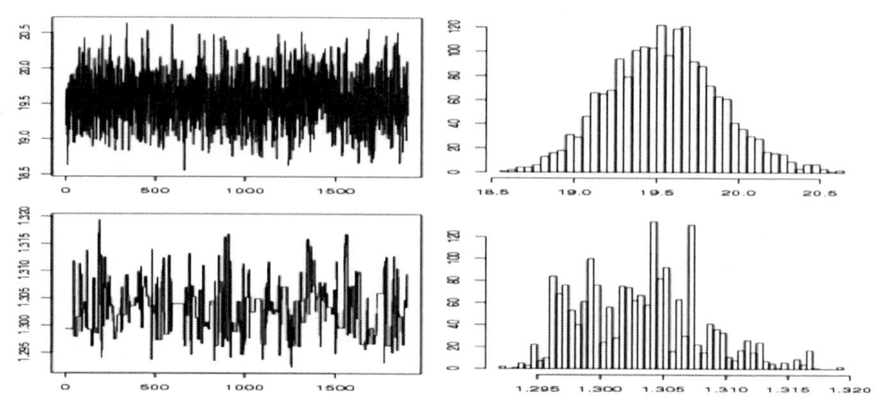

Fig. 8.14. Dataset **Menteith**: Raw plots and histograms of the σ^2's and β's based on 2,000 iterations of the hybrid Gibbs sampler (*the first row corresponds to* σ^2)

and the maximum a posteriori estimator (2.4),

$$\hat{\mathbf{x}}^{MAP} = \arg\max_{\mathbf{x}} \pi(\mathbf{x}|\mathbf{y}),$$

respectively. Note that it makes sense that the $\hat{\mathbf{x}}^{MPM}$ estimator only depends on the marginal distribution of the pixels, given the linearity of the loss function. Both loss functions are nonetheless associated with image reconstruction rather than true classification (Exercise 8.14).

The estimators $\hat{\mathbf{x}}^{MPM}$ and $\hat{\mathbf{x}}^{MAP}$ obviously have to be approximated since the marginal posterior distributions $\pi(x_i|\mathbf{y})$ ($i \in \mathcal{I}$) and $\pi(\mathbf{x}|\mathbf{y})$ are not available in closed form. The marginal distributions of the x_i's being by-products of the MCMC simulation of $\mathbf{x}$, we can use, for instance, as an approximation to $\hat{\mathbf{x}}^{MPM}$ the most frequent occurrence of each pixel $i \in \mathcal{I}$,

$$\hat{x}_i^{MPM} = \max_{g\in\{1,\ldots,G\}} \sum_{j=1}^{N} \mathbb{I}_{x_i^{(j)}=g},$$

based on a simulated sequence, $\mathbf{x}^{(1)}, \ldots, \mathbf{x}^{(N)}$, from the posterior distribution of $\mathbf{x}$. (This is not the most efficient approximation to $\hat{\mathbf{x}}^{MPM}$, obviously, but it comes as a cheap by-product of the MCMC simulation and it does not require the use of more advanced simulated annealing tools, mentioned in Sect. 6.7.)

Unfortunately, the same remark cannot be made about $\hat{\mathbf{x}}^{MAP}$: the state space of the simulated chain $(\mathbf{x}^{(t)})_{1\le t\le T}$ is so huge, being of cardinality $G^{100\times 100}$, that it is completely unrealistic to look for a proper MAP estimate out of the sequence $(\mathbf{x}^{(t)})_{1\le t\le T}$. Since $\pi(\mathbf{x}|\mathbf{y})$ is not available in closed form, even though this density could be approximated by

$$\hat{\pi}(\mathbf{x}|\mathbf{y}) \propto \sum_{t=1}^{T} \pi(\mathbf{x}|\mathbf{y}, \beta^{(t)}, \boldsymbol{\mu}^{(t)}, \sigma^{(t)}),$$

thanks to a Rao–Blackwellization argument, it is rather difficult to propose a foolproof simulated annealing that converges to $\hat{\mathbf{x}}^{MAP}$ (although there exist cheap approximations; see Exercise 8.15).

The segmented image of Lake Menteith is given by the MPM estimate that was found after 2,000 iterations of the Gibbs sampler. We reproduce in Fig. 8.15 the original picture to give an impression of the considerable improvement brought by the algorithm.

8.5 Exercises

8.1 Find two conditional distributions $f(x|y)$ and $g(y|x)$ such that there is no joint distribution corresponding to both f and g. Find a necessary condition for f and g to be compatible in that respect; i.e., to correspond to a joint distribution on (x, y).

8.2 Using the Hammersley–Clifford theorem, show that the full conditional distributions given by (8.3) are compatible with a joint distribution. Deduce that the Ising model is a Markov random field.

8.3 If a joint density $\pi(y_1, \ldots, y_n)$ is such that the conditionals $\pi(y_{-i}|y_i)$ never cancel on the supports of the marginals $m_{-i}(y_{-i})$, show that the support of π is equal to the Cartesian product of the supports of the marginals.

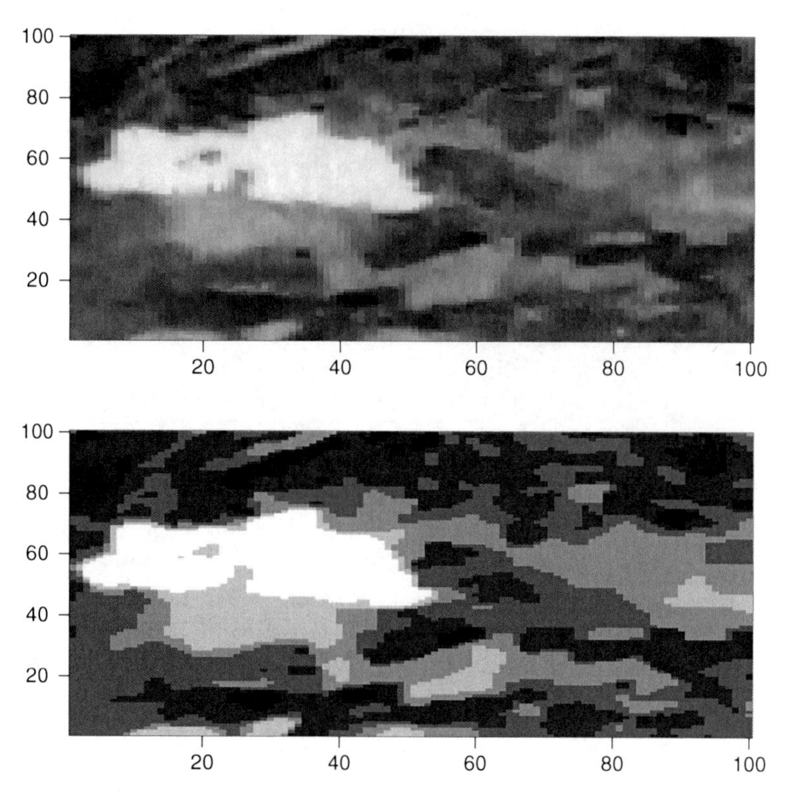

Fig. 8.15. Dataset **Menteith**: (*top*) Segmented image based on the MPM estimate produced after 2,000 iterations of the Gibbs sampler and (*bottom*) the observed image

8.4 Describe the collection of cliques C for an eight-neighbour neighbourhood structure such as in Fig. 8.2 on a regular $n \times m$ array. Compute the number of cliques.

8.5 Draw the function $Z(\beta)$ for a 3×5 array. Determine the computational cost of the derivation of the normalizing constant $Z(\beta)$ of (8.4) for an $m \times n$ array.

8.6 Show that the joint distribution (8.5) is indeed compatible with the full conditionals of the Potts model. Can you derive this joint distribution from the Hammersley–Clifford representation (8.1)?

8.7 For an $n \times m$ array $\mathcal{I}$, if the neighbourhood relation is based on the four nearest neighbors, show that the $x_{i,j}$'s for which $(i+j) \equiv 0(\text{mod } 2)$ are independent conditional on the $x_{i,j}$'s for which $(i+j) \equiv 1(\text{mod } 2)$ $(1 \leq i \leq n,\ 1 \leq j \leq m)$. Deduce that the update of the whole image can be done in two steps by simulating the pixels with even sums of indices and then the pixels with odd sums of indices. (This modification of Algorithm 8.16 is a version of *the Swendsen–Wang* algorithm.)

8.8 Determine the computational cost of the derivation of the normalizing constant of the distribution (8.5) for an $n \times m$ array and G different colors.

8.9 Use the Hammersley–Clifford theorem to establish that (8.5) is the joint distribution associated with the conditionals above. Deduce that the Potts model is an MRF.

8.10 Derive an alternative to Algorithm 8.17 where the probabilities in the multinomial proposal are proportional to the numbers of neighbors $n_{u_\ell,g}$ and compare its performance with that of Algorithm 8.17.

8.11 Show that the Swendsen–Wang improvement given in Exercise 8.7 also applies to the simulation of $\pi(\mathbf{x}|\mathbf{y},\beta,\sigma^2,\boldsymbol{\mu})$.

8.12 Using a piecewise-linear interpolation of $f(\beta)$ based on the values $f(\beta^1),\ldots,f(\beta^M)$, with $0<\beta_1<\ldots<\beta_M=2$, give the explicit value of the integral

$$\int_{\alpha_0}^{\alpha_1} \hat{f}(\beta)\,\mathrm{d}\beta$$

for any pair $0\le\alpha_0<\alpha_1\le 2$.

8.13 Show that the estimators $\hat{\mathbf{x}}$ that minimize the posterior expected losses $\mathbb{E}^\pi[L_1(\mathbf{x},\hat{\mathbf{x}})|\mathbf{y})]$ and $\mathbb{E}^\pi[L_2(\mathbf{x},\hat{\mathbf{x}})|\mathbf{y}]$ are $\hat{\mathbf{x}}^{MPM}$ and $\hat{\mathbf{x}}^{MAP}$, respectively.

8.14 Determine the estimators $\hat{\mathbf{x}}$ associated with two loss functions that penalize differently the classification errors,

$$L_3(\mathbf{x},\hat{\mathbf{x}})=\sum_{i,j\in\mathcal{I}}\mathbb{I}_{x_i=x_j}\,\mathbb{I}_{\hat{x}_i\neq\hat{x}_j} \quad\text{and}\quad L_4(\mathbf{x},\hat{\mathbf{x}})=\sum_{i,j\in\mathcal{I}}\mathbb{I}_{x_i\neq x_j}\,\mathbb{I}_{\hat{x}_i=\hat{x}_j}\,.$$

8.15 Since the maximum of $\pi(\mathbf{x}|\mathbf{y})$ is the same as that of $\pi(\mathbf{x}|\mathbf{y})^\kappa$ for every $\kappa\in\mathbb{N}$, show that

$$\pi(\mathbf{x}|\mathbf{y})^\kappa=\int\pi(\mathbf{x},\theta_1|\mathbf{y})\,\mathrm{d}\theta_1\times\cdots\times\int\pi(\mathbf{x},\theta_\kappa|\mathbf{y})\,\mathrm{d}\theta_\kappa\,, \tag{8.9}$$

where $\theta_i=(\beta_i,\boldsymbol{\mu}_i,\sigma_i^2)$ $(1\le i\le\kappa)$. Deduce from this representation an optimization scheme that slowly increases κ over iterations and that runs a Gibbs sampler for the integrand of (8.9) at each iteration.

8.16 For the Ising model, show that the distribution (8.4) can be also defined as

$$\pi(\mathbf{x})\propto\exp\left(2\beta\sum_{j\sim i}\mathbb{I}_{x_j=x_i=1}\right)$$

when the number of neighbors is constant.

8.17 Show that the joint distribution (8.4) can be obtained from the full conditionals (8.3) by virtue of the Hammersley–Clifford representation (8.1).

8.18 Show that the Ising distribution is symmetric in that inverting the color of all pixels does not change the probability (8.4).

8.19 For the Ising model, run a simulation experiment that should locate the limiting value of β above which almost all pixels are of the same color. Same question for the (negative) limiting value of β below which the image is a perfect checkerboard.

8.20 Show that the ABC algorithm implemented with $\epsilon=0$ and a distance between sufficient statistics is not approximate in that the output is truly simulated from the posterior distribution $\pi(\theta|\mathbf{x})\propto f(\mathbf{x}|\theta)\pi(\theta)$.

About the Authors

Jean-Michel Marin is Professor of Statistics at Université Montpellier 2, France, and Head of the Mathematics and Modelling research unit. He has written over 40 papers on Bayesian methodology and computing, as well as worked closely with population geneticists over the past 10 years.

Christian P. Robert is Professor of Statistics at Université Paris-Dauphine, France. He has written over 150 papers on Bayesian Statistics and computational methods and is the author or co-author of seven books on those topics, including *The Bayesian Choice* (Springer, 2001), winner of the ISBA DeGroot Prize in 2004. He is a Fellow of the Institute of Mathematical Statistics, the Royal Statistical Society and the American Statistical Society. He has been co-editor of the *Journal of the Royal Statistical Society, Series B*, and in the editorial boards of the *Journal of the American Statistical Society*, the *Annals of Statistics*, *Statistical Science*, and *Bayesian Analysis*. He is also a recipient of an Erskine Fellowship from the University of Canterbury (NZ) in 2006 and a senior member of the Institut Universitaire de France (2010–2015).

J.-M. Marin and C.P. Robert, *Bayesian Essentials with R*, Springer Texts in Statistics, DOI 10.1007/978-1-4614-8687-9,

References

Agresti, A. (1996). *An Introduction to Categorical Data Analysis.* John Wiley, New York.

Besag, J. (1974). Spatial interaction and the statistical analysis of lattice systems. *J. Royal Statist. Soc. Series B*, 36:192–326. With discussion.

Brockwell, P. and Davis, P. (1996). *Introduction to Time Series and Forecasting.* Springer Texts in Statistics. Springer-Verlag, New York.

Cappé, O., Moulines, E., and Rydén, T. (2004). *Hidden Markov Models.* Springer-Verlag, New York.

Carlin, B. and Louis, T. (1996). *Bayes and Empirical Bayes Methods for Data Analysis.* Chapman and Hall, New York.

Casella, G. and Berger, R. (2001). *Statistical Inference.* Wadsworth, Belmont, CA.

Chambers, J., Cleveland, W., Kleiner, B., and Tukey, P. (1983). *Graphical Methods for Data Analysis.* Chapman and Hall, New York.

Chen, M., Shao, Q., and Ibrahim, J. (2000). *Monte Carlo Methods in Bayesian Computation.* Springer-Verlag, New York.

Chib, S. (1995). Marginal likelihood from the Gibbs output. *J. American Statist. Assoc.*, 90:1313–1321.

Christensen, R. (2002). *Plane Answers to Complex Questions: The Theory of Linear Models.* Springer Texts in Statistics. Springer-Verlag, New York.

Congdon, P. (2001). *Bayesian Statistical Modelling.* John Wiley, New York.

Congdon, P. (2003). *Applied Bayesian Modelling.* John Wiley, New York.

Crawley, M. (2007). *The R Book.* John Wiley, New York.

Cressie, N. (1993). *Statistics for Spatial Data.* John Wiley, New York.

Dalgaard, P. (2002). *Introductory Statistics with R.* Springer-Verlag, New York.

Del Moral, P., Doucet, A., and Jasra, A. (2006). Sequential Monte Carlo samplers. *J. Royal Statist. Soc. Series B*, 68(3):411–436.

Dupuis, J. (1995). Bayesian estimation of movement probabilities in open populations using hidden Markov chains. *Biometrika*, 82(4):761–772.

J.-M. Marin and C.P. Robert, *Bayesian Essentials with R*, Springer Texts in Statistics, DOI 10.1007/978-1-4614-8687-9,

Flury, B. and Riedwyl, H. (1988). *Multivariate Statistics, A Practical Approach.* Cambridge University Press, Cambridge.

Frühwirth-Schnatter, S. (2006). *Finite Mixture and Markov Switching Models.* Springer-Verlag, New York, New York.

Gaetan, C. and Guyon, X. (2010). *Spatial Statistics and Modeling.* Springer-Verlag, New York.

Gelfand, A. and Dey, D. (1994). Bayesian model choice: asymptotics and exact calculations. *J. Royal Statist. Society Series B*, 56:501–514.

Gelman, A., Carlin, J., Stern, H., and Rubin, D. (2013). *Bayesian Data Analysis.* Chapman and Hall, New York, second edition.

Gelman, A., Carlin, J., Stern, H., Dunson, D., Vehtari, A. and Rubin, D. (2013). *Bayesian Data Analysis.* Chapman and Hall, New York, New York, third edition.

Gelman, A. and Meng, X. (1998). Simulating normalizing constants: From importance sampling to bridge sampling to path sampling. *Statist. Science*, 13:163–185.

Geman, S. and Geman, D. (1984). Stochastic relaxation, Gibbs distributions and the Bayesian restoration of images. *IEEE Trans. Pattern Anal. Mach. Intell.*, 6:721–741.

Gill, J. (2002). *Bayesian Methods: A Social and Behavioral Sciences Approach.* CRC Press, Boca Raton, FL.

Gouriéroux, C. (1996). *ARCH Models.* Springer-Verlag, New York.

Green, P. (1995). Reversible jump MCMC computation and Bayesian model determination. *Biometrika*, 82(4):711–732.

Hastie, T., Tibshirani, R., and Friedman, J. (2001). *The Elements of Statistical Learning.* Springer-Verlag, New York.

Hjort, N., Holmes, C., Müller, P., and Walker, S. (2010). *Bayesian Nonparametrics.* Cambridge University Press, Cambridge.

Hoff, P. (2009). *A first course in Bayesian statistical methods.* Springer-Verlag, New York.

Holmes, C., Denison, D., Mallick, B., and Smith, A. (2002). *Bayesian Methods for Nonlinear Classification and Regression.* John Wiley, New York.

Hurn, M., Husby, O., and Rue, H. (2003). A Tutorial on Image Analysis. In Møller, J., editor, *Spatial Statistics and Computational Methods*, volume 173 of *Lecture Notes in Statistics*, pages 87–141. Springer-Verlag, New York.

Lebreton, J.-D., Burnham, K., Clobert, J., and Anderson, D. (1992). Modelling survival and testing biological hypotheses using marked animals: a unified approach with case studies. *Ecological Monographs*, 62(2):67–118.

Liu, J. (1996). Peskun's theorem and a modified discrete-state Gibbs sampler. *Biometrika*, 83:681–682.

Lunn, D., Jackson, C., Best, N., Thomas, A., and Spiegelhalter, D. (2012). *The BUGS Book: A Practical Introduction to Bayesian Analysis.* Chapman and Hall/CRC, London, UK.

Marin, J. and Robert, C. (2010). Importance sampling methods for Bayesian discrimination between embedded models. In Chen, M.-H., Dey, D., Müller, P., Sun, D., and Ye, K., editors, *Frontiers of Statistical Decision Making and Bayesian Analysis*. Springer-Verlag, New York.

Marin, J.-M. and Robert, C. (2007). *Bayesian Core*. Springer-Verlag, New York.

McCullagh, P. and Nelder, J. (1989). *Generalized Linear Models*. Chapman and Hall, New York.

McDonald, I. and Zucchini, W. (1997). *Hidden Markov and other models for discrete-valued time series*. Chapman and Hall/CRC, London.

Meyn, S. and Tweedie, R. (1993). *Markov Chains and Stochastic Stability*. Springer-Verlag, New York.

Møller, J., Pettitt, A. N., Reeves, R., and Berthelsen, K. K. (June 2006). An efficient markov chain monte carlo method for distributions with intractable normalising constants. *Biometrika*, 93(2):451–458.

Murrell, P. (2005). *R Graphics*. Chapman and Hall, New York.

Nolan, D. and Speed, T. (2000). *Stat Labs: Mathematical Statistics through Applications*. Springer-Verlag, New York.

Pole, A., West, M., and Harrison, J. (1994). *Applied Bayesian Forecasting and Time Series Analysis*. Chapman and Hall, New York.

Pritchard, J., Seielstad, M., Perez-Lezaun, A., and Feldman, M. (1999). Population growth of human Y chromosomes: a study of Y chromosome microsatellites. *Mol. Biol. Evol.*, 16:1791–1798.

Ripley, B. (1988). *Statistical Inference for Spatial Processes*. Cambridge University Press, Cambridge.

Robert, C. (2007). *The Bayesian Choice*. Springer-Verlag, New York, paperback edition.

Robert, C. and Casella, G. (2004). *Monte Carlo Statistical Methods*. Springer-Verlag, New York, second edition.

Robert, C. and Casella, G. (2009). *Introducing Monte Carlo Methods with R*. Use R! Springer-Verlag, New York.

Rue, H. and Held, L. (2005). *Gaussian Markov Random Fields: Theory and Applications*, volume 104 of *Monographs on Statistics and Applied Probability*. Chapman & Hall, London.

Särndal, C., Swensson, B., and Wretman, J. (2003). *Model Assisted Survey Sampling*. Springer-Verlag, New York, second edition.

Spector, P. (2009). *Data Manipulation with R*. Springer-Verlag, New York.

Tanner, M. (1996). *Tools for Statistical Inference: Observed Data and Data Augmentation Methods*. Springer-Verlag, New York, third edition.

Tomassone, R., Dervin, C., and Masson, J.-P. (1993). *Biométrie: Modélisation de Phénomènes Biologiques*. Masson, Paris.

Tufte, E. (2001). *The Visual Display of Quantitative Information*. Graphics Press, second edition.

Venables, W. and Ripley, B. (2002). *Modern Applied Statistics with S*. Springer-Verlag, New York, fourth edition.

Whittaker, J. (1990). *Graphical Models in Applied Multivariate Statistics.* John Wiley, Chichester.

Zellner, A. (1971). *An Introduction to Bayesian Econometrics.* John Wiley, New York.

Zellner, A. (1984). *Basic Issues in Econometrics.* University of Chicago Press, Chicago.

Index

J.-M. Marin and C.P. Robert, *Bayesian Essentials with R*, Springer Texts in Statistics, DOI 10.1007/978-1-4614-8687-9,

Printed by Publishers' Graphics LLC
CAMZ131106.15.17.15